Fishery Science

We dedicate this book to our good friend John Blaxter, the gentleman scientist.

*His scientific excellence and creativity as well as his
personal charm and good humor have made
permanent impressions on both of us. John's
scientific contributions permeate this book, which we
hope will carry his legacy to many future generations
of fishery scientists.*

Fishery Science

The Unique Contributions of Early Life Stages

Edited by

Lee A. Fuiman

Department of Marine Science, University of Texas at Austin,
Marine Science Institute, Port Aransas, Texas, USA

and

Robert G. Werner

College of Environmental Science and Forestry,
State University of New York, Syracuse, New York, USA

Blackwell
Science

Editorial offices:
Blackwell Science Ltd, 9600 Garsington Road, Oxford OX4 2DQ, UK
Tel: +44 (0) 1865 776868
Blackwell Publishing Professional, 2121 State Avenue, Ames, Iowa 50014-8300, USA
Tel: +1 515 292 0140
Blackwell Science Asia Pty Ltd, 550 Swanston Street, Carlton, Victoria 3053, Australia
Tel: +61 (0)3 8359 1011

First published 2002 by Blackwell Science
Transferred to digital print, 2007

ISBN: 978-0-632-05661-3

Library of Congress Cataloging-in-Publication Data
Fishery science: the unique contributions of early life stages / edited by Lee A. Fuiman and
Robert G. Werner.
    p.cm.
Includes bibliographical references.
ISBN-13: 978-0-632-05661-3
ISBN-10: 0-632-05661-4
    1. Fishes--Larvae. 2. Fishes--Eggs. I. Fuiman, Lee A. II. Werner, Robert G.
QL639.25 .F57 2002
597.139--dc21

                                                                                2002074741

A catalogue record for this title is available from the British Library

Set in Times by Gray Publishing, Tunbridge Wells, Kent

For further information on Blackwell Publishing, visit our website:
www.blackwellpublishing.com

Contents

Contributors

JAMES H. COWAN, JR. *Department of Oceanography and Coastal Sciences, Louisiana State University, Baton Rouge, Louisiana, USA*

LEE A. FUIMAN *Department of Marine Science, University of Texas at Austin, Marine Science Institute, Port Aransas, Texas, USA*

G. JOAN HOLT *Department of Marine Science, University of Texas at Austin, Marine Science Institute, Port Aransas, Texas, USA*

EDWARD D. HOUDE *University of Maryland Center for Environmental Science, Chesapeake Biological Laboratory, Solomons, Maryland, USA*

CYNTHIA M. JONES *Center for Quantitative Fisheries Ecology, Old Dominion University, Norfolk, Virginia, USA*

HUBERT KECKEIS *Institute of Ecology and Conservation Biology, Department of Limnology, University of Vienna, Vienna, Austria*

KARIN E. LIMBURG *College of Environmental Science and Forestry, State University of New York, Syracuse, New York, USA*

THOMAS J. MILLER *University of Maryland Center for Environmental Science, Chesapeake Biological Laboratory, Solomons, Maryland, USA*

PIERRE PEPIN *Department of Fisheries and Oceans, St. Johns, Newfoundland, Canada*

JAMES A. RICE *Zoology Department, North Carolina State University, Raleigh, North Carolina, USA*

EDWARD S. RUTHERFORD *School of Natural Resources and Environment, University of Michigan, Ann Arbor, Michigan, USA*

FRITZ SCHIEMER *Institute of Ecology and Conservation Biology, Department of Limnology, University of Vienna, Vienna, Austria*

RICHARD F. SHAW *Department of Oceanography and Coastal Sciences, Louisiana State University, Baton Rouge, Louisiana, USA*

YOSHIRO WATANABE *Ocean Research Institute, University of Tokyo, Tokyo, Japan*

ROBERT G. WERNER *College of Environmental Science and Forestry, State University of New York, Syracuse, New York, USA*

Preface

There has been explosive growth in research on the early life of fishes over the past 30 years. A great deal of that research was undertaken in order to develop a more complete understanding of the dynamics of fish populations. It is now clear that information derived from fish eggs and larvae makes a number of unique contributions to fishery science that are crucial for accurate assessment and management of fish populations. Nevertheless, very little of this information has found its way into textbooks and college-level courses in fishery science. After several decades of intensive research on early life stages of fishes, there is now a mature body of knowledge that is ready to be summarized and distilled for students and the non-specialist audience. Our intention with this book is to improve the training of young fishery scientists by demonstrating why fish eggs and larvae are important to consider, how the characteristics of early life stages require a somewhat different research approach, and how information on early life stages can be applied and interpreted to yield unique and important insights into fish populations.

This is designed to be a supplemental textbook, to complement the material that is covered in existing textbooks and courses on fishery science. The chapters are written in a didactic and readable style by university instructors who are also respected specialists in larval fish research. The first 10 chapters focus on the concepts that are traditionally covered in such courses. Each chapter ends with a list of references that provide more information on the material covered. A selection of case studies follows, each of which demonstrates several specific applications of early life-history information to a fishery problem. The case studies represent three very different aquatic environments: the open ocean, a lake, and a river. The final chapter provides a starting point for learning more about the methods for working with fish eggs and larvae. Throughout the book we attempted to provide a balanced coverage of marine and freshwater environments and geographical regions, but uneven availability of information and personal biases were inevitable.

We hope that this book will be useful not only for instruction of introductory fishery science courses, but also as a core text for graduate-level courses in larval fish biology. It may also serve well as a reference that brings to a wider fishery science audience an understanding of the unique contributions of early life stages.

We gratefully acknowledge the strong support of the Early Life History Section of the American Fisheries Society, which generously provided funding and encouragement for production of this book. We are grateful to our colleagues who enthusiastically agreed to contribute chapters and did a fine job of understanding and following our plan. We also appreciate the assistance of Darrel Snyder, who offered thoughtful advice, and Heather Alexander, who help with the graphics. Finally, we would like to acknowledge the thoughtful support of our wives, Linda and Jo, who made this effort much more pleasant than it otherwise would have been.

L.A.F.

R.G.W.

Chapter 1
Special Considerations of Fish Eggs and Larvae

Lee A. Fuiman

1.1 Introduction

Fishery science attempts to understand the dynamics of fish populations with the goal of optimizing some human benefit or value, such as the yield to a commercial fishery, hours of pleasurable angling, or conservation of a species, population, or ecological community. Virtually all management goals center on adult fishes and so the traditional tools of fishery science were forged in an understanding of the biology and ecology of adult fishes. The critical role that early life stages play in the dynamics of fish populations is easily recognized when we consider that every fish taken by commercial trawlers and long-liners, each trophy fish landed by the patient and skilled angler, and all of the spawners in the relic population of an endangered species were, at one time, inconspicuous embryos and larvae, and that only a tiny fraction of their cohort survived the first few months of life. With this realization it becomes clear that fishery science demands a broader understanding of fishes, one that encompasses the early life stages.

This chapter highlights some of the important differences between early life stages of fishes and the more familiar juveniles and adults. It provides a general introduction to the developmental changes that take place during early life and some of the ways in which early life-history traits are interrelated. A comprehensive description of all developmental changes or the breadth of variation among species is beyond the scope of this chapter. Rather, the intent is to impart an appreciation that morphology cannot be considered static during this period of life. With this dynamic morphology comes continually changing behavioral and physiological capabilities which alter the nature of a young fish's ecological interactions. Recognizing the fundamental differences between the life of an embryo or larva and that of an adult is a first step toward appreciating the unique contributions of early life stages to fishery science.

1.2 The life cycle and the nature of early life

The life history of a fish can be divided into five primary periods: embryo, larva, juvenile, adult, and senescent. The first four of these form a cycle and only those individuals that are especially adept at survival, or just lucky, become senescent (Figure 1.1). Each of these life-history periods can be characterized by one or two dominant physiological processes that

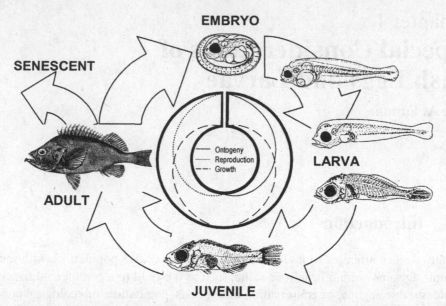

Figure 1.1   Generalized life cycle of a fish, using *Sebastes marinus* as an example. Major periods and the corresponding dominant intrinsic processes are shown. The circular chart at the center shows that ontogenetic changes (solid line) begin at fertilization of the egg and gradually diminish toward the juvenile period. Growth (dashed line) also begins at fertilization and diminishes toward adulthood. Reproduction (dotted line) begins when gonads differentiate and diminishes in senescence. Drawings from Bigelow & Welsh (1925).

largely determine the changes in morphological structure, physiological capabilities, behavioral motivation, and ecological role of an individual at that time of life. The embryonic period is a time of ontogeny, a complex set of changes that include rapid proliferation of cells, differentiation of new tissues, and reorganization or loss of existing ones. Since the embryo receives all of its energy from the maternal investment of yolk, the total weight of the embryo does not increase (it may actually decrease). This apparent lack of growth, as measured by total weight, disguises the conversion of yolk into new metabolically active biomass. In contrast, growth is the dominant process during the juvenile period. At the onset of the juvenile period, nearly all organs are present and the fish has the appearance of a small adult. Growth during the juvenile period can increase a fish's dry weight 1000–1 000 000 times, while differentiation is mostly confined to the reproductive organs. Between the embryonic and juvenile periods is the larval period, an especially dynamic interval when ontogeny continues and biomass begins to increase, sometimes by a factor of 10–1000 (Figure 1.2). Together, growth and ontogeny bring about major changes in structure and function over a brief span of time. Reproduction is the dominant process of the adult period. Growth continues at a reduced rate or ceases altogether. The post-reproductive senescent period is characterized by degeneration, which takes the form of reduced growth rate and spawning frequency, changes in external appearance, and endocrine dysfunction.

The embryonic and larval periods have important ecological and evolutionary functions. For many species, they represent an effective means of dispersal that can extend the range of a population and mix the gene pool. This is accomplished passively in species that have

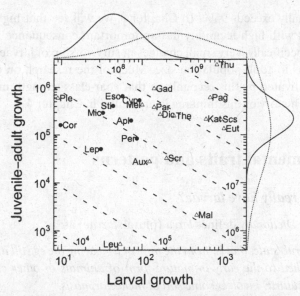

Figure 1.2 Biomass increase during major periods of the life cycle. The dry weight of larvae increases by one to three orders of magnitude before metamorphosis and three to six orders of magnitude after metamorphosis. Dashed lines show lifetime growth, expressed as a multiple of egg dry weight. Data points identify means for genera, based on 38 species in 23 genera from 14 families. Filled symbols represent freshwater species; open symbols are marine species. Abbreviations for genera are: Apl, *Aplodinotus*; Aux, *Auxis*; Cor, *Coregonus*; Cyp, *Cyprinus*; Dic, *Dicentrarchus*; Eso, *Esox*; Eut, *Euthynnus*; Gad, *Gadus*; Kat, *Katsuwonus*; Lep, *Lepomis*; Leu, *Leuresthes*; Mal, *Mallotus*; Mel, *Melanogrammus*; Mic, *Micropterus*; Pag, *Pagrus*; Par, *Paralichthys*; Per, *Perca*; Ple, *Pleuronectes*; Scr, *Scomber*; Scs, *Scomberomorus*; Sti, *Stizostedion*; The, *Theragra*; Thu, *Thunnus*.

nearly neutrally buoyant, planktonic eggs and early larvae, such as most marine fishes. Even the more robust larvae of freshwater species are feeble swimmers in comparison to the water movements of streams and rivers. This is not to say that dispersal is entirely passive. There are many examples of embryos or larvae timing their exposure to dispersing currents, such as using light intensity to initiate hatching from eggs in a nest, synchronous emergence of larvae from the gravel of a stream bed, or swimming vertically to achieve tidal stream transport. Lacustrine fish larvae often migrate actively from incubation sites along the shore out to the limnetic zone, and the later larval stages of coral reef fishes can swim across currents toward reef habitats.

The early life stages of fishes are sometimes referred to as ichthyoplankton and are typically considered inert particles. While this approach is convenient for examining large scale physical processes, such as oceanographic currents, it is a mistake to dismiss fish larvae as passive animals or to consider their ecology simple or unimportant. They are interactive components of the ecosystem. They can, for example, temporarily reduce local zooplankton populations so that the potential for competition for food is heightened (see Chapter 8). With the exception of reproduction, they face the same challenges that adult fishes experience. These challenges are made more severe by the larva's small size, incomplete state of development, and the fact that their size, structure, and behavioral and physiological capabilities are changing rapidly as they attempt to survive. It is no wonder that mortality to the juvenile period usually exceeds 95%. In Chapter 3, we will see that high mortality during early life combined with high fecundity has an important consequence for understanding fish populations. Specifically, very small changes in survivorship of larvae can translate into very large variations in adult population size. Much of the research on early life stages of fishes has been motivated by this recognition that year-class strength may be determined during early life. A history of this thinking is presented in Chapter 4.

1.3   Developmental traits and patterns

1.3.1   Do fishes really have larvae?

The *Oxford English Dictionary* defines *larva* (plural: *larvae*) as:

> *"an insect in the grub state, that is, from the time of its leaving the egg till its transformation into a pupa; applied to the early immature form of animals of other classes, when the development to maturity involves some sort of metamorphosis."*

Our question now involves metamorphosis, which is defined as:

> *"change of form in animals and plants, or their parts, during life … ; especially in entomology, a change or one of a series of changes which a metabolous insect undergoes, resulting in complete alteration of form and habit."*

Whether fishes have larvae hinges upon the presence of a "complete alteration of form and habit." There are no known fishes that have a post-hatching stage analogous to the insect pupa, during which both form and habit change so dramatically and apparently abruptly. Rather, metamorphosis in fishes is more like that of amphibians, a gradual change in form. When changes in habit occur at metamorphosis, they are more subtle than the transition from an aquatic to a terrestrial existence in amphibians. Figure 1.3 shows clearly that fish larvae are quite different from their adult counterparts.

In practice, the completion of metamorphosis defines the boundary between the larval and juvenile periods and is determined on the basis of outward appearance (but see Box 1.1). Many species with pelagic larvae undergo a relatively subtle change in habit or habitat as metamorphosis finishes, such as leaving the plankton to become associated with a substrate (coral, rock, or bivalve reefs or vegetation). Settlement is sometimes used as a synonym for metamorphosis, even though these two terms refer to different changes, one a change in habit, the other a change in form. Despite these more common habitat shifts, there are splendid examples of fishes with changes in form or habit that approach those of amphibians and even insects. One of the most familiar and impressive is the metamorphosis of flatfishes. Flatfish larvae are similar to other fish larvae in most ways (Figure 1.3t) until the end of the larval period when one eye migrates over the top of the head to the other side. At the same time, body pigmentation becomes asymmetrical, they begin to swim with a noticeable tilt, and finally take up a benthic habit. This transition is not abrupt, it takes many days to a few weeks. Other dramatic metamorphoses include the transformation of sea lampreys (*Petromyzon marinus*) from a benthic, filter-feeding ammocoetes larva to an adult ectoparasite of other fishes, and the transparent, ribbon-like leptocephalus larva

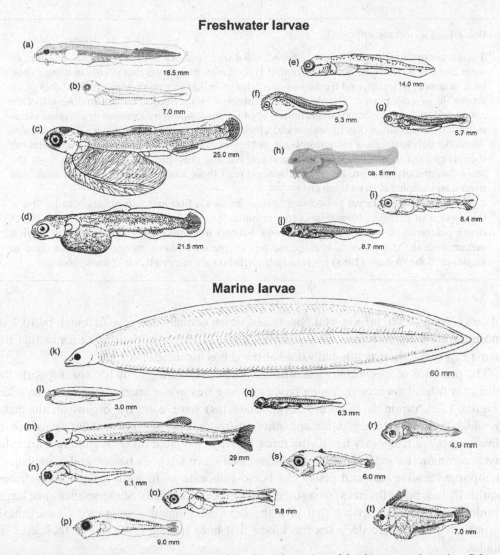

Freshwater larvae

Marine larvae

Figure 1.3 Representative stages of larvae for selected species of freshwater and marine fishes. Sources are given in parentheses: (a) lake sturgeon, *Acipenser fulvescens* (Jude 1982); (b) bloater, *Coregonus hoyi* (Auer 1982); (c) chinook salmon, *Oncorhynchus tshawytscha* (Kendall & Behnke 1984); (d) lake trout, *Salvelinus namaycush* (Fish 1932); (e) northern pike, *Esox lucius* (Gihr 1957); (f, g) common carp, *Cyprinus carpio* (Nakamura 1969); (h) brown bullhead, *Ameiurus nebulosus* (Armstrong & Child 1962); (i) largemouth bass, *Micropterus salmoides* (Conner 1979); (j) yellow perch, *Perca flavescens* (Mansueti 1964); (k) American eel, *Anguilla rostrata* (Schmidt 1916); (l) Japanese anchovy, *Engraulis japonicus* (Mito 1961); (m) Atlantic herring, *Clupea harengus* (Ehrenbaum 1909); (n) haddock, *Melanogrammus aeglefinus* (Dunn & Matarese 1984); (o) walleye pollock, *Theragra chalcogramma* (Matarese *et al.* 1981); (p) Atlantic cod, *Gadus morhua* (Schmidt 1905); (q) striped bass, *Morone saxatilis* (Mansueti 1958); (r) Atlantic croaker, *Micropogonias undulatus* (Hildebrand & Cable 1930); (s) bluefin tuna, *Thunnus thynnus* (Collette *et al.* 1984); (t) California halibut, *Paralichthys californicus* (Ahlstrom *et al.* 1984). Drawings reproduced with permission of Great Lakes Fishery Commission (a, b), American Society of Ichthyologists and Herpetologists (c, n, s, t), Muséum d'Histoire Naturelle Genève (e), National Science Museum Tokyo (f, g), Syracuse University Press (h), Kyushu University (l), Estuarine Research Federation (q).

Box 1.1   Two metamorphoses?

The metamorphosis of fish larvae is considered a true vertebrate metamorphosis, like that of frogs and toads. There is, however, another type of metamorphosis that occurs in some fishes, such as lampreys, trouts, and freshwater eels. This so-called second metamorphosis takes place during the juvenile period. The changes brought about by the second metamorphosis are often associated with a diadromous migration. One of the most familiar examples of a second metamorphosis is smoltification in salmonids, which prepares the freshwater parr for life at sea. Anguillid eels undergo a metamorphosis as they migrate into estuaries before undertaking their migration to oceanic spawning grounds. For the eels, the changes resulting from the second metamorphosis are much less dramatic than those that occur in the earlier transition from a leptocephalus larva to an elver.

   The presence of a larval period in a species implies a first metamorphosis because this is the process of changing from a larva to a juvenile. Species that lack a larva, those with direct development, have no first metamorphosis. Second metamorphosis is independent of first metamorphosis and, as such, is a second type of metamorphosis rather than the second in sequence. John Youson (1988) provided a thoughtful summary of fish metamorphoses.

---

of eels (Figure 1.3k), tarpon, and bonefishes, which actually decrease in length by 10% or more during metamorphosis. So distinct are these larvae from their adult forms that the name *Leptocephalus* was originally used as the genus for these species.

The presence of specialized structures that exist only during early life also supports the idea that fishes have larvae. Young larvae of some freshwater species (for example, pikes [Figure 1.3e], cyprinids [Figure 1.3f, g], characins) have adhesive organs on the head by which they attach to vegetation and other substrates. Marine larvae exhibit a particularly diverse array of temporary larval structures. Greatly elongated fin spines and rays are relatively common (for example, some flatfishes [see Figure 1.3t], sea basses, and goosefishes). Temporary head spines and plates can be so elaborate as to form a helmet (tilefishes, squirrelfishes, butterflyfishes) or a suit of armor (ocean sunfish). Some species have large, fluid-filled subdermal spaces that give the larva an inflated appearance (goosefishes). Especially fantastic are deep sea fish larvae that have trailing intestines or their eyes on stalks.

These larval forms that have striking metamorphoses or ephemeral structures represent one extreme in a broad spectrum of developmental programs in fishes. At the other extreme are species with direct development, where the larval stage is greatly reduced or missing and most of their development occurs before they become free-living individuals. In these species, including such evolutionarily diverse groups as sharks and surfperches (Embiotocidae), most of the developmental changes take place inside large eggs or the parent, and the offspring is nourished through its period of ontogeny by a large yolk supply, some sort of maternal–fetal connection, or even siblicide. Although technically very different, mouthbrooding fishes, such as some catfishes and cichlids, are ecologically equivalent in that their embryos hatch into larvae but the larvae are retained in a parent's mouth often into the juvenile period. The vast majority of species have a developmental program somewhere between these extremes, in which eggs incubate in the water and free-living larvae, which appear different from juveniles to varying degrees, change form, habit, or both as they grow.

1.3.2 Basic anatomy and development

Eggs

It is convenient in fishery science to refer to embryos as eggs. Because the actions of embryos are limited, most fishery-based investigations need not be directly concerned with the changes to the embryo that take place within the confines of the chorion. Substituting "egg" for "embryo," however, can lead to some confusion for those concerned with the production of eggs and the measurement of fecundity. A useful distinction can be made between ova or oocytes – internal, unfertilized cells in any stage of oogenesis – and eggs – ovulated cells after being expelled from the body and/or fertilized. These definitions do not cover all possibilities, but will suffice for many purposes of fishery science.

Fish eggs have a relatively small suite of permanent and distinctive features that can be used for identification. The most useful traits include: habit (for example, demersal, planktonic, encased in gelatinous material, in a nest, attached to vegetation or other substrate); overall size and shape; sculpturing or ornamentation of the chorion; width of the perivitelline space; amount, texture, and color of yolk; presence, number, and size of oil globules or droplets; and pigmentation on the body, yolk, or oil globules. These traits usually vary little within species or over the course of the incubation period. As ontogeny progresses, additional and more specific features appear that aid identification.

Most eggs are spherical with an apparently smooth chorion lacking ornamentation (Figure 1.4). Non-spherical eggs occur in diverse groups and may be ovoid (for example, anchovies, parrotfishes) to pear-shaped (gobies). The chorion of some species has obvious sculpturing but in others patterns can only be seen with the aid of scanning electron microscopy. Some species have threads, spikes, stalks, or other projections on the chorion that are used for attachment or to reduce their sinking rate. Eggs of bony fishes are generally between 0.7 and 7 mm in diameter. Oviparous sharks and rays have much larger eggs, each with a tough, leathery case and often with tendrils for attachment to the substrate. Within species of bony fishes, egg diameters vary by only a few tenths of a millimeter, some of which is due to maternal influences and some is due to environmental conditions, such as water hardness or salinity. Variations due to maternal investment are directly related to the size of the embryo and newly hatched larva, which can have important implications for survival as the larvae grow (see Chapters 3 and 4). Salinity and water hardness are inversely related to egg diameter by their influence on the width of the perivitelline space. The size of the embryo is not affected. Among species, the embryo (including yolk) may fill the chorion, leaving a negligible perivitelline space (Figure 1.4), or it may occupy as little as 50–60% of the egg diameter. Yolk commonly ranges from transparent and colorless to various degrees of translucent yellow, although greenish and reddish hues are known. Oil globules may be absent, numerous and small, or few and larger. They provide a final source of energy during the larval period, although they also provide a small degree of buoyancy. These characteristics of embryos do not vary much within a species.

Ontogeny begins with activation of the oocyte when sperm penetrates the chorion through the micropyle. Activation triggers delamination of the membrane that encloses the egg, dehiscence of cortical alveoli around the periphery of the yolk, and rapid uptake of water from the surrounding environment, resulting in the formation of the fluid-filled perivitelline space and a reduction in specific gravity of the egg (Figure 1.4). At the same

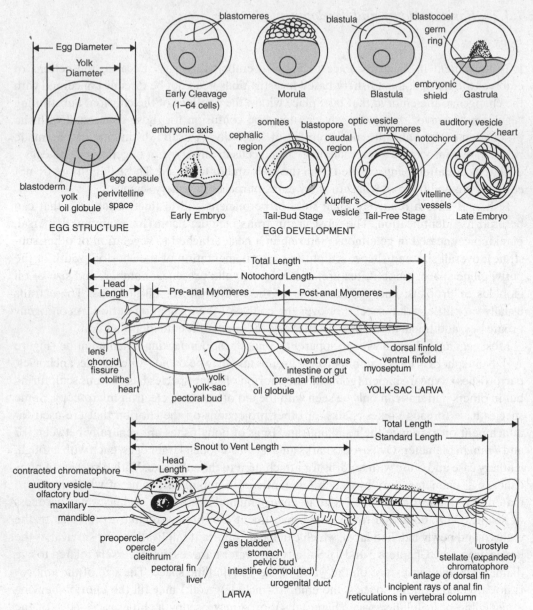

Figure 1.4   Anatomy of teleost embryos and larvae (from Hardy *et al.* 1978).

time, the micropyle closes to prevent polyspermy. These changes, which are sometimes referred to as "water-hardening," may take place even if fertilization fails to occur. Indeed, even some initial cleavages can occur in the absence of fertilization, making it difficult to distinguish fertilized and unfertilized eggs early in their development. The perivitelline space and hardened chorion provide a barrier for the embryo against some degree of mechanical and chemical damage.

Fertilization is the merging of nuclei from the egg and sperm and is required for full, normal development. The sequence of events follows the general scheme of vertebrate embryology: cleavage, morula, blastula, gastrula, neural crest, and so forth (Figure 1.4).

As the end of the embryonic period approaches, the fin folds, somites, heart, optic and auditory vesicles with otoliths, and chromatophores become apparent. The embryo increases in length to the point that it entirely surrounds the yolk, with its tail meeting or slightly overlapping its head.

Hatching results from the combined action of hatching enzymes (chorionase) that degrade the inner layer of the chorion and physical forces produced by the embryo's vigorous movements. Hatching enzymes are produced by clusters of unicellular hatching glands usually located on the head, anterior portion of the yolk sac, or pharyngeal region. Therefore, the embryonic or incubation period* ends only after these glands become functional and the appropriate trigger stimulates their secretions. Under normal conditions, hatching may be triggered when the embryo reaches a size where its metabolic demand for oxygen exceeds the rate at which oxygen diffuses across the chorion and perivitelline space. Indeed, hypoxia is one of the best-known triggers for hatching, whether it is brought about experimentally by reducing the oxygen concentration of the surrounding water, chemically inhibiting oxygen uptake of the embryo, or elevating the metabolic demand for oxygen. The reverse response has also been observed; hatching is delayed in hyperoxic conditions. Demersal eggs deposited in areas of reduced water flow may be at risk of suffocation, so accelerated hatching is adaptive because it frees the embryo to seek a more oxygen-rich environment. Interestingly, this response to reduced oxygen can be observed in species that are never likely to experience reduced oxygen levels, such as coral reef fishes that produce planktonic eggs.

Temperature, light, and pH also affect production and activity of hatching enzymes. In the few species studied experimentally, more hatching enzyme was produced, and it degraded the chorion faster, at warmer temperatures. Likewise, more eggs hatched during the light phase of a diel cycle than the dark phase, and hatching success was reduced under conditions of constant darkness. This response may require that a photoreceptive organ, such as the eye or pineal gland, be functional before hatching can occur unless the hatching gland cells respond directly to light or through a dermal light sense. In contrast, there are many species in which hatching naturally takes place in hours of darkness. This does not rule out the possibility that light, or lack of it, regulates production of hatching enzyme in those species as well. Finally, mechanical disturbance can accelerate hatching. This happens under laboratory conditions when a container of eggs at an advanced stage of development is transported. The cause of this response has not been examined, but it may be some degree of stress, like the other external agents that promote early hatching.

Delayed hatching is another adaptive response to environmental conditions. Several species deposit eggs out of the water and therefore out of reach of aquatic predators. For example, grunion (*Leuresthes tenuis*) and capelin (*Mallotus villosus*) deposit eggs at the high tide line so that the eggs spend much of the incubation period in moist sand or gravel. Other species, such as inanga (*Galaxias maculatus*) and the splashing tetra (*Copeina arnoldi*) attach eggs high on vegetation and other substrates that become exposed to air periodically. The chorion and perivitelline fluid protect the embryos from desiccation and collapse

---

* The term "hatching period" is sometimes incorrectly used to refer to the incubation period. The hatching period is the span of time over which a batch of eggs hatches, from the first to the last. The incubation period extends from fertilization until hatching.

under their own weight in air, but hatched larvae have no such protection. The high oxygen tension while the eggs are in air probably reduces the production of hatching enzyme which would destroy the protective chorion. Such eggs do, however, hatch soon after they are reimmersed. Extremes of pH, both high and low, can also increase the incubation period by inhibiting the activity of hatching enzymes. This can have important consequences in poorly buffered freshwater systems (see Chapter 7).

Environmental conditions also regulate the rate of embryonic development and, there-fore, the time it takes for hatching glands (and everything else) to appear. This has a more substantial impact on the incubation period than the minor adjustments caused by varia-tions in the production or activity of hatching enzymes. Temperature, the most potent envi-ronmental regulator of fish physiology, follows a strong, negative relationship with incubation period. The precise relationship is species-specific and only roughly linear within the natural range of temperatures. Therefore, the product of temperature and the duration of incubation, usually expressed in units of day-degrees or degree-days, is roughly constant within a species and is a convenient tool for first approximations of the time from fertilization to hatching. The temperature term in the day-degree formulation sometimes incorporates a non-zero baseline or "biological zero," which is extrapolated from empirical data and usually assumes values less than 0°C. Over a broad range of temperatures, the relationship between incubation period (I) and temperature (T) is not linear (Figure 1.5). Several mathematical models have been applied, including the simple power function

$$I = \alpha \cdot T^{\beta} \tag{1.1}$$

where α and β are empirical constants.

Other environmental variables, including salinity, oxygen, and light, may also influence development rate. This is usually concluded from observations of effects on the incubation period, but in most cases the size and state of development at hatching are also affected.

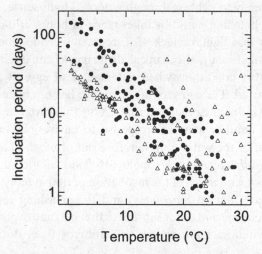

Figure 1.5   The effect of temperature on the duration of the incubation period of fish embryos. The rela-tionship appears linear because the ordinate is logarithmic. Eggs deposited in freshwater (filled symbols) have longer incubation periods for a given temperature because they are larger. Data points represent 16 freshwater and 22 marine species.

For example, low salinity, low ambient oxygen, or high light intensity can result in a shorter incubation period and smaller, less developed larvae at hatching. It is possible that some portion of this response is due to the environmental effects on the activity of hatching glands, as described above, rather than the rate of development. Because environmental conditions can cause embryos to hatch at different stages of development, the duration of the incubation period is not a good measure of development rate. Regardless of the mechanism, the magnitude of the effects of salinity, oxygen, and light on incubation period is small in comparison to the effect of temperature.

There is a strong, species-specific component to the duration of the incubation period. Even at a fixed temperature, incubation periods vary greatly among species, sometimes more than they do within a species over the entire range of tolerable temperatures. For example, at 14°C, the incubation period for brook trout (*Salvelinus fontinalis*) is 30 days, for plaice (*Pleuronectes platessa*) it is 10 days, and for striped mullet (*Mugil cephalus*) it is 3–4 days. This interspecific variation in incubation period is highly correlated with egg size. Larger eggs have longer embryonic periods than smaller eggs. In the above example, the trout has the largest egg (4 mm diameter), followed by the plaice (2 mm) and mullet (1 mm). Although a species' egg size does vary among populations and individuals, the amount of intraspecific variation does not seem to have a large impact on the duration of the incubation period.

Larvae

Variation in the stage of development at hatching among species and in response to environmental conditions within species highlights an important, but unappreciated point: hatching is not strictly a developmental event; it is a physiologically regulated process that marks a major transition in the ecology of a fish. A larva emerging from an egg is able, for the first time, to move in space, to interact with biotic and abiotic components of the environment, and to actively influence its survival in varied ways. The ontogenetic changes that begin in the embryo continue after hatching so that additional physiological and behavioral capabilities accrue for some time. In addition to ontogenetic changes, the larva's size increases rapidly, with concomitant improvements in performance. Together, ontogeny and growth allow a larva's ecological interactions to become increasingly sophisticated and diverse.

The very small size and apparently featureless form of most fish larvae (Figure 1.3) is probably responsible for them being largely ignored by fishery scientists for so long. Identification of species or even genera is a difficult challenge that relies on taxonomic characters that are very different from those used on adult fishes. Newly hatched larvae can range in morphological complexity from very poorly developed animals that look scarcely fish-like, to nearly complete miniatures of the adult ("precocial" larvae; for example, in direct development). Identification becomes easier as development progresses and more distinctive features appear.

Larvae that are poorly developed at hatching, so-called "altricial" larvae, are usually only a few millimeters long and transparent (for example, Figure 1.3f, l). The large yolk sac, which still contains oil globules that were present in the embryo (although some may have coalesced), protrudes below the anterior half of the body. The larva's head may be bent downward over the front of the yolk sac and attached to it. The eyes are formed but they lack pigment and therefore are not functional. Jaws are absent and there may not even be an oral opening. Gills may be absent, so the larva relies on cutaneous respiration facilitated by a circulatory plexus at one or more locations on the body surface, such as the yolk syncitium and bases of the future dorsal and anal fins. Bones, including vertebrae, are not ossified. Nearly all myomeres are formed; the few that may appear after hatching develop at the caudal end. A median fin fold fringes the body from a point on the mid-dorsal surface behind the head, around the urostyle, and anteriorly to the anus. A preanal fin fold extends forward from the anus. No fin rays are present and the caudal region of the median fin fold bears a superficial resemblance to the isocercal tail of some adult gadoid fishes because the urostyle is aligned with the rest of the notochord and the fin fold is symmetrical around it. Paired fins (pectoral and pelvic) are absent. Few, if any, chromatophores are present. Given this simple structure, identification is often based on general shape (length relative to depth), size at a given stage, position of the anus, numbers of myomeres, character of the yolk, and circumstantial data (location and date of collection, local fauna).

The preceding paragraph describes hatching at a very early stage of development. Many species progress somewhat beyond this point by the time they emerge from the egg and some will pass many of the subsequent stages as embryos. As ontogeny proceeds, the head separates from the yolk sac and aligns with the body axis. An opening forms at the mouth before the jaws appear. Gills begin to develop and paddle-shaped pectoral buds appear, although they lack fin rays. The yolk sac gradually diminishes but oil globules remain until all yolk has been absorbed when they, too, are metabolized. It is important to realize that larvae normally begin feeding before all yolk is absorbed. This is the period of mixed feeding. The stage of development when gas first appears in the swim bladder varies greatly among species. Once gas appears, the bladder remains inflated, although there may be a diel cycle in the volume of gas present.

Development of many ecologically important features, including fins, sense organs, skeleton, and external pigmentation, occurs gradually and over a large portion of the larval period. The caudal fin is often the first fin to show signs of differentiation when the urostyle, the final segment of the vertebral column, turns upward (shown in Figures 1.3r, s and 1.4). The term "flexion" is frequently used to refer to this stage of development. Soon afterward, hypural bones begin to form immediately below the upturned urostyle, and primordial fin rays appear as striations in the adjacent median fin fold. A slight constriction of the median fin fold develops in the region of the caudal peduncle as the outline of the median fins begins to take shape. Bony elements appear at the junction of the median fin fold and the body where the future dorsal and anal fins will be located, followed closely by their primordial fin rays. Rays are added to the fins sequentially, so that species that ultimately have a large number of rays in one or more fins take a long time to complete fin development. As the rays develop, sections of the median fin fold that lie between the final fins disappear and the margins of the median fins are refined. Meanwhile, the pectoral bud enlarges and begins to develop fin rays. Pelvic buds appear late as small outgrowths at the base of the preanal fin fold. This is a common sequence of fin development, but the sequence varies from species to species. Scales are among the last external features to be completed. They typically appear first on the sides of the caudal peduncle very late in the larval period and spread forward. Complete formation of fin rays is commonly used to designate the end of the larval period and the beginning of the juvenile period (complete metamorphosis), however, complete squamation (coverage of scales) may be a more accurate endpoint. The late appearance of scales means that they cannot be used to age larvae as they are for adults (see Chapter 2) and it may account for the non-zero intercept observed in plots of scale radius on body length for juvenile and adult fishes.

Because of their critical role in survival, it is surprising that the sensory systems have such a prolonged period of development after hatching. A great many species hatch with poorly developed eyes. The eyeball and lens are formed but the retina lacks pigment to intercept light. Before all yolk and oil is absorbed, melanin forms in the eyes, making them very conspicuous black spots. When eyes reach this stage prior to hatching, as in some freshwater and anadromous species such as salmonids, the embryos are sometimes referred to as "eyed eggs." With very few known exceptions, the first photoreceptors to appear in the retina are cones. Their number increases rapidly, resulting in an increase in the number of cones per degree of visual angle. There is some evidence that a few rods may be present soon after hatching, but when present, their number remains low for most of the larval period. Rods typically begin rapid proliferation toward the end of the larval period. Developmental changes in visual system function will be described more thoroughly in Section 1.4.1. The lateral line system begins with a few neuromasts on the body surface in embryos. The number of these free or superficial neuromasts gradually increases throughout the larval period in specific patterns on the head and trunk. Lateral line canals begin to form very late. Development of the chemosensory system is not as well known. In the few species that have been studied, chemoreceptor cells in the nares appear to develop quickly. By the time of settlement in some reef fishes, the density of chemoreceptor cells is as high as in adults of many species.

Ossification of the skeleton progresses slowly. The outlines of some bones, such as the pectoral girdle and jaws, become visible relatively soon after hatching. Differential staining of cartilage and bone shows that most skeletal elements are cartilaginous when they first appear and that they become ossified later. Bones of the pectoral girdle, jaws, and gill arches calcify relatively early. Vertebral centra ossify later and more gradually, sometimes finishing after the fish has assumed a juvenile external appearance.

Pigmentation on the head and body surface generally increases during development. Melanophores are most obvious because they persist after fixation and storage in common preservatives. Chromatophores bearing more colorful pigments have been described, but only occasionally. Melanophores often spread over upward facing surfaces, such as the dorsal surface of the cranium and trunk. One can speculate that this arrangement may provide some degree of countershading while maintaining transparency along a horizontal line of sight, or it may shield sensitive tissues from ultraviolet radiation. By the time a larva approaches metamorphosis, its size has increased greatly making its body more opaque. Transparency is no longer an option for avoiding predators. Pigment extends onto the lateral body surfaces in species-specific patterns.

Differences in structure are not the only features that set larvae apart from juveniles and adults. Larvae have a different shape, which is often quantified in terms of body proportions. Eyes of larvae are relatively large, as they are in young of many other vertebrates. The head and tail are generally smaller proportions of total length than they are in adults. Species that are deep bodied as adults often have slender larvae. To reach adult proportions from these initial sizes, various body parts must grow at different rates. This allometric growth is a distinctive feature of the larval period. Since juveniles have the adult appearance (shape), their growth is more or less isometric. Isometric growth maintains constant body proportions as all parts increase in size at the same rate. The transition to isometric growth is one criterion that has been used to signify completion of metamorphosis, although a careful analysis is required to determine when this transition occurs. Analyses of the pattern of relative growth (allometric and isometric) along the body of larvae show that there may be smooth growth gradients that are responsible for the orderly changes in shape that lead to metamorphosis.

1.3.3   Variety and patterns in early life history

Surveying the variety of egg and larval traits displayed by fishes brings to light some interesting and complicated relationships that tie early life-history traits to environmental characteristics, reproductive strategies, and population dynamics. These relationships are intertwined in such complex ways that cause and effect are difficult to distinguish. In addition, the trends are "noisy" and based on interspecific comparisons so that exceptions are not difficult to find. Nevertheless, these general relationships provide a useful context for understanding the diversity of early life-history traits and the ways in which early life stages are important to the practice of fishery science.

Egg size

The size and number of eggs produced, the duration of their incubation period, and their location in the environment vary tremendously among species and all of these traits have important consequences for survival. Egg size – principally weight and, secondarily, diameter (because of variations in the width of the perivitelline space) – is positively related to the duration of the incubation period and the size of the larva. The greater amount of yolk in large eggs provides more total energy for growth and ontogeny before the larva requires exogenous nutrition, resulting in larger larvae at hatching and at first feeding. This yields several vital benefits:

(1)   larvae emerge with a better repertoire of behavioral and physiological capabilities than less developed larvae from smaller eggs;
(2)   they are more resistant to starvation because weight-specific metabolic rates are lower and bodily energy stores are greater; and
(3)   the larval period is shorter.

One disadvantage of large eggs is that the embryonic period, during which they are unable to fend for themselves, is prolonged. Trade-offs involving egg size are discussed below (see *Patterns and strategies*) and the importance of larval stage duration to mortality and recruitment will be discussed more thoroughly in Chapters 3 and 4.

Planktonic vs. demersal

The broadest distinction between classes of fish eggs is whether they are planktonic or demersal, that is, where they occur in the environment. Demersal eggs that are not attached to a substrate sink to the bottom because their low water content, which is generally between 55% and 85% of their wet weight, increases their specific gravity. By comparison, planktonic eggs have a very high water content (>90% of their wet weight), which makes them buoyant, or nearly so. The difference in water content can be seen in the high transparency and sometimes wide perivitelline space of planktonic eggs. The yolk of demersal eggs is optically more dense and the chorion may be thicker than that of planktonic eggs. Demersal eggs have a larger diameter than planktonic eggs, sometimes in excess of 15 mm. Planktonic eggs range from about 0.5 to 5.5 mm in diameter, but usually are 0.7–2.0 mm. This difference in size between planktonic and demersal eggs is amplified when the comparison is based on dry weight because the larger, fluid-filled perivitelline space of planktonic eggs is neglected.

Species that produce small eggs do not have the advantage of large, well-developed larvae at hatching. Rather, they improve their chances of having offspring survive by producing more eggs. Fecundity generally is inversely related to egg size, although it is constrained by adult body size and influenced by other factors, including age and diet of the female. As a result, fecundity in species with planktonic eggs can be 10^5–10^6 eggs, whereas species with demersal eggs generally produce less than 10^5 eggs (one exception is large sturgeon, with a fecundity of almost 10^6 demersal eggs that average about 3.2–3.5 mm in diameter).

Species that produce planktonic eggs release them somewhere in the water column, rarely close to the substrate. This separates the eggs from benthic predators and facilitates their dispersion. Conversely, species with demersal eggs often spawn very close to a substrate so that their eggs are all in relatively close proximity to one another throughout the incubation period. Even species that scatter demersal eggs do so over a restricted area. Most demersal eggs receive some degree of protection from predators during the vulnerable embryonic period. This protection ranges from falling into interstices of a gravel bottom or being attached to vegetation to being deposited in a nest then abandoned or being guarded or carried by a parent. Planktonic eggs rely on transparency and perhaps dispersion itself for protection from predators.

Freshwater vs. marine

The distinction between demersal and planktonic eggs has an ecosystem component: planktonic eggs are almost exclusively found in the marine environment and freshwater fishes generally spawn demersal eggs. Accordingly, differences between freshwater and marine eggs and larvae (Table 1.1) parallel the differences between demersal and planktonic eggs. Marine fish eggs generally are smaller than freshwater eggs. Although the difference in modal diameter is not great, the distribution of egg diameters for freshwater species is strongly skewed toward larger eggs. Most freshwater species have eggs from 1.3 to 2.8 mm in diameter, and some are much larger. Eggs of most marine species are between 0.9 and 1.4 mm in diameter. The size difference is greater when expressed as dry weight because planktonic eggs, which have a higher water content, dominate in the marine environment. This difference in egg size accounts for the longer incubation periods of freshwater species at similar temperatures (Figure 1.5). Since embryos do not gain weight, the difference in egg weight can be seen in the 10-fold difference in the mean dry weight of larvae at hatching (Table 1.1). The larger freshwater larvae generally are at a more advanced state at

Table 1.1 Comparison of traits of freshwater and marine teleost fish eggs and larvae.

Early life-history trait	Freshwater	Marine
Egg diameter (median, mm)	1.70[a]	1.02[a]
Egg buoyancy	Mostly negative[a]	Mostly positive[a]
Incubation period (days)	10.9 ± 0.27[a]	7.0 ± 0.33[a]
Hatching length (median, mm)	5.40[a]	2.87[a]
Hatching dry weight (μg)	359.7 ± 72.8	37.6 ± 6.4
Metamorphic dry weight (mg)	9.3 ± 1.6	10.8 ± 0.95
Larval duration (days)	20.7 ± 1.1	36.1 ± 1.1
Metabolic rate ($\mu l\,O_2\,mg^{-1}h^{-1}$)	2.8 ± 0.4	5.9 ± 0.4
Ingestion rate ($\mu g\,\mu g^{-1}day^{-1}$)	0.46 ± 0.09	0.57 ± 0.07
Growth rate ($\mu g\,\mu g^{-1}\,day^{-1}$)	0.18 ± 0.02	0.20 ± 0.01
Growth efficiency	0.32 ± 0.03	0.29 ± 0.3
Instantaneous mortality (day^{-1})	0.16 ± 0.04	0.24 ± 0.02
Expected larval mortality (%)	94.7	99.9
Starvation risk	Lower	Higher
Larval mortality	Density-independent	Density-dependent
Stage for recruitment regulation	Juvenile period	Larval period
Recruitment variability	Lower	Higher

Values (means ± 1 SE, unless stated otherwise) are adjusted for differences in temperature, where appropriate. Data are from Houde (1994), except where indicated by superscript "a". Houde omitted salmonids, sturgeons, and ictalurid catfishes from calculations for freshwater fishes because of their unusually large, demersal eggs. Other calculations (with superscript "a") are based on data for 42 freshwater species from 21 families and 42 marine species from 34 families, derived from various published sources.

hatching. Metamorphosis is complete at about the same size in both environments. Therefore, the amount of growth and ontogeny that takes place between hatching and metamorphosis and the duration of the larval period are considerably greater for marine fishes. Calculations based on a sample of 22 species in 15 genera from six families show that marine larvae increase in dry weight between 130- and 660-fold during the larval period, whereas freshwater larvae (16 species, eight genera, six families) increase only by a factor of 30–80 (Figure 1.2). The average duration of the larval period in marine fishes (calculated from the increase in dry weight and temperature-adjusted growth rates) is 36 days compared to 21 days for freshwater fishes. These developmental differences are complemented by differences in the relative importance of density-dependent and density-independent mechanisms of mortality during the larval period, the time of life when recruitment is most strongly regulated and the potential for recruitment variability is greatest (Table 1.1, Chapter 3).

Parental care vs. independent offspring

Parental care of early life stages has obvious effects on the relationships outlined so far for demersal eggs. Fecundity is reduced and egg size is usually larger in species that provide a moderate or high degree of care for their eggs and larvae. The magnitude of these differences is directly related to the degree or duration of parental care. The majority of caregiving species produce large, yolky eggs that have all of the characteristics described for such eggs: a long incubation period, large and precocial larvae at hatching, and a brief larval period. Some nest guarders, however, have small, highly transparent eggs that are attached to a substrate, and the eggs may even appear to be buoyant if not for their hold-fast. Although they are demersal, these are essentially attached planktonic eggs. They have a shorter incubation period and hatch as small, altricial larvae. Their parental care ceases at hatching and the larvae disperse from the nest into a long, planktonic phase before they settle and undergo metamorphosis. Examples are found among the damselfishes, gobies, blennies, and darters.

Patterns and strategies

One of the early attempts to understand this diversity in early life-history traits was formu-lated in the 1940s by Sergei Kryzhanovskii (also Kryzhanovsky) at the Research Institute of Morphogenesis at Moscow University. Kryzhanovskii recognized the importance of the embryonic and larval periods to adult populations and believed that two factors, predation mortality and oxygen availability, are of overriding importance to survival. He reasoned that the habitat in which eggs are released defines the respiratory conditions and predation potential for the early life stages and explains the diversity of eggs, larvae, and reproductive styles found in fishes. With extensive knowledge of the reproductive habits of many species, especially those in freshwaters, and the morphology and physiology of their eggs and larvae, he devised an ecological classification based on the spawning habitat and the degree of parental care provided. This system was expanded and renamed as a classification of repro-ductive guilds by Eugene Balon in 1975. Such ecological classifications are particularly use-ful for understanding the habitat requirements of early life stages, a matter we will discuss more thoroughly in Chapter 7.

Many early life-history traits vary in parallel with well known relationships among repro-ductive and demographic parameters, such as age at maturity, spawning frequency, fecun-dity, and survivorship, which have been interpreted in the context of evolutionary ecology. The traditional paradigm of *r*- and *K*-selection defines a two-sided continuum of strategies. On one side are species that are able to discover and quickly exploit resources that are unpredictable or ephemeral in space or time. Ideally, these so-called *r*-strategists are char-acterized by effective dispersal and a high rate of population increase (*r*). Most of the com-mercially important species of the oceans, with their high fecundities, small planktonic eggs, and altricial larvae, are nearer this end of the continuum. The food supply for their offspring is scattered in patches and the large number of propagules improves the chances of at least some offspring locating these patches. At the other extreme are *K*-strategists, species that are adept at competing for spatially and temporally stable resources. *K*-strategists typically have small batches of large eggs, precocial young, and are often pro-vided parental care. Some fish species that tend toward this strategy are important to sport or subsistence fisheries and aquaculture, such as sharks, some catfishes (for example, *Clarias gariepinus, Ictalurus punctatus*) and tilapias (species of *Oreochromis, Sarotherodon, Tilapia*).

Analyses by Kirk Winemiller and his colleagues have expanded the traditional paradigm in a way that distinguishes three life-history strategies, rather than two. These strategies are defined by the relative magnitudes of fecundity, survivorship from fertilization until first reproduction, and age at maturity (Figure 1.6). The periodic strategy maximizes batch

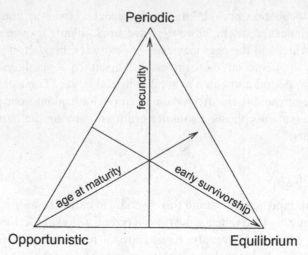

Figure 1.6 Adaptive surface of fish life-history strategies based on trade-offs among reproductive and demographic traits (after Winemiller & Rose 1992). The opportunistic strategy is typified by early maturity, which results in a shorter generation time but a smaller body size. This constrains fecundity and egg size, and reduces adult survivorship. The equilibrium strategy is marked by the high early survivorship achieved through large eggs and parental care. Larger eggs produce stronger, precocial larvae with a shorter larval stage duration. The high fecundity of the periodic strategy results in small eggs, poorly developed larvae, and a long larval stage, which reduces early survivorship.

fecundity but with costs in terms of delayed maturation and low and variable early survivorship. The reduced survivorship is a result of smaller, more altricial eggs and larvae that receive no parental care. This tactic functions well when resources, such as larval food supply, vary regularly (perhaps seasonally) or are distributed in large scale patches. The abundance of eggs or larvae is a poor predictor of future year-class strength for these species, and early life mortality plays a pivotal role in population dynamics. This is the dominant strategy among commercial fish stocks as well as some large tropical freshwater species, such as pacu (*Colossoma* sp.) and tigerfish (*Hydrocynus* sp.). The opportunistic strategy maximizes the rate of population increase by emphasizing early maturity and frequent spawning (that is, high reproductive effort). As a result of the smaller body size at maturity, fecundity is low, eggs are small (reflecting low early survivorship), and adult survivorship is also low. This strategy is favored when resources vary unpredictably on small temporal or spatial scales. These species, exemplified by anchovies, silversides, and a variety of small tropical freshwater species, can repopulate quickly after major disturbance to the environment. The third strategy, the equilibrium strategy, is much like the *K*-strategy of the traditional paradigm. It optimizes early survivorship through the production of a small number of large eggs, precocial offspring, and parental care. These species, including viviparous fishes and mouthbrooding cichlids, thrive in the presence of suitable habitat and stable environmental conditions.

1.3.4   *Developmental progress and intervals*

Early stages of several species appear so different from their familiar adult forms that upon their discovery they were assigned to their own genus. Some of these names are still used

Table 1.2 Modern intervals (roman text) for early stages of fishes and the criteria (italic text) used to separate them.

Balon (1975b)	Ahlstrom *et al.* (1976)	Snyder (1976)	Hardy (1978)
Eleutheroembryo	Yolk-sac larva	Protolarva	Yolk-sac larva
Yolk absorption	*Yolk absorption*	*First median fin ray*	*Yolk absorption*
Protopterygiolarva	Preflexion larva	Mesolarva	Larva
First median fin rays	*Urostyle flexion*	*Adult median fin ray complement and pelvic buds or fins*	*Adult fin ray complement*
Pterygiolarva	Flexion larva	Metalarva	Prejuvenile
	Adult caudal fin ray complement		
	Postflexion larva		

informally when referring to these distinctive forms (acronurus, ammocoetes, kasidoron, leptocephalus, querimana, tholichthys, vexillifer). Perhaps this is why there have been many attempts to divide the developmental period into discrete, named intervals. Or, maybe it is an extension of the practice of classical embryology to identify "normal stages." Regardless of its origins, the topic of terminologies for intervals of fish development continues to be controversial and there exists no single, widely accepted classification system. The principal value of these classification systems is to divide the range of ontogenetic variation into smaller, more tractable pieces for constructing taxonomic keys and other tools for identifying specimens. It is common to use one set of characters for distinguishing larvae during an early interval of development, then switch to a different set of characters to identify the same species later. One of the first widely used terminologies was developed for salmonids and included: egg, alevin, fry, fingerling, parr, smolt, adult, and kelt. These terms are still in use but are only applied to trouts and salmons. There are at least four modern terminologies for intervals of development (Table 1.2). Their popularity varies with geographic region of the investigator and ecological realm of the fish. A particularly ambiguous term that should be avoided at all costs is "postlarva."

Outlining the fish life cycle at the start of this chapter (Figure 1.1) automatically created intervals: embryo, larva, juvenile, adult, and senescent. These are based on changes in the dominant physiological processes that define a fish's changing capabilities and requirements. The criteria dividing these categories – hatching, metamorphosis, maturation, and senescence – are significant life-history transitions. To be of importance to fishery science – as opposed to, say, embryology or developmental biology – further subdivision must reflect natural intervals in which ecological interactions take a new course. Most terminologies devised for fish development are based on morphological changes that were presumed to reflect changes in a fish's ecological role. The most common developmental milestones used in developmental classifications are hatching, yolk absorption, and fin development. There is no doubt that hatching is a major transition. It frees the larva to make effective behavioral responses to environmental stimuli. Yolk absorption, or more accurately the transition from endogenous to exogenous nutrition, is also a critical developmental and ecological transition for a fish larva. Thus, the period between hatching and yolk absorption is ecologically distinct and the term "yolk-sac larva" is often applied (Table 1.2). Terminologies born out of classical embryology use yolk absorption as the criterion to separate the embryonic and larval periods, so that the term "embryo" applies long after hatching. This can cause some confusion for those unfamiliar with the differences among terminologies.

The fin-development criteria used in various systems include upward flexion of the urostyle, the first appearance of caudal fin rays, completion of median fins, and other traits. These milestones were selected with the expectation that they mark a new level of swimming performance. While this seems reasonable, subsequent studies of swimming performance in larvae have not shown distinct changes when these structures appear. This does not mean that the intervals defined by fin development are not useful, only that their contribution to understanding the ecological processes relevant to fishery science is limited. It can be argued that subdividing the larval period beyond recognizing yolk absorption is not necessary for a useful understanding of the role of eggs and larvae in population-level processes. The fact that developmental intervals more refined than egg, larva, and juvenile are rarely mentioned in subsequent chapters in this book supports this contention.

There are two schools of thought on the fundamental model for the time course of fish development. One view is that development is gradual, that changes accumulate continually, but not necessarily at a constant rate. The alternative view is that development is accomplished by periods of dramatic change in body structure and physiology, followed by longer intervals of little or no change. While the latter idea, proposed as "saltatory ontogeny" by Eugene Balon, is interesting, direct evidence is equivocal. Since fishery science is an especially quantitative endeavor, incorporating developmental trends into population models is easier if a larva's ontogenetic progress is expressed numerically on a continuous scale of measurement, rather than representing it as discrete, qualitative steps. Models that imply gradual development seem to produce reasonable results for purposes of fishery science.

Experimental studies and observations of field-caught larvae show that the appearance of specific structures and levels of performance during the development of a species are more closely tied to the size of a larva than its age or other convenient measures of "biological time." The importance of this observation cannot be overstated. Experimentalists, working with larval fishes under controlled environmental conditions, often use age as a means of identifying a larva's developmental progress. Comparisons based on age are only valid under identical environmental conditions. Size, on the other hand, integrates a larva's environmental experience, whether constant or fluctuating, and it is much easier to measure on field-caught larvae than age. Thus, size is a useful measure of developmental state, but what about the progress of development? Development follows a logarithmic progression with respect to body size. The rate of developmental change is most rapid early and continually slows toward metamorphosis. For example, a 2-mm difference in length of larvae soon after hatching represents a greater change in form and function than a 2-mm difference in length near metamorphosis. Therefore, a logarithmic scale of length appears to be an appropriate, if not convenient, scale for development within a species. Because species differ in the size of their larvae at comparable stages of development, it is necessary to adjust the logarithmic scale accordingly when it is desirable to compare species. This can be done simply by scaling a species' developmental period to one logarithmic cycle

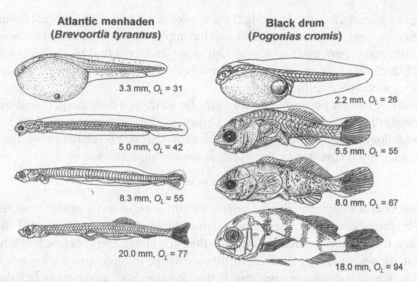

Figure 1.7   Selected larval stages of a clupeid and a sciaenid demonstrating a two-fold difference in length at comparable stages of development. The ontogenetic index, O_L, adjusts for these differences (from Fuiman & Higgs 1997, drawings from Joseph *et al.* 1964 and Mansueti & Hardy 1967, reproduced with permission of Kluwer Academic Publishers, American Society of Ichthyologists and Herpetologists, and University of Maryland).

so that metamorphosis always assumes a value of 1.0 or 100%. Mathematically, this is accomplished by selecting a logarithm base equal to the species' size at metamorphosis, or equivalently:

$$O_L = \frac{\log L}{\log L_{\text{juv}}} \tag{1.2}$$

where O_L is the ontogenetic index, a numerical representation of the ontogenetic state of a larva of length L, and L_{juv} is the length at which that species completes metamorphosis. This results in a continuous measurement scale that assigns similar numerical values to individuals of different species but similar developmental state. For example, Atlantic menhaden (*Brevoortia tyrannus*) are roughly twice as long as black drum (*Pogonias cromis*) at comparable stages of development. This includes metamorphosis (complete squamation) at lengths of approximately 45 and 22 mm, respectively. Using the ontogenetic index, their similar poorly developed state at hatching takes on values of 31% and 26%, respectively. Their median fins are complete at about 79% and 71%, respectively (Figure 1.7). This quantitative scale for ontogeny is new and has proven useful when it has been applied.

1.4   Development and performance

Survival of individuals at any stage of life requires adequate levels of performance against a variety of ecological challenges, such as obtaining food, evading predators, and locating and remaining in a suitable habitat. Meeting these challenges is especially difficult during early life because eggs and larvae are so small and incompletely developed, and body size and structure determine performance levels. Ontogeny and growth, the defining processes during early life, make this an especially dynamic period during which new capabilities arise and performance levels improve rapidly.

The effects of size on performance are more familiar, well documented, and quantitatively modeled than the effects of ontogeny. Diverse studies of animal physiology and functional morphology in many groups of organisms show that size is related to most measures of performance in a common way, according to the scaling relationship

$$\text{Performance} = \alpha \cdot \text{Size}^\beta \tag{1.3}$$

a power function in which size may be length, weight, area, or some other measure, and α and β are empirical constants. The scaling relationship depicts different rates of change in performance for smaller, younger individuals than for larger and older ones. Only in the unusual situation where $b = 1$ does performance change in direct proportion to a change in size. One familiar scaling relationship is that for standard metabolism in fishes. The exponent of body weight (b) for juvenile and adult carp (*Cyprinus carpio*, 25 g to 3.5 kg wet weight) is 0.85. This describes adult carp as having a higher standard metabolic rate than juveniles but a lower weight-specific metabolic rate. Generally, exponents of a scaling relationship vary little among species during the juvenile and adult periods. The consistency of exponent values underscores how fundamental and general the effects of size are for a given measure of performance.

Strictly speaking, scaling relationships apply to individuals that differ only in size and not structure or shape. Embryos and larvae are characterized by intense structural and shape change as a result of ontogeny and allometric growth, so we cannot expect early life stages to fit even the strongest scaling relationships derived for older fishes. Returning to the carp example, the weight exponent (b) for larvae (Figure 1.3f, g) ranging from 2 mg to 3.8 g is 0.98, much higher than the value of 0.85 for larger carp. The exponent of approximately 1.0 indicates that the weight-specific metabolic rate of larvae is independent of body size (weight), an observation that has been confirmed for several other species. The empirical relationship between standard metabolic rate and weight for larvae is not really a scaling relationship. Developmental relationship is a better term because the data include the combined effects of scaling (changes in size), ontogenetic changes in structure, and allometric growth. The differences between scaling and developmental relationships mean that we cannot extrapolate our knowledge of juveniles and adults to earlier stages or vice versa. For now, developmental relationships must be derived empirically. Despite the more complex processes that produce them, developmental relationships are just as useful as scaling relationships for understanding the mechanisms responsible for changes in performance and for modeling those changes. The following examples demonstrate the nature and magnitude of effects of ontogeny and growth on selected, ecologically important measures.

1.4.1   Sensory systems

Visual acuity is a measure of performance that determines the distance at which a visually feeding fish can see its prey, all other conditions being equal. Acuity is defined by the

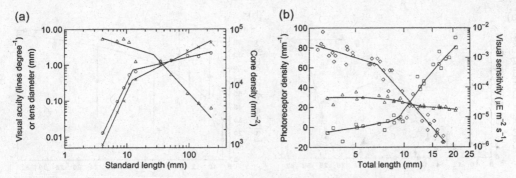

Figure 1.8 The effect of changes in eye morphology on visual performance. (a) Lens diameter (○) increases and cone cell areal density (△) decreases from the beginning of the larval period, resulting in rapidly increasing visual acuity (×) in the roach (*Rutilus rutilus*). (b) Cone cell lineal density (△) decreases slightly but rod cell lineal density (□) increases rapidly late in the larval period of red drum (*Sciaenops ocellatus*), to improve visual sensitivity (◇). Redrawn from data of Zaunreiter *et al.* 1991 and Fuiman & Delbos 1998.

diameter of the lens and the density of cone photoreceptors in the retina and can be calculated from measurements made on histological sections. Acuity of the cyprinid fish known as the roach (*Rutilus rutilus*) improves rapidly in the earlier stages of retinal development but the rate of improvement declines in later stages (Figure 1.8a). The driving force for the improvement is the rapid growth of the eye, and in particular the lens, which is more positively allometric (steeper slope) in larvae than in juveniles. At the same time and in opposition, the areal density of cones decreases because the expanding retinal area outpaces the differentiation of new cones, although the difference is less for larvae because cone differentiation is quicker than in juveniles. Such changes in acuity may constrain a larva's ability to find food.

Visual sensitivity, the minimum light intensity necessary for a particular visual task, is another measure of visual performance that improves in two phases. In red drum (*Sciaenops ocellatus*, similar in appearance to Figure 1.3r), improvement during the first phase is slower and not until rod photoreceptors begin to differentiate does the second phase begin, which brings a much more rapid improvement in sensitivity (Figure 1.8b). Ontogenetic improvements in sensitivity, such as this, may be an important determinant of vertical distribution patterns of larvae (see Section 1.5.3).

Successfully evading a predator's attack requires a larva to detect the predator and to initiate a well-timed response of suitable magnitude and in the appropriate direction. The gradual development of sensory systems after hatching has a direct impact on the ability of a larva to perceive the threat, to initiate its evasive response, and to do so at the appropriate moment. There is a general trend for increasing responsiveness to attacks as larvae develop. For most species studied to date, responsiveness increases in proportion to developmental progress. That is, there is a linear relationship between responsiveness and the logarithm of body size. This improvement coincides with increases in the peripheral end organs of the visual and lateral line systems, although reliance on a particular sensory system varies from species to species. Atlantic herring (*Clupea harengus*, Figure 1.3m) larvae provide an excellent example of a developmental bottleneck for sensory systems.

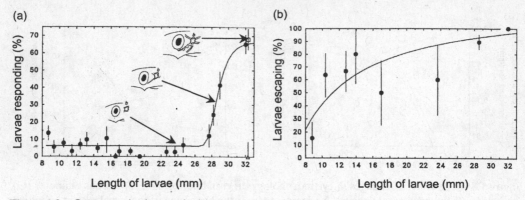

Figure 1.9   Ontogenetic changes in (a) responsiveness and (b) effectiveness of responses of Atlantic herring larvae to attacks by a predatory fish. Drawings of the developmental stage of the auditory bulla (*b*, posterior to the eye) and lateral line canal system show the role of these structures in initiating escape responses. Data from Fuiman (1989), drawings from Blaxter & Batty (1985).

This species and its close relatives rely on an air sac adjacent to the ear, called the auditory bulla, and its connection to the lateral line canal system on the head to provide them with an acute sense of underwater sounds and water movements. The bulla does not become functional until it fills with gas when herring larvae reach a length of about 25 mm. In laboratory experiments, only 6% of herring larvae less than 24 mm long (no gas in bulla) responded to attacks by larger herring. In contrast, 65% of 32-mm larvae (complete canals) responded. In addition, larger larvae initiated responses earlier in the attack sequence than younger stages, indicating improving sensitivity to predatory attacks. This sensory bottleneck is especially curious because the ability of herring larvae to escape successfully improves substantially during this period (Figure 1.9). Behavioral information such as this provides a more thorough understanding of the mechanisms of predation mortality, which is arguably the most important source of mortality during early life (Chapter 3).

1.4.2   Swimming performance

Changing size and structure have strong and varied influences on swimming performance. The gradual appearance of fin rays and reduction in the median fin fold are obvious changes to a làrva's swimming apparatus. In many species, the caudal fin displays an exceptionally high degree of positive allometry during the larval period which results in a rapidly increasing propulsive surface area. Highly allometric growth also occurs in the region of the caudal peduncle, signaling disproportionate growth of locomotor muscle mass. The muscles themselves undergo important changes, starting with a single dominant fiber type and only later developing the two-gear system of red and white muscle. These ontogenetic changes, together with actual scaling effects, result in gradually improving swimming performance as larvae develop.

The routine swimming speeds of larval fishes, as well as their burst speeds, increase approximately linearly with body length among species (Figures 1.10 and 11.31). Generally, routine speeds increase by a factor of four to five or more from a length of 5 mm to a length of 20 mm. Therefore, length-specific speeds of larvae are about one to three body lengths

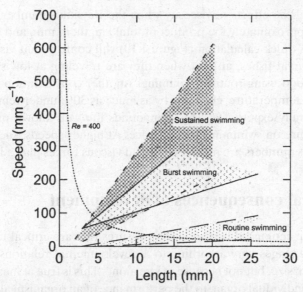

Figure 1.10 Ranges of swimming speeds of fish larvae relative to their length in different types of swimming. The dotted line represents the range of conditions for which the Reynolds number (*Re*) is equal to 400. Water viscosity and its natural variations can have a measurable effect on swimming when *Re* is less than 300–450. Data for routine and burst swimming are from freshwater and marine species (from summaries by Fuiman & Webb [1988], Miller *et al.* [1988] and Williams *et al.* [1996]). Data for sustained swimming are from coral reef fishes (Stobutzki & Bellwood [1994] and Fisher *et al.* [2000]).

per second ($BL\,s^{-1}$). Burst speeds increase by a factor of two to three over the same range. On a length-specific basis, burst speeds are highest for small larvae, commonly 15–20 $BL\,s^{-1}$ at 5 mm, with individual values recorded as high as 66 $BL\,s^{-1}$. This far surpasses the traditional rule of thumb of 10 $BL\,s^{-1}$ for maximum speed that is accepted for adult fishes. Recent work on young reef fishes shows that larvae have surprisingly high endurance at relatively high sustained swimming speeds. Just prior to settlement, some species are capable of sustained speeds of 20–60 $BL\,s^{-1}$. All of these general trends contain a good deal of variation, part of which can be attributed to differences in morphology at a given size. As swimming performance improves during development, larvae are able to forage over larger areas and they become better able to escape from predators, as well as plankton nets.

A less familiar, but very important, influence on swimming in larvae is the interplay between body size, swimming speed, and hydrodynamics. Small, slowly moving larvae experience hydrodynamic conditions in which water viscosity has a major impact on their motion, and inertia has little effect. For larger or faster fishes, inertial forces dominate and viscous effects are minor. The small size and large amount of growth that larvae exhibit take them from a viscosity-dominated regime to an inertia-dominated one in a short span of time. The effects of viscosity on small fish larvae can be seen in their routine swimming movements. Very small larvae come to rest immediately after they stop beating their tail. So, in order to travel appreciable distances, they swim continuously and vigorously. Larger larvae, juveniles, and adults, are able to save energy by incorporating glides into their swimming. Although definitive experiments have not been done, this hydrodynamic transition must have important consequences for a larva's energy budget. Experiments suggest that larvae escape these effects of viscosity when the Reynolds number (a hydrodynamic parameter that is approximately the product of total length, in mm, and speed, in $mm\,s^{-1}$) exceeds 300–450. A quick calculation (Figure 1.10) will confirm that viscosity is a unique consideration for larval fishes, at least when they are traveling at low speeds, as they do when cruising for food using routine swimming. Another consideration is that viscosity is inversely related to temperature, changing by as much as 30% under temperature changes a larva might naturally experience. This compounds the well known physiological (Q_{10}) effects of temperature on swimming performance. At higher speeds, such as when fleeing predators, Reynolds numbers are much higher and viscous forces play a minor role.

1.5   Ecological consequences of development

A great many of the ecological performance measures that are critical to the survival and growth of early life stages vary according to a developmental relationship. That is, performance varies with size, but not because of size alone. This is true at many levels, from the performance of an individual organ, to the performance of an organism alone or in its interactions with other organisms. The following examples illustrate the ways in which ontogeny and growth (development) can influence the abilities of a fish in various ecological contexts.

1.5.1   Food and feeding

The great importance to fishery science of a larva's transition from endogenous nutrition to foraging in the environment was recognized a century ago by fishery scientists in Norway, among whom Johan Hjort is generally credited because of his pioneering publication of 1914, *Fluctuations in the Great Fisheries of Northern Europe*. One of Hjort's ideas, later called the Critical Period Hypothesis, recognized that larvae can survive for only a brief period without food after their supply of yolk and oil globules is gone, and that natural abundances of their food vary greatly in time and space. Therefore, starvation could be an important source of mortality during the larval period and this could translate into large variations in year-class strength. Hjort's ideas have given rise to several important hypotheses, which will be outlined in Chapter 4.

Starvation is a more serious risk for larval fishes than juveniles or adults because of their high weight-specific metabolic energy demand (see Section 1.4). The length of time a fish can survive without food, its starvation resistance, is governed by the rate of energy expenditure (metabolic rate) and the amount of energy stored in the tissues. Recall that the metabolic rate of larvae is directly proportional to their weight ($W^{1.0}$), and that the weight exponent for juveniles and adults is lower (for example, $W^{0.85}$). Assuming relative constancy of body composition, a fish's energy stores are directly proportional to its weight. Therefore, larvae have smaller reserves on which to draw when food is scarce but the ability to withstand starvation improves with growth (Figure 1.11). When food is withheld from larvae, they reach a point of no return at which starvation is irreversible. The point of no return varies among species and, as expected, is related to body size and temperature. At temperatures of 5–10°C, larvae may reach irreversible starvation after 20–35 days, but at 25–30°C it may only take 4 or 5 days.

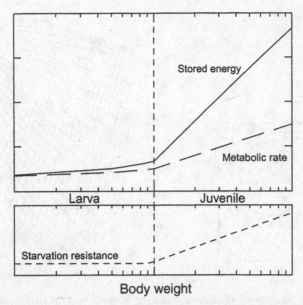

Figure 1.11   Changes in stored energy and metabolic rate with growth during the larval and juvenile periods. Stored energy is directly proportional to body weight through both periods (although the logarithmic abscissa distorts this trend). Metabolic rate is proportional to body weight for larvae (that is, weight-specific metabolic rate is constant), whereas weight-specific metabolic rate decreases in the juvenile period. The diverging trends result in increasing starvation resistance, the ratio of stored energy to metabolic rate in units of time.

The risk of starvation diminishes through the course of development not only because of lower weight-specific metabolism and increasing energy reserves, but also because of improvements in sensory and swimming performance. Older larvae are better equipped to locate more distant food supplies. Growth also allows larvae to select larger, more energy-rich prey while retaining the ability to feed on the smaller, more abundant prey. This increasing diet breadth may be critical to their ability to continue growing. The highly allometric growth of fish larvae leads to a particularly rapid increase in gape size, so much so that some species can cannibalize members of their own cohort that are 70–90% of their own length. By comparison, the maximum size of prey for piscivorous adults is generally 30–50% of their length.

1.5.2   *Predator detection and escape*

As mentioned above, burst swimming performance improves with development and this translates into increasingly effective escape responses (Figure 1.9b). Many investigations have measured the predator's capture success and arrived at a common negative trend relating capture success to larva size. These results on different species and sizes of larval prey attacked by various juvenile or adult predators can be summarized by a single curve according to which capture success decreases rapidly as prey become larger relative to the predator, implying that the sole or primary cause of the change in capture success is a matter of scaling (Figure 4.7).

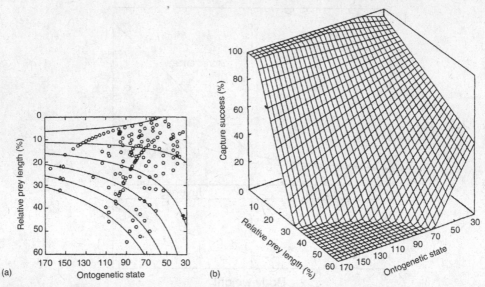

Figure 1.12 Capture success of predators attacking various species of fish larvae as a function of ontogenetic state and relative prey size (larva length divided by predator length) (from Fuiman 1994).

This final example of a developmental relationship (between capture success and relative prey size) assumes that developmental differences among species at a common size are trivial, a point that is often forgotten when the empirical results follow such obvious trends. Recall that body size is a suitable measure of developmental state within a species, but also that species differ in their size at a given developmental stage. An 8-mm Atlantic menhaden is still early in its developmental program, while a black drum at the same size is quite well developed (Figure 1.7). Therefore, in the capture-success relationship (Figure 4.7), prey:predator length accounts for scaling effects but the ontogenetic effects are confounded with size in this multi-species data set. The fact that the data appear to follow the curve so closely suggests that scaling is the principal determinant of capture success, but it may also reflect a limited or biased selection of larva types in the data set. Since we know that ontogenetic changes have important effects on aspects of performance that ultimately determine capture success, such as sensory and locomotor performance, the general model for capture success can be improved by adding a variable that describes ontogenetic state, such as O_L. Doing so results in a biologically more complete model that yields better insight into the mechanisms of change in capture success for larvae and improved quantitative predictions (Figure 1.12). Nevertheless, body size alone does tell us a great deal about how a larva will be affected by many of the mechanisms that influence its survival.

1.5.3   Habitat shifts vs. sensory development

Physical separation of life-history stages is a common strategy that is usually mentioned in regard to spawning migrations, which place the eggs and larvae in a different habitat from the adults. Diadromous fishes provide an extreme example. Benefits generally attributed to this strategy include: a better match with life-stage-specific habitat requirements, reduced competition among age classes, and reduced cannibalism. Migrations or habitat shifts are also common within the early life of many species. As we will see in Chapter 7, these movements may serve the same ecological functions as in other periods of life. Their timing may be determined by one or more developmental events that provide the skills necessary to locate or succeed in a new habitat.

Atlantic herring hatch from demersal eggs and spend their larval period in offshore waters. They move into coastal waters around the time of metamorphosis. Their prolonged, low responsiveness to attacks by larger fishes (Section 1.4.1) poses interesting questions, especially in light of the fact that their evasive responses improve early, long before responsiveness increases (Figure 1.9). Why does responsiveness remain so low for so long, given that surviving an attack depends on responding? Why has there not been natural selection for earlier functionality of the auditory bulla or another sensory system to initiate a response? One possibility is that herring larvae may not experience much predation pressure from other fishes. The abundance of planktivorous fishes offshore is probably considerably lower than it is in coastal habitats. Thus, the timing of the movement of herring into coastal waters coincides with, and may be constrained by, the development of a specific sensory system that is needed for their survival in that environment. This does not mean that predation is not an important source of mortality for herring larvae. Invertebrate predators, principally medusae, are abundant and voracious consumers of fish larvae offshore. Herring larvae do not require functional auditory bullae to respond to these predators; they respond by touch or by sensing nearby water movements.

Various ontogenetic events have been implicated in habitat shifts of several other species. Like its relative, herring, Atlantic menhaden larvae (Figure 1.7) also move inshore after the auditory bullae and other sense organs reach a high level of functionality. Movements in the vertical direction have also been associated with ontogenetic changes. Young rockfish (*Sebastes diploproa*) and red sea bream (*Pagrus major*) larvae live in shallow seas until differentiation of the retina improves their visual sensitivity, when they move to depths of 500 and 150 m, respectively. At a finer level, vertical distribution patterns may be a result of ontogenetic constraints on vision and feeding performance in larval fishes, as evidenced by concordance between light sensitivity and field observations on the vertical distributions of reef fish larvae. Vertical distribution is also influenced by swim bladder inflation and resulting changes in buoyancy. Transition from pelagic larvae to epibenthic juveniles coincides with, or is preceded by, major changes in the lateral line system in flatfishes and dentition in a grouper.

This leads us to the complex matter of settlement in coral reef fishes, which involves both long distance horizontal movement and transition from a pelagic to a bottom-associated lifestyle. Ninety-five percent of the families of coral reef fishes have pelagic larvae and the duration of this phase can range from about 1 week to more than 3 months. Eggs and larvae are dispersed from the reef occupied by their parents, often far from any reefs, and weeks later must find a reef on which to settle and complete metamorphosis. Settlement takes place over a very narrow range of sizes for a given species, indicating that competency to settle is determined by their developmental state. Intriguingly, at least some species are able to delay settlement and metamorphosis, perhaps when there is no suitable habitat nearby. During this delay their growth rate slows significantly, otherwise metamorphosis would probably begin and they would lose their pelagic body form. It is not clear which developmental events are critical for settlement. Those related to locating a reef over a distance of kilometers to tens of kilometers (sensory performance) and swimming to it seem paramount. Recent investigations have demonstrated the extraordinary swimming capabilities of pre-settlement-stage reef fishes. They are clearly not at the mercy of prevailing currents and we must hesitate applying the term ichthyoplankton. In addition, they seem to be able to orient their swimming in a consistent direction. Research attention is now turning to the cues that guide them to a reef. The most promising candidates appear to be chemical and acoustical signals. Settlement, or recruitment to a reef, is considered an important regulator of local species abundance and community structure under the Recruitment Limitation Hypothesis (Chapter 4).

Despite the relatively strong swimming performance of later larvae, few larval stages are able to swim against the strong flows of tidal currents. This challenge is faced by many species that enter estuaries from offshore and coastal spawning areas, as well as those spawned within the estuary but advected out of the system as eggs or young larvae, only to return later. Some species that enter estuaries from the sea appear to use selective tidal stream transport. In a tidal environment, where currents reverse once or twice each day, the velocity of the current is always very low in the boundary layer immediately adjacent to the bottom. Larvae can progress into the estuary by entering the water column on flood tide then sheltering in the boundary layer when the flow reverses. This requires a sense of the direction of flow, landward or seaward, or at least a circatidal rhythm in activity. Early observations of tidal stream transport were made on young eels (*Anguilla anguilla*) entering a Dutch estuary. The elvers were distributed throughout the water column on flood tides and near or on the bottom during ebb tides. Similar observations have been made since that time on other species, including flatfishes and sciaenids. Laboratory experiments indicate that the change in vertical position and behavior may involve detection of one or more stimuli, possibly olfactory cues in dissolved or particulate organic matter, that are associated with seaward flow of estuarine water. This may prevent younger larvae, with less developed sensory systems, from orienting with respect to certain stimuli.

1.6   Summary

The early life of fishes is distinct from the more familiar juvenile and adult periods because of two intrinsic processes, ontogeny and growth, that give rise to profound changes in body structure and size. Not only are embryos and larvae different from their later forms, but the degree of difference changes continually. These changes influence the behavioral and physiological performance of early life stages, which in turn alter the nature of ecological interactions. This chapter has examined a few representative performance measures and their associated morphologies to demonstrate the ways in which developmental changes work their way from morphology to ecology. In practice, it is better to avoid examining one organ system or one performance measure in isolation. A more comprehensive approach, when possible (Figure 1.13), will provide a better basis for understanding a species' requirements and limitations during early life.

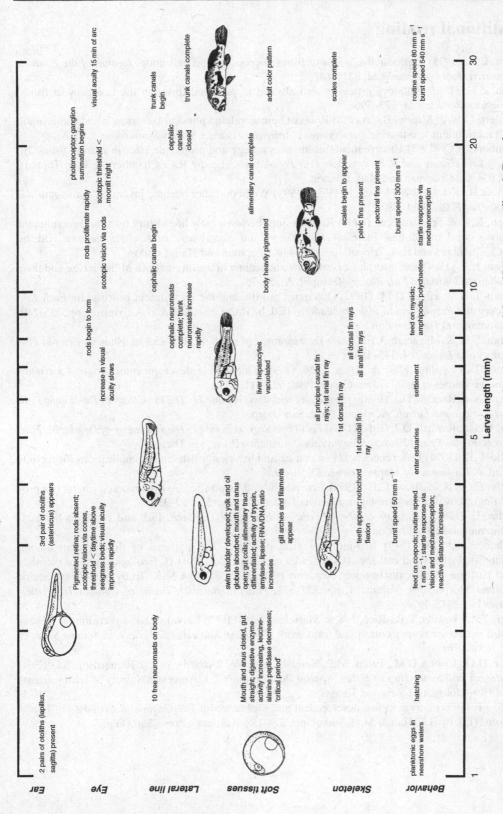

Figure 1.13 Composite of developmental changes in body form, function, and ecology in red drum (*Sciaenops ocellatus*). Data from numerous sources. Drawings from Holt *et al.* (1981) and Pearson (1929).

Additional reading

Balon, E.K. (1975) Reproductive guilds of fishes: a proposal and definition. *Journal of the Fisheries Research Board of Canada* **32**, 821–864.

Balon, E.K. (1981) Saltatory processes and altricial to precocial forms in the ontogeny of fishes. *American Zoologist* **21**, 573–596.

Boehlert, G.W. & Mundy, B.C. (1988) Roles of behavioral and physical factors in larval and juvenile fish recruitment to estuarine nursery areas. *American Fisheries Society Symposium* **3**, 51–67.

Chambers, R.C. (1997) Environmental influences on egg and propagule sizes in marine fishes. In: *Early Life History and Recruitment in Fish Populations*. (Ed. by R.C. Chambers & E.A. Trippel), pp. 63–102. Chapman and Hall, London.

Copp, G.H., Kovác, V. & Hensel, K. (Eds) (1999) When do fishes become juveniles? *Environmental Biology of Fishes* **56**, 1–280.

Cowen, R.K. & Sponaugle, S. (1997) Relationships between early life history traits and recruitment among coral reef fishes. In: *Early Life History and Recruitment in Fish Populations*. (Ed. by R.C. Chambers and E.A. Trippel), pp. 423–449. Chapman and Hall, London.

Fuiman, L.A. (1994) The interplay of ontogeny and scaling in the interactions of fish larvae and their predators. *Journal of Fish Biology* **45**(suppl. A), 55–79.

Fuiman, L.A. & Higgs, D.M. (1997) Ontogeny, growth, and the recruitment process. In: *Early Life History and Recruitment in Fish Populations*. (Ed. by R.C. Chambers & E.A. Trippel), pp. 225–249. Chapman and Hall, London.

Fuiman, L.A. & Magurran A.E. (1994) Development of predator defences in fishes. *Reviews in Fish Biology and Fisheries* **4**, 145–183.

Fuiman, L.A., Poling, K.R. & Higgs, D.M. (1998) Quantifying developmental progress for comparative studies of larval fishes. *Copeia* **1998**, 602–611.

Hoar, W.S. & Randall, D.J. (Eds) (1988) *Fish Physiology, Volume 11. The Physiology of Developing Fish. Part A. Eggs and Larvae*. Academic Press, San Diego.

Hoar, W.S. & Randall, D.J. (Eds) (1988) *Fish Physiology. Volume 11. The Physiology of Developing Fish. Part B. Viviparity and Posthatching Juveniles*. Academic Press, San Diego.

Houde, E.D. (1994) Differences between marine and freshwater fish larvae: implications for recruitment. *ICES Journal of Marine Science* **51**, 91–97.

Houde, E. D. & Zastrow, C.E. (1993) Ecosystem- and taxon-specific dynamic and energetics properties of larval fish assemblages. *Bulletin of Marine Science* **53**, 290–335.

Kamler, E. (1992) *Early Life History of Fish: An Energetics Approach*, Fish and Fisheries Series 4. Chapman and Hall, London.

Kryzhanovskii, S.G. (1949) Eco-morphological principles and patterns of development among minnows, loaches and catfishes (Cyprinoidei and Siluroidei). Part II. Ecological groups of fishes and patterns of their distribution. Academy of Sciences of the USSR. Trudy of the Institute of Animal Morphology, Volume 1, pp. 237–331. *Fisheries Research Board of Canada Translation Series* No. 2945. 1974.

Miller, T.J., Crowder, L.B., Rice, J.A. & Marschall, E.A. (1988) Larval size and recruitment mechanisms in fishes: toward a conceptual framework. *Canadian Journal of Fisheries and Aquatic Sciences* **45**, 1657–1670.

Moser, H.G., Cohen, D.M., Fahay, M.P., Kendall, Jr., A.W., Richards, W.J. & Richardson, S.L. (Eds) *Ontogeny and Systematics of Fishes*. Special Publication No. 1, American Society of Ichthyologists and Herpetologists. Lawrence, Kansas.

Webb, J.F. (1999) Larvae in fish development and evolution. In: *The Origin and Evolution of Larval Forms*. (Ed. by B.K. Hall & M.H. Wake), pp. 109–158. Academic Press, San Diego.

Chapter 2
Age and Growth

Cynthia M. Jones

2.1 Introduction

This chapter introduces the concepts of age and growth and the special considerations that are necessary to estimate them in fish larvae. Later chapters show how age and growth information for early life stages is used to develop an understanding of mortality, population analysis, and cohort identification, processes that are vital to the persistence of populations. These processes are termed the vital rates and elementary characteristics of the population. Vital rates are measures of the processes that increase or decrease the abundance and biomass of the population. The processes that increase a population are birth and growth, while the process that decreases a population is death. The difference between the number of births (hatchings) and deaths in a population determines whether the population grows or declines. The number and timing of hatchings, along with survival, give the population its age structure. Similarly the number of young that survive to reproduce themselves determines the population's reproductive capacity and ultimately, its persistence. The age-specific growth rates determine the population's biomass. In human populations, this science of vital rates is termed demography and in non-humans it is population ecology.

The study of fish population dynamics is one of the subdisciplines of applied population ecology. Much of the research in fishery science and fish ecology has focused on the characteristics of adult fishes and their vital rates. Even so, researchers have long understood that the survival of the youngest fishes determines the future abundance of the population. This early period has been the most difficult to study, in part because of the difficulty in measuring survival and growth during this period and in understanding the biotic and abiotic effects on survival. The development of new techniques for measuring age and growth over the past 30 years has spurred renewed interest in determining the fundamental factors that influence survival and growth of young fishes. This is termed analysis of survivors.

The length of early life is short relative to the potential life span of the fish, yet during this brief time the fate of the cohort is established. To predict the fate of the cohort, we must be able to measure survival accurately during this period. The most important breakthrough in measuring survival in larval fishes was development of the daily ageing technique. Before this technique was developed, scientists could only investigate survival during this period by measuring changes in abundance-at-size (size–frequency distribution). However, the change in the average size includes not only growth – where growth may also be influenced by density – but is also compromised by the addition of new hatchings that maintain a low average size. In measuring abundance-at-size alone, without knowing age, growth will appear slower, and mortality greater, than would be evident by following age-specific cohorts. Thus, change in size alone often will not allow us to predict survival or growth. To measure survival and growth accurately, we need to measure age.

Age-specific estimates of survival have been commonplace in adult fishes since the early 1900s. During that time scientists found that the bony parts of temperate fishes produced yearly marks termed annuli. With these annual marks, fishery scientists were able to track the decline in numbers of a cohort over time. Because larval fishes are less than a year of age, they are too young to form these annual marks. Only with the discovery of daily marks in the 1970s were scientists able to age larval fishes and develop age-specific measures of survival and growth.

The growth of individuals also influences population dynamics. Growth rates influence survival and, upon maturation, fecundity. As we will see in Chapters 3 and 4, fishes that grow faster are more likely to survive because they reach a large size sooner, are able to swim faster, can escape predators better, and are better able to capture food. For adult fishes, individuals that grow faster are sexually mature sooner, thus they can potentially reproduce longer, and are more fecund because fecundity typically increases with size.

2.2   History of the study of age and growth in larval fishes

The study of age and growth is the foundation on which our knowledge of fish population dynamics is built. Even though the use of age and growth data is fundamental to the modern practice of fishery science, much of our knowledge of fish age and growth has developed only within the past 150 years. By the mid-1800s, people observed wide fluctuations in the abundance of fishes from year to year. These fluctuations in fish abundance drove fluctuations in price. The fluctuations in price created boom and bust cycles in the economies of cities that were dependent on fisheries. Even so, prior to the late 1800s marine fishes were considered an inexhaustible resource and little effort was directed toward predicting fish abundance. Demographic or age-based studies of fishes began with collapses of nearshore populations in the northeastern and northwestern Atlantic Ocean in the late 1880s, including the Lofoten cod and French sardine. Sadly, this story would repeat itself a century later in the northwestern Atlantic cod fishery even after tremendous increases in the use of age-based data and in the use of models that relied on these age data.

Before modern methods of ageing were developed, fishery scientists used size differences as a proxy for age. C.G.J. Petersen, a Danish biologist, plotted the frequency of different sized fish and used such plots to identify age classes, with each mode of size indicating a separate year class. Although these plots were useful for determining the age of a group of fish, the age of individuals could not be determined. Determination of individual age awaited another discovery, annual marks similar to tree rings on the bones of individual fish. These marks were first seen in cross-sections of pike vertebrae as early as 1759. In the last decade of the 1800s they were also seen in carp scales and in plaice otoliths (see *Scaling Fisheries* by T.D. Smith for further discussion). Because the discoveries of annuli in scales and otoliths came at an opportune time, they were used widely to determine age of individual fish. Other scientists used data on size-at-age and abundance-at-age to assess growth rates and survival – all based on individually determined age from the analysis of bony structures. It soon became apparent that the larval stage was the critical period for survival in many fish populations. In 1914, Johan Hjort championed the hypothesis that the larval period was critical to determining the ultimate abundance of recruits for a given year class (see Chapter 4). He developed this hypothesis for marine fishes because they produce hundreds to millions of eggs each spawning season. From this multitude, only a handful survives to reach the juvenile stage. Small changes in the proportion of larvae surviving to the juvenile stage result in large changes in the abundance of recruits to the fishery. Therefore, it becomes clear that, if the factors affecting survival and growth during the larval period could be clarified, the fate of fish populations could be better predicted. However important the larval period, little progress in predicting larval-cohort survival could be made until daily age could be determined as convincingly as annual age was in adult fishes.

In the 1970s Gregor Pannella, a geologist, noted the presence of daily increments in the otoliths of adult temperate and tropical fishes. Otoliths from adult tropical fishes had daily increments that showed a fortnightly pattern that was related to fish behavior during lunar cycles. The discovery of daily increments soon encouraged fish ecologists to look for daily patterns in the otoliths of larvae and juvenile fishes. Prior to this discovery there was no reliable method of ageing individual larvae or juveniles. Larval fish biologists eagerly embraced daily ageing, and in the early 1980s a burgeoning series of papers appeared, which validated this technique in many species of freshwater and marine larvae. During the 1980s, scientists first reared fishes under optimal and suboptimal conditions to confirm the constancy of increment formation, and noted when daily increments could not be seen. Under suboptimal conditions increments sometimes were so closely spaced that individual increments could not be accurately counted under a light microscope. The increments were still present, but they could only be viewed with a scanning electron microscope. In some instances, for example, in herring (Figure 1.3m) and turbot, daily increments may not be formed during part of the larval period. Nonetheless, validation studies have shown that daily increment formation occurs in the overwhelming majority of young fishes that survive to metamorphosis. During the 1980s scientists also used daily ageing to calculate hatching dates, and these dates were used to impute environmental effects on growth and survival (see, for example, Section 11.11). This decade of scientific advance clarified the value of using larval fish otoliths as the basis for estimating age-specific survival and growth and rendered these processes capable of being measured in much the same age-based way for larvae as they were for adults.

When demographic data became available for adults in the beginning of the last century, these advances spawned the development and application of mathematical models such as virtual population analysis (VPA), the Beverton–Holt Yield-Per-Recruit Model, and W.E. Ricker's Stock–recruitment Model in fishery management in the 1950s. These models concentrate on the characteristics of adults – survival, growth, and fecundity – but, as will be discussed in Chapter 9, are sensitive to variations in survival during the first year of life. Development of demographic models that included rates specific to larval life awaited the daily ageing breakthroughs. Today, the newest models use early life data that include detailed information of age- and size-specific processes. The best known of these are the individual-based models. On the horizon there are further breakthroughs in the use of spatially explicit models that rely on the time-keeping properties of otoliths. These new age-based spatial models will sharpen our appreciation not only of the importance of the larval period, but just how much survival and growth may vary in different habitats.

2.3   Determining the age of larval fishes

2.3.1   Using otoliths, scales, and other hard parts

In temperate waters, the seasonal cycles of light and water temperature result in optical banding in the hard parts that reflect the seasonally driven growth of adult fishes. In tropical waters, the seasonal cycles that cause banding are the result of rains, salinity changes, and other environmental factors. For adult fishes, age is determined by counting these yearly bands in the hard parts. In adults, several hard parts are routinely used for ageing, including otoliths, scales, fin rays, and spines. Whereas scales and other hard parts are used to age adults, the age of larvae and juveniles is determined by counting daily increments only in otoliths. Scales and other hard parts are not formed prior to hatching and have limited use for ageing larvae. Complete calcification and full formation of fin rays and spines do not occur until metamorphosis is nearly complete (see Chapter 1). Hence, these structures cannot be used to age larvae. Likewise, full formation of scales over the entire body does not occur until metamorphosis, which precludes their use as reliable ageing structures in larvae. In contrast, otoliths are enduring calcified structures that are present before hatching. Thus, if we are interested in the larva's response to temporal events, we must use otoliths. They not only give age, but also provide a retrospective history of growth.

Daily marks can be seen in the otolith microstructure of most young fishes reared under a wide range of conditions (Figure 2.1). These daily marks begin to form at hatching or within a week or so after hatching. Otoliths begin forming from centers of crystallization as the embryo develops in the egg, but not all otoliths are formed by hatching. Fishes have three pairs of otoliths, the sagittae, lapilli, and asterisci. Typically the sagittae and lapilli are present at hatching and the asterisci form later (Figure 1.13). Soon after the onset of crystallization, marks form as a result of the differential deposition of calcium carbonate (typically in the form of aragonite) and a high molecular weight protein called otolin. Differential deposition of the calcium carbonate occurs in an endogenously driven diurnal pattern. The initiation of daily marks may begin prior to hatching, at hatching, or at yolk absorption depending on the species. Daily marks are more distinct during certain parts of larval life. Often the daily marks are very clear initially and are uniformly spaced. The clarity of daily marks may vary thereafter, especially when otoliths begin to take their adult, species-specific shape. Clarity of the daily marks can degrade as the larva approaches metamorphosis, when new centers of crystallization (accessory primordia) form at the edge of the largest otolith, typically the sagitta, and at this point it is difficult to maintain an accurate count of daily bands (Figure 2.2). For fishes whose sagittae have accessory primordia, the lapilli are used instead for ageing. Nonetheless, daily marks continue to form throughout the first year or two of life (and probably thereafter), but as the otolith gets larger and somatic growth declines the marks narrow too much to be distinct.

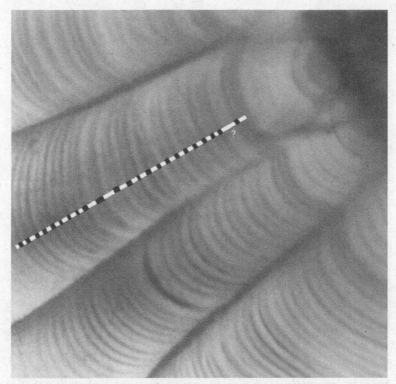

Figure 2.1   Daily marks can be seen on the sagitta of a 120-day-old Atlantic croaker, *Micropogonias undulatus*. A daily mark is composed of a pair of light and dark bands indicating the alternating pattern of protein- and calcium-rich layers. The white question mark indicates an area where we are uncertain about the count. There is a faint band and this may be another daily increment.

The daily increment technique is widely used now to obtain ages, hatching dates, and age-specific growth in larval and juvenile fishes. The age of the larva is estimated by counting the daily bands from the hatching mark (or the birthmark in live bearers) to the edge of the otolith. When the hatching mark is ill defined, age can be calculated by counting from the time of yolk absorption (usually marked by a well-defined change in microstructure) to the otolith edge if the mean age at yolk absorption is known. Once the larva's age is known, then the hatching date is estimated by subtracting the age from the date of capture. Otolith microstructure also permits growth to be measured from the width of the daily bands when the relation between otolith size and fish size is predictable.

Changes in the otolith microstructural patterns also mark major transitions in early life, such as hatching, yolk absorption, and metamorphosis. Similar to adult fishes, whose growth slows upon sexual maturity and is marked by a narrowing of annuli, larval life events are marked in the otolith by changes in the clarity and width of daily bands. Under normal conditions, the width of the bands reflects the fish's somatic growth, but otolith growth and fish growth can become decoupled. Even so, microstructural patterns persist to mark life-history events. In fishes with indirect development, the period of endogenous nutrition is associated with clear, even bands. The period immediately after yolk absorption is often marked by a narrowing of these bands as the larva relies solely on exogenous food at this time. In a sim-

Figure 2.2   Accessory primordia develop on the periphery of the otoliths and they interrupt the regular formation of daily marks. Although precise counts of daily increments can be made up to the accessory primordium, thereafter there is no clear path along which to count.

ilar manner, changes in habitat and feeding are reflected in the microstructural patterns. For reef fishes, the transition between the pelagic pre-settlement phase of life and the reef-associated post-settlement phase is marked by a change in the microstructure. For marine spawning, estuarine-dependent fishes, the transition to the estuarine nursery ground is often marked by wider daily increments. For amphidromous fishes, the transition to a different salinity regime is marked by dramatic changes in physiology and morphology. When these changes affect somatic growth, they also affect otolith microstructure (see, for example, Section 11.11). Moreover, salinity changes are also registered in otolith chemistry.

Preparing, handling, and viewing otoliths

The methods of collection, removal, and storage of otoliths in young fishes are modified from those for adults. The modifications are due to the small size of larvae, incomplete calcification of the larva's skeleton, and to the small size of larval otoliths. Otoliths are easily seen and removed in very young fishes because larvae are relatively transparent, but once removed otoliths are easy to lose. In very young fishes, otoliths do not need much preparation other than adhering the otolith to a slide with a mounting medium. As larvae approach metamorphosis otoliths increase in size and may require more extensive preparation, including grinding and even sectioning. Several good manuals and papers are available that present the methods of removal, storage, and preparation and a few are listed at the end of this chapter.

Otoliths must be prepared so that the daily increments are visible throughout, from the edge to the core. Once daily increments are visible, they must be read in a standardized way so that knowledge of the specifics of rearing or capture will not influence the judgment of the reader. If the reader knows that the larvae were caught late in the season, he may be influenced by this knowledge to count more subtle microstructural features as daily increments than the reader would without this knowledge. This type of subtle influence should be avoided whenever possible. One of the methods used to avoid prejudicing the readings is to cover any information on the slide and to select the slides for reading in a randomized order. Otoliths are typically read several times to determine the precision of the readings (see discussion below). Standard practice is to take a time break between readings and to perform the subsequent readings with the same random selection procedures. This practice insures that otoliths will not be read in the same order as before and the readings will be independent of each other.

Otolith reading has a high potential for accuracy but a similarly high potential for error. Beginning scientists are disconcerted to learn that there is a great deal of subtle understanding, not just mechanics, in reading daily increments. The skill of the reader has a lot to do with the accuracy and precision of the readings. First it is important to establish criteria for what and where increments will be counted. It is best to work with an expert who has aged larval fishes and has been trained on known-age fishes. Although some parts of the otoliths have clear increments, other parts do not and have subtle shadings of microstructure. People who age fishes use a variety of techniques to interpret ages. For instance, in adults one expects to see a gradual decrease in width of the bands that mirrors the slowing of somatic growth. A very wide band may indicate that an annulus has been missed. Because larval fishes have daily variability in somatic growth, the width of the daily bands is intrinsically more variable. But again, an abnormally wide band may indicate that a daily increment has been missed. One of the best methods for learning to age larval fishes is to work with known-age fishes reared for the purpose of training. The daily bands on known-age fishes will differ from those seen in field-caught larvae. Typically laboratory-reared larvae have lighter, less distinct bands. Nonetheless, they do demonstrate the growth that the species is capable of, and the method of preparation that provides the most accurate readings. Readings of known-age otoliths can be used to determine the age of deposition for the first increment and to compare true age to increment counts. The counting techniques can be modified if counts grossly under- or overestimate true age.

Part of the mechanics of ageing larval fishes is to have good microscopes available for this research. Otoliths contain little organic tissue so the same techniques used to look at tissues will not work as well with otoliths. Otoliths are composed of calcium carbonate and should be handled more like a geological specimen – a rock – than a biological one. Because they are composed of crystals they can refract light, and they can be read best using polarized light. Under polarized light, they sparkle through the surrounding tissue and are easier to retrieve. It is best to use a compound microscope with good optical properties. Otoliths of larval fishes are small. In newly hatched fishes, they can be as small as 10 μm. Given such small size, otolith microstructure is observed under the microscope using either transmitted or reflected light and the characteristics of the bands, whether seen as translucent or opaque, will depend on the light used. The bands themselves are made up of layers rich in calcium carbonate alternating with layers richer in protein. The difference in composition shows up as different opacity under the microscope. A pair of light and dark bands comprises a daily increment. Otoliths of larval fishes must be read under adequate magnification to discern bands that can be as narrow as 1–2 μm; typically magnification is 400–1000×. The appropriate magnification will depend on the species (for example, higher magnification is used with flatfishes, which have relatively small otoliths), and whether there are numerous sub-daily changes in microstructure. The appropriate magnification may also change depending on which part of the otolith is examined. It may be easier to read the bands near the core than nearer the edge, but this will also depend on the species, age, and ontogenetic stage. Similarly, the clarity of bands will be different depending on the growth rate. The otoliths of fishes that grow very slowly or very fast are each difficult to read.

Precision of increment counts

The quality of our estimates of survival and growth will depend on the precision and accuracy of the age readings. Precision is the measure of how close the repeated readings of an otolith are to one another. Perfect precision is obtained when repeated readings of each otolith are the same. Otoliths are rarely this clear and interpretable and normally precision will be lower. The level of acceptable precision will depend in part on the age of the fish. In very young fishes, say less than a week, readings of the same otolith that are different by two bands would be off by some 30% whereas, for a fish that is 30 days old, this same difference is less than 10% and is a less egregious error. There are routine procedures that can be used to measure precision. One procedure is the percent agreement test, which is based on the mean difference between repeated readings on an individual otolith and the mean of those readings expressed as a percentage of the mean (sometimes referred to as average percent error or APE). These tests have drawbacks. The percent agreement will depend on the ages sampled and these tests do not reveal any pattern in the disagreements that might exist. Two different readers may have the same APE, yet one may consistently attribute an older age to each individual. This difference will not be detected by an APE calculation. Another common pattern is to have more difficulty precisely ageing older fishes, which may only be apparent as a decrease in overall precision. Rather than use percent agreement tests, scientists have recently been using symmetry tests. Symmetry-test statistics are based on the idea that if repeated readings were perfectly precise, the results will all fall on the diagonal of the matrix of first vs. second readings (Table 2.1). The first reading would be equal to the second and all off-diagonal cells would be zero. Symmetry tests not only indicate the level of agreement between readings, but also whether there is a pattern in the disagreements. This is revealed when the off-diagonal cells are other than zero. When the upper triangle of the matrix is filled, then the second reading was less than the first, and so forth. Of course any patterns that exist can be seen by graphing the repeated readings to observe the agreement between readings directly.

Validating age procedures

Although precise measurement of age is desirable, accuracy is fundamental to the validity of our age readings. Age readings are accurate when they give us true age. Careful scientists make every effort to validate their age readings through a variety of techniques. In adult

Table 2.1 The symmetry test is one method that can be used to estimate bias or imprecision in counting daily marks. This test can be used to compare one scientist's reading precision when repeatedly counting a set of otoliths, called within-reader precision, or to compare the repeatability between the readings of two scientists for a given set of otoliths, called between-reader precision.

Second reading (age in days)	First reading (age in days)									
	27–32	33–38	39–44	45–50	51–56	57–62	63–68	69–74	75–80	81–86
27–32	1									
33–38	2	4	1							
39–44		3	2	1						
45–50		2	4							
51–56	1			5	4					
57–62			1	1	3	3				
63–68					1	3	4		1	
69–74					1	1				
75–80									1	
81–86						2		1		2

fishes, age estimates are validated directly through rearing known-age fishes in the laboratory, carrying out mark-and-recapture experiments, or indirectly, through radiochemical means (for example, by radioisotope dating) and marginal increment analysis. Although these methods work well enough for adults, most cannot be used to validate daily ages in larval fishes. Of the techniques mentioned above, only known-age fishes, and less often, mark-and-recapture methods are used in larvae. The radiochemical tests, such as $^{210}Pb : {}^{226}Ra$ or ^{14}C dating, can only be used with long-lived fishes and cannot be used to determine ages of fishes less than 1 year old. These radiochemical methods rely on the natural decay of radioactive elements to act as a chronometer. Depending on the half-life of the radioisotope, it can take several years for the daughter element to be formed in sufficient quantity to be measurable.

Marginal increment analysis of adult fishes relies on the seasonal cycle of growth as reflected by the otolith to demonstrate the annual pattern of band formation. The pattern of daily increment formation and the small size of the increments on a larval otolith make this technique impractical for most species. Only a few scientists have used daily marginal increments to validate bands in larval fishes. Fish must be sampled throughout 24 h, and the width of the edge increment must be measured. To validate daily bands, the width should increase during 24 h until another band is completed. Use of this validation technique requires a fast growing otolith in which the edge increment can be seen clearly.

Mark–recapture techniques have been widely used to validate age in hard parts of adult fishes and recently this technique has been applied to larval fishes. A dye is injected into a sample of adult fish and this dye becomes incorporated into the hard parts. The fish is released and later recaptured after a time at large. Upon recapture, its hard parts are removed and examined for growth subsequent to the chemical mark that identifies the age at the fish's release. This technique is valuable when sufficient adults can be recaptured over their entire life span. This technique has more limited use during the larval period. It can be useful during this period for species that have low mortality and low dispersal. The technique is problematic, however, for fishes with high fecundity and high mortality during the larval period.

Three problems are encountered in using mark–recapture methods in larval fishes. First, marked fish die from natural mortality and thus, are unavailable for recapture. To conduct a mark–recapture experiment, enough fishes must be marked so that a sufficient number survive and are recaptured. This is a problem in freshwater fishes that experience mortalities of several orders of magnitude during the larval period and even a greater problem for marine fishes whose survival through this period may be on the order of one in a million, as we will see in the next chapter.

Second, larvae are typically too small for individual dye injection, and the number of larvae that must be dyed too great. Thus the dye technique used in adults is modified for larvae wherein a large number of small larvae are immersed in dye instead of individual dye injections. Mass marking with chemicals, such as alizarin complexone, has made this technique feasible. Third, fishes can be advected out of the release area and then are not available for recapture. If the larvae disperse at all or mix with wild, unmarked larvae, then great numbers must be captured to obtain the rare marked larva. Millions of fishes must be marked if there is to be any hope of recapture. This is not a problem in lotic habitats, but it is a problem for lentic, estuarine, and marine habitats. When there is advective transport of larvae, the sampling area for recapture must be large enough to include the area to which the fishes are transported. Logistically this becomes unmanageable when transport is over a wide area. To overcome potential transport, fishes can be resampled quickly while they are still in the general release area. One advantage of using mark–recapture techniques for validation of daily increments is that losses from mortality or advection do not have to be quantified. It is only necessary to have sufficient recaptures to show that the number of increments between release and recapture equals the number of days at large. If it does, then increments have been laid down each day.

Ageing procedures can be validated directly by rearing known-age fishes in the laboratory. It is hard to rear sufficient numbers of fishes to adulthood and beyond to validate their annuli over a range of ages. Rearing is especially hard for those adults that have long potential life spans. Hence, laboratory rearing of known-age adult fishes is uncommon as a validation technique. This technique is commonly used, however, to validate age (and growth) of the early life stages. Hardy larvae are easily reared in the laboratory from hatching to several months of age. It poses greater problems, often insurmountable, for larvae of pelagic marine species that are delicate and difficult to rear. Nonetheless, daily increment formation has been validated for the larvae of many species and scientists now commonly assume that the microstructure they see are daily increments, even for species in which daily formation of increments has not been validated. Even though this is becoming common practice, daily increments should be validated for a species whenever possible, if for no other reason than to familiarize oneself with the ontogenetic changes in otolith microstructure and to assess the best counting path and otolith preparation.

Beyond the difficulty of rearing fish larvae, validation experiments entail other considerations. Periodically throughout the experiment, fishes are sacrificed and their otolith

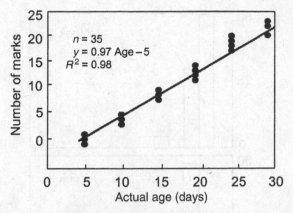

Figure 2.3   The accuracy of putative daily marks is tested against known-age larvae that have been reared in the laboratory. In this example, the age of fish, reared from birth in the laboratory, is regressed against the number of daily marks that have been counted. We accept "daily" counts as accurate estimates of age when the slope of this regression line is not significantly different from 1.0.

increments are counted. The slope of a line will equal 1.0 when these increment counts are regressed on known age (Figure 2.3). (Statistically, we will fail to reject the null hypothesis that the slope $= 1.0$.) In most species tested, increment formation is daily even under fairly stressful conditions. Some species are sensitive to unfavorable conditions, however, and fail to form daily increments. Notable among these are the flatfishes. One argument made against this technique is that laboratory conditions are very different from those of the field and we stretch our inference frame when we extrapolate to the field. Increments from field-captured fishes are usually more distinct compared to laboratory-reared fishes and this difference in increment clarity is usually attributed to the diel variations of temperature in the field and the higher daytime light levels. Nonetheless, there is good corollary evidence from the field that, under most circumstances, larvae do in fact form daily increments. This corollary evidence comes from:

(1)   the few mark–recapture experiments performed with larvae and juveniles;
(2)   following the progressive changes in mean age of field-caught larvae produced from species with short spawning seasons; and
(3)   from matching the earliest hatching dates calculated from increment counts to the known starting date of the spawning season.

2.3.2   Alternate methods of determining age

Age can be determined by alternate methods besides the analysis of hard parts. The most widely used alternate method in adult fishes is length–frequency analysis. Length–frequency analysis has proven useful for ageing fish species that have a limited spawning season and fast early growth. The technique is straightforward. All life stages of the population are sampled and the frequency is plotted against fish length. Under the optimum circumstances each age group will appear as a separate mode of length. For adults, this method estimates the growth of year classes by following the size obtained by each mode of abundance-at-length.

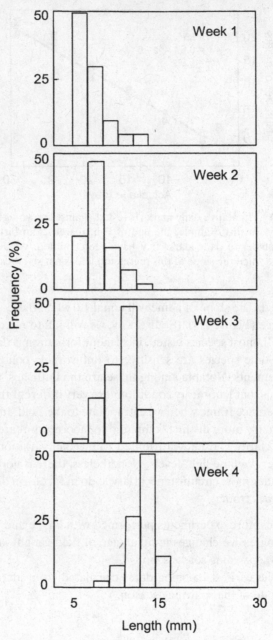

Figure 2.4 Length–frequency analysis can be used to follow age and growth of young fish when the spawning season is composed of discrete spawning events. These plots illustrate a unimodal peak in spawning that results in an identifiable age mode that shows a weekly increase in size. In 4 weeks, these larvae have grown almost 10 mm.

The technique becomes problematic when growth slows at older ages and when peaks merge. It is also difficult to use for larval fishes. When a species has a short spawning season, the abundance and growth of the newest cohort can be followed through the first year of life if sampling occurs frequently (Figure 2.4). Such frequent sampling is uncommon.

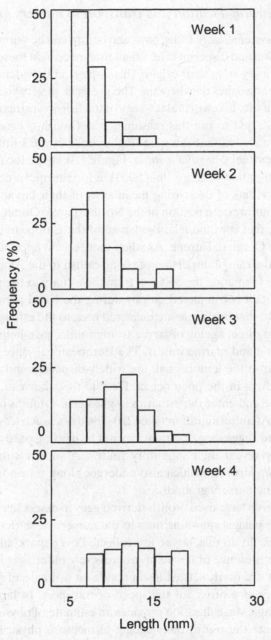

Figure 2.5　In contrast to Figure 2.4, these plots illustrate continuous production of appreciable numbers of hatchlings that obscures any age peak. When hatchlings are constantly added to the population, the size increase of an age cohort cannot be tracked by modal progression of length without knowledge of age. Although the analysis of length frequencies is often used to determine age of adult fishes, it is difficult to use for larvae and juveniles for this reason.

Beyond this problem, this method cannot be used to estimate daily ages of individuals or the survival of weekly cohorts of larvae because spawning often occurs over several weeks. During this time, an unknown number of new and small larvae are added to the population and distinct peaks between weekly cohorts do not occur (Figure 2.5).

2.3.3 Clarifying spawning location and transport to nursery grounds

Larvae from major river systems, large lakes, bays, and oceans can be sampled spatially to pinpoint the spawning location and dispersal rate. When fish are born in these environments, the current transports them away from their origin. This dispersion is evidenced by the progression of size and age of young fishes downstream. The presence of very young larvae indicates proximity to the spawning site. Likewise, relative age of the fish downstream indicates the rate of dispersion. One of the first to use this reasoning was Danish oceanographer Johannes Schmidt, who in 1922 concluded that the spawning grounds of the European eel (*Anguilla anguilla*) and the American eel (*Anguilla rostrata*, Figure 1.3k) were located in the Sargasso Sea, which required a migration of more than 5000 km for European eels. This conclusion was based on his observations of decreasing mean sizes of their larvae (leptocephali) the closer they were to the southwestern portion of the North Atlantic Ocean. He also estimated, from hydrographic data, that it would take the larvae of the European eels 2 or 3 years to travel across the Atlantic Ocean to Europe. Another, more recent application of this method has been to understand the rate of dispersion of young catfish in the Amazon River. It takes 5 months or so for adults to migrate the 3300 km from the estuary to their spawning location upstream. The return circuit is completed quickly during the larval period. Using size as a proxy for age, it is thought that the larvae are delivered back to the estuary in 13–20 days.

Several applications of direct ageing of larvae from otoliths have helped to clarify transport of larvae in estuarine and marine waters. The European eel, once again, provides an excellent example. Because the leptocephali are widely dispersed and are in low density, it is difficult to collect them in the open ocean. Despite this, they can be easily collected after they metamorphose and enter the estuaries as glass eels. Otoliths obtained from glass eels show that the trans-Atlantic migration takes 220–284 days, much less than the 2–3 years estimated by Schmidt and suggesting that they may not be drifting passively. Trace-element analysis of otoliths may reveal their migratory routes. A similar story can be told for the Japanese eel, *Anguilla japonica*, which also undergoes long transport periods of about 170 days to their estuarine nursery grounds.

More generally, scientists have used otolith-derived ages to assess the length of time that transport takes from the pelagic spawning area to the nursery or settlement grounds for a variety of marine species. To do this, larvae are collected over space and time throughout the area of interest. The presence of larvae of progressively older ages indicates the speed of transport. Fishes that are newly settled also provide an estimate of pelagic stage duration, which in turn serves as a proxy for the speed of transport. In larvae that are transported passively, the pelagic stage duration provides an estimate of physical transport speed by currents. To complicate the use of daily ageing to measure physical transport, recent research on reef fishes has shown that larvae can actively swim to their settlement sites. In this case, analysis of age can also provide insights into the relative contribution that swimming provides over passive transport. Similarly, daily ageing can provide an estimate of larval retention over banks and other features in oceans, lakes, and rivers.

The newly developing field of otolith microchemistry relies on the otolith as an accurate chronometer to track early life migrations. Otolith microchemistry provides the signal that indicates the fish's exposure to different water masses, while the otolith microstructure provides the timing of this exposure based on daily increments. Juvenile or newly settled fishes can be collected and their otoliths sectioned to reveal a retrospective, dated history of their lives. With microsampling devices or lasers, periods of their early life can be sampled with highly sensitive, analytical instruments, such as inductively coupled plasma mass spectrometry (ICPMS) to determine the elemental and isotopic composition of the otolith. The trace-element composition during these few days of the fish's life is then classified according to the known properties of the water mass, or of the signature the water mass has imparted to fishes of known exposure. Although still in its infancy, analyses of otolith microchemistry may add immeasurably to our understanding of how fishes use habitat at specific times in their early life. We will learn more about this technique in Chapter 6.

2.3.4   Problems with using age information to clarify survival in early life

The classic comparison of vertebrate survival was made by Raymond Pearl in 1928 and was based on the numbers of organisms alive at a given age (Figure 2.6). By his simple illustration he showed that fishes and humans had very different patterns of survival. Human survival is high until they become senescent, whereas very few fishes survive to become juveniles, but after that time survival is relatively high. This simple approach was followed by other techniques that relied even more heavily on obtaining accurate estimates of age throughout the fish's lifetime. The fundamental technique that underlies our modern models is the method of life-table analysis. Life tables are still used in entomology, even in the larval stages. In fishery science, however, life tables are less recognizable; their features are retained only in VPA, which will be discussed more thoroughly in Chapter 5. The life table is built by recording observational data on absolute abundance and fecundity for each age (Table 2.2). Other rates such as proportion dying, instantaneous mortality, survival, and

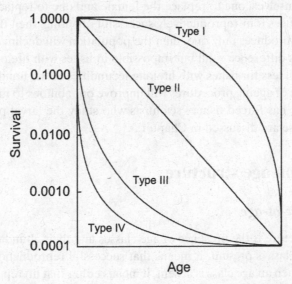

Figure 2.6   Raymond Pearl developed survival curves to illustrate the time course of survival for populations of different organisms. Fishes typically exhibit a survival curve in which huge mortalities occur in early life (Type III or IV). Redrawn and modified from Pearl (1928).

Table 2.2 Life and fertility tables are used to estimate a population's potential for growth. The vital rates are calculated for each age, x. The per capita rate of increase, R_0, is then calculated by multiplying the age-specific birth rate (hatchings), b_x, by the standardized abundance of survivors, l_x, and then taking the sum. For this example, the per capita increase, R_0, is 2.3, the generation time is 2.7 years, and the intrinsic rate of growth, r, is 0.3. Thus the population will probably increase.

Age, x	Abundance, n_x	Age-specific survival rate, s_x	Standardized survival rate, l_x	Standardized death rate, d_x	Age-specific birth rate, b_x	Age-specific average reproductive rate, $F_x = l_x b_x$
0	1 000 000	0.0001	1.000000	0.9999	0	0
1	100	0.6	0.000100	0.00004	5000	0.5
2	60	0.5	0.000060	0.00003	10 000	0.6
3	30	0.5	0.000030	0.000015	20 000	0.6
4	15	0.5	0.000015	0.000007	20 000	0.3
5	8	0.25	0.000008	0.000006	30 000	0.24
6	2	0	0.000002		30 000	0.06
						$R_0 = 2.30$

fecundity are then calculated and added to the table. Further calculations of age-specific reproduction are made and used to estimate the population replacement rate. Life tables have not been used often in fishery science because they demand very accurate measures of abundance-at-age and this has been all but impossible to obtain for larval fishes, the life stage in which over 90% of the lifetime mortality occurs.

The age-based approaches that are used with adult fishes prove to be problematic for larvae. This is clear when we think about the process that yields a population that is stable in abundance. In such a population, each female needs only produce two offspring which then reproduce themselves, one to replace the female and one to replace the male parent. If each female produces four reproductively successful offspring, then the population doubles. If each female produces only one, then the population will decline by one half. Being able to measure this difference is all but impossible in fishes with lifelong fecundities of a thousand eggs, much less for fishes with lifelong fecundities in the hundreds of millions of eggs. No refinement of ageing procedures will improve our abilities to make estimates with such precision. This has forced fishery scientists who study the larval period to use other approaches and these are discussed in Chapter 5.

2.4   Population age structure

2.4.1   *Abundance-at-age*

Population age structure is the presence of age classes and their abundance in the population. When an age class is present, it means that successful reproduction occurred during that time period. When an age class is absent, it means either that no reproduction occurred or that the young produced did not survive. When an age class is present, its abundance can be used to indicate whether conditions were conducive to good survival, thus giving high abundance. One commonly used method of measuring survival is to sample all ages over

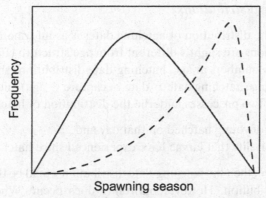

Figure 2.7 An idealized distribution of hatching would follow a bell-shaped curve (—) that measures the abundance of newborns from daily sampling throughout the spawning season. This would
result in a bell-shaped curve for abundance-at-age. This abundance-at-age curve cannot be used as a
catch curve, as it gives initial negative mortality (there are more old fish than young fish). By season's
end, early-spawned larvae have experienced greater mortality than those recently hatched. This
results in a skewed curve (---) of abundance when measured from the hatching dates of survivors
taken shortly after spawning has ceased.

a short period. This produces a snapshot of the population at a specific time. Measured
this way, the age structure incorporates the number born in a cohort and the mortality that
has occurred to them. Alternately, if we follow a given cohort over time, we do not measure
age structure but rather we produce a survival curve. Typically in adult fishes, the snapshot
method is used to estimate the survival from population age structure. The decline in numbers at age will reflect survival only if equal numbers of young are born each year. For adult
fishes this problem is skirted by assuming approximately equal hatchings and by having
enough year classes so that departures from equal hatchings average out over the total
range of ages.

The use of a snapshot of population age structure is far more problematic with larvae
than with adults. Rather than being measured on yearly time frames, larval age structure is
measured on daily time frames. As in adults when the age structure is interpreted from a
single time period, the age structure indicates both production of young and survival at age.
Unlike adults, the daily production of eggs is not equal from one day to the next, or even close
to equal throughout the spawning period. This age structure cannot be used as a proxy for
following daily-cohort survival as with an adult catch curve. To illustrate simply, the daily
production of eggs increases from the beginning of the spawning period to a peak and then
declines to the cessation of spawning. In the simplest case, this would follow a bell-shaped
curve (Figure 2.7). Depending on the actual statistical distribution of abundance-at-age,
you might see more older fish than younger fish. In this example it is obvious that you are
not seeing true mortality, for a negative mortality would indicate the spontaneous generation of new life – when there are 100 3-day olds, something is wrong when you see 200 4-day
olds the next day. Data from the field do not usually produce such an obvious example,
although occasionally they do, and scientists can mistakenly believe that they can obtain
estimates of mortality from these data. In larval fishes, a catch curve is not a proxy to show
mortality and must be used with great caution.

2.4.2 *Hatching-date distributions*

In the early life stages the distribution of hatching dates is as informative as age structure. Hatching-date distributions are slightly different from age structure (Figure 2.8). Whereas age-structure plots show numbers at age, hatching-date distributions show numbers by the estimated date of hatching (hatching date = date of capture − age) and is the transpose or mirror of age structure. Two processes underlie the distribution of hatching dates:

(1)   the actual number of young hatched on that day and
(2)   the cumulative mortality that larvae have experienced since hatching.

If no mortality occurred, then the hatching-date distribution would be the larval-production or egg-production distribution. However, mortality does occur. When viewed as daily cohorts (all the larvae born on the same day), the older the cohort becomes, the fewer larvae remain. In species with low mortality, the hatching-date distribution will not be very different from the egg-production distribution. In species with high mortality, such as pelagic fishes, the hatching-date distribution will differ significantly from that of egg production. The abundance of older larvae on given hatching dates will underestimate their true initial abundance. Thus, in species with high mortality the hatching-date distributions must be corrected by an independent measure of mortality during the larval stage. Lacking this correction, other more intensive methods for estimating hatching-date distributions can be used as shown in Chapter 5 with egg-production methods or by intensive sampling of the larval phase (see Campana & Jones 1992).

The number of hatchings and the subsequent mortality are of paramount interest to the fish ecologist. The fish ecologist seeks to tie these processes to density-dependent and density-independent chronological events, so that the persistence of the population can be understood. Hatching-date distributions have been used to study the seasonal and interannual variation in the survival of larval northern anchovy, *Engraulis mordax* (similar to Figure 1.31). Survival of northern anchovy is affected by transport and by prey availability. Survival from

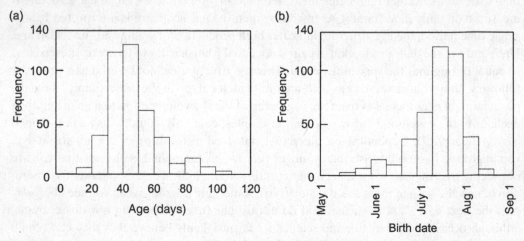

Figure 2.8   (a) A distribution of ages for a sample of larvae. (b) Distribution of hatching dates for these same fish. The hatching-date distribution is a mirror image of the age distribution.

hatching through the larval stage was not attributed to a single cause, but differed from year to year. Similarly, hatching-date distributions can be used, albeit with caution, to clarify the effect of catastrophic events such as storms, droughts, or precipitous temperature changes on the survival of larvae. For example, periodic sampling may reveal that fishes from an early spawn suddenly disappear from the samples taken after a cold snap and only larvae born after the event appear in the sample. Likewise, the hatching dates of some cohorts may be over-represented, which would indicate that events occurred that contributed to their greater survival (see Section 11.11). In this way, hatching dates give us a tool to explore the consequences of environmental conditions for survival through the larval period.

2.4.3   Age–length keys

Age–length keys are used in fishery science as a way to impute age to the entire catch or sample by determining age from a smaller subsample. The method is far simpler and less prone to error for larval and juvenile fishes than for adults. Estimates of the total abundance of larvae can be made from samples of larval density that have been obtained by random sampling with trawls and nets. A statistically chosen subsample of fish is then obtained from the nets or trawls for ageing. Age, length, and weight relationships in the subsample accurately reflect the entire sample. The errors in estimating these relationships include measurement errors of age, length, and weight and the improper choice of mathematical model to fit these relationships. Additionally, there may be errors in the estimates that come from incorrect sampling. With careful sampling, measurement, and analysis, however, scientists can develop good estimates of age, length, and weight for larvae and juveniles. In contrast, the estimation of age–length keys for adults involves additional steps and sources of error. For adults, the catch is measured in weight, not abundance, length, or age. Weight must be converted to abundance, then a subsample must be taken for length. Further, another subsample is taken for ageing. Once age has been determined for the subsample, the proportion of age at length from the sample is expanded to the entire sample. Because more indirect sampling is done, there are more sources of error in these estimates.

Age–length keys are rarely used during the larval period. Direct estimates of size-at-age are important because of the greater relative variability in age-at-length for larval fishes. For example in Atlantic croaker (*Micropogonias undulatus*, Figure 1.3r), a marine sciaenid, a fish of 20 mm standard length can be 30 or 60 days old, depending on whether it hatched early or late in the spawning season. The proportion of size-at-age is not constant. In this example, an age–length key is inappropriate and the scientist needs to determine the age structure by ageing subsamples of larval fish throughout the sampling period.

2.5   Growth

2.5.1   Characteristics of growth

During its lifetime a fish will grow by several orders of magnitude in length and weight (Figure 1.2). For example, marine fishes such as the large sciaenids hatch at 2–3 mm and grow to over 1 m in length (almost a 1000 fold increase). Young perciform fishes grow from

10–20 µg at first feeding to 1–2 mg 30 days later. Similarly, bluefin tuna (Figure 1.3s) hatch at less than 1 mg and grow to over 500 kg in weight (eight orders of magnitude). During the larval period, marine fishes may grow on the order of one-tenth to one-half of their size per day. Changes in length and weight occur rapidly during the period of the life when fishes are sexually immature, and slowly when they reach maturity.

Growth during the larval period is influenced by both exogenous and endogenous factors. Exogenous factors include food, temperature, oxygen, salinity, and the interaction with other organisms. Endogenous factors include genetics, maternal contribution, and the fish's previous growth history. Of the exogenous (or environmental) factors, food availability and temperature have the greatest influence on growth. The size spectrum of available prey determines the quantity and quality of food. As larval fishes grow, they trade off the abundance and ease of capture of small prey for the caloric value of scarcer, larger, and higher quality prey. Just after yolk absorption, the larva's smaller gape limits them to small food particles. As larvae grow, their mouth becomes larger and they can eat larger prey items when available, but can still prey on smaller organisms. Larval growth depends on encountering and capturing prey of the right size.

The amount of food needed by larvae and the availability of this food are also influenced by temperature. Temperature directly affects the growth rate of larvae. Typically fish growth rates respond to increasing temperature in a non-linear fashion. A general pattern is for growth rates to be slow at colder temperatures, reach a maximum at intermediate temperatures, and fall with temperatures warmer than optimal. Environmental temperature also has a profound effect on the seasonal availability of prey. David Cushing developed a hypothesis that good survival of larval fishes occurs when fish spawning coincides with the seasonal bloom of prey – termed the Match/Mismatch Hypothesis. In temperate climates, waters undergo a seasonal overturn that promotes nutrient release into sunlit surface waters, and thereby triggers a series of seasonal blooms of phytoplankton followed by zooplankton. This seasonal cycle is driven by temperature. So beyond the direct effects of temperature on metabolism and growth rates, secondary temperature effects control the timing and availability of prey. The Match/Mismatch Hypothesis will be revisited in Chapter 4.

Endogenous factors also affect growth rates, as all fishes are not born equal. Some larvae are born with the right combination of genes to meet the challenges of the environment into which they are born. Beyond genes, the quality and amount of yolk for each larva is dependent on the vitality and age of the mother. A good supply of yolk gives a larva a better chance to survive. Larvae that are born bigger or have quicker growth are more likely to continue to be bigger and grow faster. The size that a larva achieves depends, in part, on the size that it has already obtained. Faster growing, more competent larvae survive better. When environmental conditions or competition is severe, such differences are exacerbated and only the fittest larvae survive to metamorphose. Sometimes luck wins out and those larvae that happen to be advected into a patch of high quality food that supports quick growth may survive.

Cumulative effects of exogenous and endogenous factors result in increasing variability in size as larvae age, a phenomenon known as growth depensation. At hatching, there is little variation in the size of individuals within a spawn, whereas large differences are present within a matter of weeks. Growth depensation produces cohorts based on larva size. Once formed, these size cohorts can persist well into the juvenile period. Although fishes, especially adults, have the capacity for compensatory growth, which reduces variability in size-at-age, size differences in the larval period are important to survivorship because most mechanisms of mortality are size-dependent. Continuation of size variability into later stages of life can profoundly affect life-history traits such as age at maturity, fecundity, and survival.

2.5.2 Size relations and body proportions

The relation of body proportions has wide use in adult fishes. It is used to indicate the condition of a fish, and differences in body proportions can also be used to identify subpopulations of a species. When body proportions remain the same, the ratio of body depth to length or head length to body length, for example, remains constant. When these ratios are constant, growth is referred to as isometric. When the growth of one body part is faster or slower than that of another, the growth is referred to as allometric. For fishes with isometric growth, the fish's biomass is a cubic function of its length

$$W = \alpha \cdot L^3 \tag{2.1}$$

where W is weight, L is length, and α is a parameter to be estimated. For fishes with allometric growth, the fish's biomass is a power function of its length:

$$W = \alpha \cdot L^\beta \tag{2.2}$$

where β is the power coefficient ($\neq 3$) and all other symbols are the same as above. These relations have limited use for larval fishes, but are widely used in the study of adults, where, within a species, weight may be less or more at a given length. As we have seen before, understandings drawn from adult fishes are not directly applicable to larvae. The relation of body proportions may be constant in some fishes that have precocial larvae or may change dramatically for fishes having altricial larvae. The length–weight relation is little used for altricial larvae because larvae have highly variable body proportions between hatching and metamorphosis. The changes in morphological proportions and the relationship between length and weight are developmental rather than a result of changes in condition. In contrast, for adults the relationship between length and weight provides insight into the condition of the fish.

The relation between otolith and body size is one of the most useful and critical to the study of the larval period. If otolith size is tightly coupled with body size, then the size of the otolith measured from sequential daily increments, can be used to calculate the historical size of the larva at a prior time in its life. For the most part, otolith size does reflect body size. Once this is demonstrated then otolith metrics, such as the maximum diameter or radius, can be used to reconstruct an individual fish's growth history (Figure 2.9). This history yields individual measures of growth rate. When individual growth rates are known, then they can be used to investigate the presence of size-selective mortality, the impact of environmental conditions on growth, and latitudinal gradients in larval growth, among others. It is a powerful technique, but it must be used wisely. Several studies have shown that otolith size and body size can become decoupled. The otolith to body-size proportion may change depending on the health of the larva or its growth rate. One study found that slower growing larvae had heavier otoliths because even though the fish were barely growing,

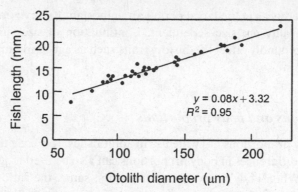

Figure 2.9   Otolith size must be closely related to fish size for the scientist to use otolith size as a measure of prior size-at-age. Such a close relation is illustrated here. Note, however, that the relation begins with 8-mm or larger fish and may provide inaccurate projected sizes with otoliths smaller than shown (70-µm in this illustration).

a new, albeit narrow, daily increment was added each day to the otolith. Whether or not this decoupling persists among the few fish that survive to adulthood in the wild remains unanswered. Given the knowledge that otolith growth can become decoupled from body growth, retrospective studies of growth should be done with this concern in mind.

2.5.3   Spatial aspects of growth

The models of growth that we commonly use in fishery science and fish ecology rely on the assumption that these fishes are part of a unit stock. The unit stock concept means that there is so much migration between subpopulations and habitats that a fish selected in one place will represent all fish in the population. We tacitly rely on this assumption when we calculate growth rates and ignore the potential differences that we might see spatially. For many populations of fishes, adults do migrate widely and this assumption may hold. This may also hold true for marine species that are estuarine-dependent and those that spawn offshore to produce pelagic larvae that, in turn, replenish several subpopulations. In this case there is no genetic adaptation of growth to local conditions. There are, however, scenarios for early life when mixing may be more restricted. One can think of species whose young are born and retained in their natal habitats, thereby showing a high degree of philopatry even in the face of routine adult migration.

The best known example of a spatial influence on growth rates is the hypothesis of countergradient variation in growth developed by David Conover and his students. Studies were done on estuarine and estuarine-dependent populations of mummichog (*Fundulus heteroclitus*), Atlantic silverside (*Menidia menidia*), and striped bass (*Morone saxatilis*). In one of the most extensive studies, mummichog were captured from several subpopulations for each of two subspecies that occur along a north–south latitudinal gradient. These fish were bred, their progeny tested in the laboratory, and then the progeny were bred and the second generation was also tested in the laboratory. By testing a second generation, the genetic effect was isolated from the maternal effect that was potentially present in the progeny of the wild-caught fish. Wild fish in the northern part of the range have a much shorter growing season than those in the southern part. Nonetheless, juveniles in both areas are the same size at the end of the growing season. Hence, northern larvae must grow faster. This growth-rate difference could be attributed to faster growth at lower temperatures by the northern fish as a result of temperature adaptation. In the laboratory, however, northern fish actually grew faster than southern fish at higher temperatures and grew similarly at lower temperatures. So the differences in growth were not due to temperature adaptation, but were genetic in origin and probably the result of selection pressures. The selection pressure is probably on the size that must be accomplished prior to overwintering in estuaries. This topic will be addressed again in Chapter 3. This research underscores the importance of clarifying the effect that spatial heterogeneity has on the traits and vital rates we measure.

2.5.4 Biochemical methods of estimating growth

Several biochemical methods have been used in adult fishes to measure growth and condition. Biochemical methods for estimating growth differ from methods based on size changes. Biochemical methods use tissue components, such as ribonucleic acid (RNA), deoxyribonucleic acid (DNA), proteins, and lipids to indirectly measure increases in somatic tissues that result in size change. The advantage of these methods is in their ability to measure immediate past growth and condition, thus also measuring the potential for growth in the immediate future. Most of the methods that are used for adults are impractical for use on larvae, given the size of larvae and the difficulty in extracting sufficient amounts of tissue from each individual. Two methods have been modified for use with larvae, RNA/DNA ratios for growth and lipid analysis for condition.

Estimates of recent growth can be obtained from the analysis of the RNA/DNA ratios in larval fish tissues. Unlike otolith-based estimates of growth, these biochemical estimates provide a quantifiable measure of growth that is the result of just the previous day or two. The concept behind this method is straightforward. The amount of DNA in cells is constant and the amount in a sample reflects the size of the sample (number of cells) being analyzed. The amount of RNA is variable and it reflects the amount of recent protein synthesis. When protein synthesis is high, RNA levels will be elevated and the RNA/DNA ratio will also be high. Conversely, when protein synthesis is low because of a recent starvation event, then the RNA levels will be depressed and the RNA/DNA ratio will be low. When RNA/DNA ratios are high, proteins are available for tissue growth and high ratios correlate with recent somatic growth and the potential for future growth. The ratio of RNA/DNA varies ontogenetically and by growth rate. In a study of larval red drum, *Sciaenops ocellatus*, 10-day-old larvae had ratios of 4–7 for faster growers and ratios of 2–4 for slower growers. By 18 days of age, faster growers had ratios of 3–5 and the slower growers had ratios of 1–3. At the point of no return (indicating irreversible starvation), Japanese sardine (*Sardinops melanostictus*) have an RNA/DNA ratio of 1.2–1.3. The case study for this species discussed in Chapter 11 demonstrates the use of RNA/DNA ratios for assessing the role of larval starvation in population dynamics.

Although not truly a measure of growth, condition factors reveal the fish's potential for continued growth in the near future. Fishes in good condition are relatively fat and well fed. Their future prospects for continued growth are good. In adults, the relation between length and weight is used to assess condition. As described above, condition indices based on length and weight relations do not necessarily reveal condition in larval fishes. Recently, the constituent lipid content of larval fishes has been used to assess condition. Among the lipids analyzed are triacylglycerol and sterol or polar lipid. Triacylglycerol is an indicator of physiological state and its ratio to sterol reflects the larva's condition.

As discussed in Chapter 1, larvae are more prone to starvation than juveniles or adults. They have smaller body mass (stored energy in Figure 1.11) available for protein catabolism during periods when there is insufficient food to meet their minimum metabolic needs. Similarly, they will respond quickly when prey densities are high. Thus, larvae are good candidates for biochemical measures of growth and condition as long as the techniques can be applied to organisms of small size, often less than 20 mm in total length. By the time fishes have gotten to the juvenile period, they are better able to physiologically integrate environmental fluctuations as they have adequate reserves to draw upon. Still, these biochemical methods are useful for assessing growth and condition.

2.5.5   Modeling growth as a change in size

Growth is typically measured as the change in length or weight with age. Over their lifetime, fishes become larger and heavier. Unlike mammals, most fishes grow throughout their lives and are said to have indeterminate growth. This and the presence of a larval form make them unique among the vertebrates. Fishes are also ectotherms and their growth is strongly influenced by the temperature of their environment, as well as by their food supply. Because their growth is so strongly influenced by their environment, size alone is not a good indicator of age and the relation between size and age varies over the fish's lifetime. For adults, growth typically slows after sexual maturity and the growth curve can become asymptotic when size is measured as length or weight. At sexual maturity, energy must now be used to produce gametes and can no longer be invested solely in somatic growth. Hence, growth slows. The rate of growth also depends on whether length or weight is measured. Growth in length slows at a younger age than does growth in weight. In larvae and juveniles growth may slow temporarily and therefore care must be used in modeling growth in early life. Growth is not only a function of age and sexual maturity, but is determined by heredity, physical state of the mother, prior growth, and ultimately by luck.

The simplest measure of an individual's growth over a period of time, $\Delta t \ (= t_2 - t_1)$, is the difference in length, L, or weight, W, termed *absolute growth*: $L_{t_2} - L_{t_1}$ or $W_{t_2} - W_{t_1}$. The rate of change in size, or *absolute growth rate* (AGR), is more instructive than the simple measure of size difference and, where growth is linear, these models are

$$\text{AGR} = \frac{L_{t_2} - L_{t_1}}{t_2 - t_1} \quad \text{or} \quad \text{AGR}' = \frac{W_{t_2} - W_{t_1}}{t_2 - t_1} \tag{2.3}$$

These linear models of growth provide a first-order estimate of growth and, in a surprising number of species, give a reasonable fit during the larval period. These simple measures are just the total change in size that has occurred during a specific time period. For example, the AGR might be 4 mm per week. For an adult, this growth rate would not be impressive, but for a 5-mm larva this would be an impressive rate of growth. Based on the importance of current body size, there are obvious advantages in expressing the AGR in proportion to the growth already achieved as the proportional growth rates, g:

$$g = \frac{L_{t_2} - L_{t_1}}{L_{t_1}(t_2 - t_1)} \quad \text{or} \quad g' = \frac{W_{t_2} - W_{t_1}}{W_{t_1}(t_2 - t_1)} \tag{2.4}$$

Although calculation of specific growth improves our ability to compare growth between fish in different stages, when we use this metric we assume that growth is linear. Linear growth measures may provide a good fit over short periods, but they fail to model growth over longer periods or the entire lifetime. In periods when growth is accelerating, an exponential model provides a better fit to the actual size increase and, when the time period (Δt) is small, it can be modeled as the instantaneous growth rate

$$G = \frac{\ln L_{t+1} - \ln L_t}{\Delta t} \quad \text{or} \quad G' = \frac{\ln W_{t+1} - \ln W_t}{\Delta t} \tag{2.5}$$

where G is instantaneous growth rate measured in length and G' is instantaneous growth rate measured in weight.

The models given here can be used to obtain growth rates to use for purposes of description or comparison, but not for prediction. The models for instantaneous growth yield the equations

$$L_t = L_0 \cdot e^{G(t-t_0)} \quad \text{and} \quad W_t = W_0 \cdot e^{G'(t-t_0)} \tag{2.6}$$

when rearranged and evaluated starting at time $t = 0$. However, fishes do not grow exponentially over their lifetime. We need better models if we are to predict future growth. For prediction, we use more sophisticated models that describe growth over the entire life span.

Lifetime models of growth try to fit observed data of size-at-age to the pattern of size from hatching to death. The general pattern of absolute growth is slow when fishes are small, fast when fishes are at an intermediate size, and slow again as fishes become large and begin to reproduce. This pattern yields an S-shaped or sigmoidal curve. Several models of lifetime growth have become popular in fishery science. Some of these models are accurate over the larval and juvenile period, while others are not. The von Bertalanffy growth function is the most commonly used growth function for adults and the model in length is

$$L_t = L_\infty \left(1 - e^{-K(t-t_0)} \right) \tag{2.7}$$

where L_t is length at time t, L_∞ is mean asymptotic length, K is the growth coefficient, and t_0 is the hypothetical age when $L = 0$.

This growth model is used widely because of its incorporation into the Beverton–Holt Yield Model. Despite its wide use, it does not accurately represent growth for larvae and juveniles. Several other growth functions are used in conjunction with stock assessments, such as the Richards seasonal growth model or the generalized model of John Schnute (see Seber & Wild 1989, for a detailed discussion of these models). However, these growth functions miss the initial slow growth that is important when predicting growth through the larval period and they are of little value to people studying the early life of fishes.

In contrast, the Laird–Gompertz growth function is a model that fits the larval period well and it is useful for predicting size throughout life. The Laird–Gompertz growth equation is

$$L_t = L_0 \cdot e^{(g_0/\alpha)\cdot(1-e^{-\alpha t})} \tag{2.8}$$

where L_t is length at time t, L_0 is length at hatching ($t = 0$), g_0 is specific growth rate at hatching, and α is the rate of exponential decay of the specific growth rate. This growth curve is useful when growth is not symmetrical around the point of inflection. It is fit with non-linear methods (see below) to estimate the parameters L_0, g_0, and α.

Care must be taken when modeling growth during the larval period. Larvae become more competent as they mature and this competence is demonstrated by more powerful swimming and better predator avoidance. As they swim more strongly, they are capable of avoiding the nets that capture younger larvae or smaller members of their age group. Thus, their catchability changes. When sampling gear catches only the smaller fish of a daily cohort, the measure of size-at-age is biased downward and will not represent the population's true growth. Additionally, gear avoidance can leave a false impression of asymptotic growth. Although growth may slow or speed seasonally, or as different transitions in form are reached, overall growth is continuous throughout life. When a plot of size against age appears asymptotic in the larval period, gear avoidance should be suspected. When gear avoidance is suspected, the age distribution can be truncated to eliminate the ages with inaccurate measures of size-at-age. Otherwise, the scientist must use independent estimates of gear avoidance by age-specific size to correct the inaccurate distribution, although these corrections exist only in rare instances.

2.5.6   Fitting growth curves

Growth is modeled differently depending on how size-at-age data are collected. The simple case of calculating growth over a limited period of time was presented above. Other approaches to modeling growth are used commonly in the fishery science literature and these cover population growth curves. In both adults and young, a population growth curve is produced when the size-at-age of capture is measured. This is the growth curve most commonly used. Ideally fish are obtained for this growth curve using a statistically sound sampling plan that insures that the fish in the sample are representative of the population under study. The sampled fish are measured for size and their otoliths or other hard parts are extracted and read for age. Only one measure of size-at-age is obtained for each fish in the sample. Because each measurement of size-at-age is independent, curves are fitted to data using either linear or non-linear least squares, which is the most commonly used curve-fitting technique.

We measure size and age from the survivors that are left in each age group from all fish that were born initially in the cohorts. Surviving fish are often those that have grown fastest. Hence, this type of growth curve includes both the measure of growth and any size-selective mortality that may have occurred. This growth curve has many important uses. When applied to adults, it forms the basis for yield calculations. It is often used to model the average growth of cohort members and then to estimate cohort biomass. For this growth curve,

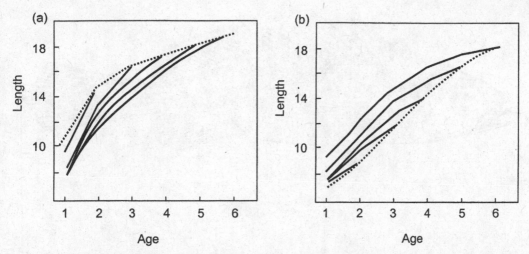

Figure 2.10   William Ricker realized that growth could be measured in two ways: as size and age at capture (....) or as back-calculated size-at-age (—), as illustrated here. He termed the latter the true growth rate of the population. If fishes are subject to size-selective mortality, either the slower growing fishes survive (a) – seen commonly for adults – or faster-growing fish survive (b) – as hypothesized for larvae and juveniles. These back-calculated growth rates will differ from those obtained from size- and age-at-capture data ((a) redrawn from Ricker 1969).

all individuals are assumed to be randomly selected from the population and the size-at-age of one individual is considered independent of others.

If no size-selective mortality has occurred, then the growth curve will accurately reflect the growth of individuals within the population. In adult fishes, however, size-selective mortality is commonplace. In exploited populations the fastest growing fish are typically harvested in greater proportion and earlier in life than the slower growing fish of a cohort. This results in Lee's Phenomenon. When Lee's Phenomenon occurs fish from the oldest cohort will be smaller at age 1 when compared to fish from younger cohorts at age 1 (Figure 2.10). The presence of this phenomenon indicates that the population was subjected to size-selective mortality.

Sometimes a scientist is interested in the growth of individuals rather than the growth of the population. In this case, individuals are sampled randomly from the population as discussed previously. Separate growth curves are built for each individual. An individual's growth curve is produced when the size-at-age is measured retrospectively from hard parts. First, a relation is established between fish size and the size of the ageing structure. Second, using this relation the individual's length is reconstructed from the size of the ageing structure at each prior age (Figure 2.11). This results in a growth curve for each individual. William Ricker understood correctly that this is the true growth rate but did not present more than a logical argument. As yet, no mathematical framework or derivation has been developed to compare the relation between individual growth and population growth.

Retrospective or individual growth curves typically rely on a linear relation between growth of the ageing structure and the somatic growth of the fish, as does the population growth curve. Assumptions of the method are that the relation between the ageing structure and body size is consistent over the entire life span from hatching until the final size

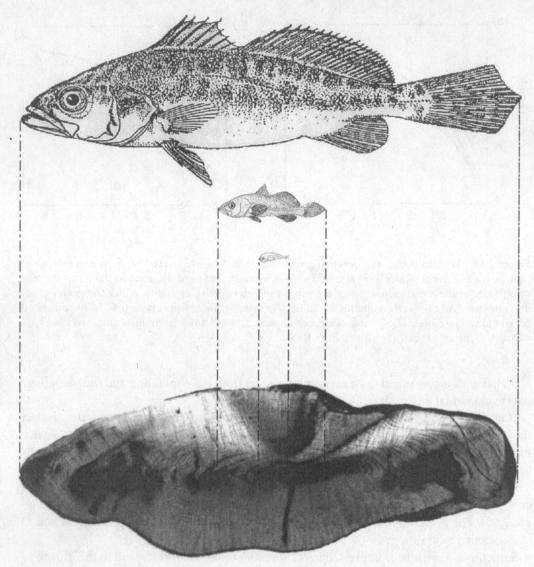

Figure 2.11   This illustration shows the relationships between fish size and otoliths size. Otolith size can be measured as a linear distance (such as diameter or radius), volume, or weight.

measured, and that the relation between the size of the ageing structure and fish size remains coupled under all conditions of growth. These assumptions have been problematic. Much has been written about the proper way to build the relation between the ageing structure and the fish's size and about the problem with growth decoupling. Additional readings on this subject appear at the chapter's end.

Because individual growth curves are built upon retrospective data from each individual, they cannot be modeled with the same methods that are used for population growth curves. The scientist cannot use techniques that assume independence, such as ordinary least squares, that are used to fit population growth curves. Each retrospective estimate of size is not independent from the others calculated for an individual fish. The size that a fish can become is correlated with the size it has already reached. Thus, we must use techniques that incorporate this autocorrelation, such as repeated-measures models. Without taking auto-correlation of growth into consideration, estimates of growth for individual growth curves can be grossly distorted. Only a few studies have recognized the importance of using proper methods to fit individual growth curves, such as those of R. Christopher Chambers and Thomas Miller, and Cynthia Jones. Unfortunately, the literature is replete with improperly applied growth curves derived from repeated measurements on individual fish.

The distortion from improperly fitting individual growth curves is greater in larvae and juveniles than adults. The number of repeated measurements for a larva or juvenile is their daily age. Thus it is not uncommon to have 30 to over 100 repeated measures for a larva or juvenile. The advantage of an individual growth curve for larvae and juveniles is that it yields a true measure of the average growth for individuals that is not comprom-ised by size-specific mortality. Moreover, we are most frequently interested in the average individual's growth rather than the population's growth during the larval period. Once the average individual's growth is estimated, we can use statistical methods such as analysis of covariance or residual analysis to estimate the effect of biotic and abiotic factors on growth.

2.5.7   The relation between growth and mortality

In much of the literature on larval fishes, characteristics that enhance survival have been size related. Thomas Miller and his colleagues have demonstrated that in fish larvae the faster growing members of a cohort have a better chance of survival. Larger larvae avoid starvation and predation better than smaller larvae. The relation between instantaneous growth rate in weight, G', and mortality, M, has been used to illustrate the application of sized-based ecosystem theory to the early life dynamics in fishes. D.M. Ware demonstrated mathematically and conceptually that, for a larval cohort to gain biomass, G' must be greater than M over the larval period. For any given cohort, the abundance is greatest at the appearance of the cohort and declines thereafter. Moreover, M can be represented as an exponential decline in numbers that is a function of weight and growth

$$\frac{N_t}{N_0} = \left(\frac{W_t}{W_0}\right)^{-M/G'} \tag{2.9}$$

where N is abundance and all other symbols are as defined previously (see p. 178 in Houde 1997). As the abundance of the cohort declines, the individual fish that survive gain weight. These two processes, G' and M, result in a maximum biomass during the life of the cohort. This maximum biomass typically occurs in adults as their growth rate slows after reaching sexual maturity. A minimum biomass may also occur in the larval stage when mortality is high, but individual growth rates are not high enough to overcome this early mortality. Small changes in abiotic and biotic conditions, such as changes in temperature or prey abundance, will alter G' and M. By altering G' and M the biomass minimum will be changed. Depending on the timing of this change, the numbers of young fish that survive to metamorphose can be dramatically altered. The relation between G' and M has also been used to provide a length-based measure of mortality expressed in abundance decline

$$\ln N = \frac{M}{G'} \cdot \ln L \qquad (2.10)$$

solving for N

$$N = e^c \cdot L^{-M/G'} \qquad (2.11)$$

where c is a constant of integration. Edward Houde has used these relationships to investigate the factors that alter the M/G' ratio in striped bass in Chesapeake Bay and, thereby, to understand the factors affecting recruitment. Although this ratio provides a good conceptual framework to understand the importance of growth rates as they affect survival, growth and mortality are highly variable during early life and are notoriously difficult to measure accurately. More about M/G' will be provided in Sections 3.4.2 and 7.5.

2.6   Summary

The larval period is profoundly important in the life of fishes. During this time the new cohort is reduced dramatically to the rare survivor that goes on to reproduce and add to the population's persistence. In freshwater fishes only one in 100 may survive and in marine fishes even fewer, perhaps one in a million. Within the first few months of life, the proportional increases in length and weight are greater than for all of the rest of the fish's life. From hatching to death, fishes may increase in size by as much as five to eight orders of magnitude. Fishes accomplish their first 10-fold increase in size within the first 30 days; they may grow as much as one-tenth to one-half of their size per day. Arguably, this period is the most important one in the life of fishes.

As we have seen, research on the age and growth of fish larvae provides unique types of data. Much of this is due to the presence of daily increments in otoliths. These increments provide high temporal resolution for age and growth estimations. Being able to make 30 to more than 100 repeated measurements of growth for an individual yields a measure of average growth that is more accurate than is possible for adults. The daily resolution can also be used to pinpoint the spawning location, dispersal rate, and transport to nursery grounds. Changes in otolith microstructure can serve as markers of major transitions in early life, such as hatching, yolk absorption, metamorphosis, and settlement. Otolith microchemistry can give clues to a fish's exposure to different water masses, which may add immeasurably to our understanding of how fishes use habitat at specific times in their early life. Hatching-date distributions can reveal seasonal and interannual variations in survival. They can even be used, albeit with caution, to clarify the effect of catastrophic events or especially beneficial conditions on the survival of larvae. Age–length keys, long used for adult fishes, can be used to impute age to a sample of larvae more simply and with less error than to adults. These unique types of data give us new insight into the processes that regulate survival through early life. Thus, the study of age and growth in larval fishes gives us tools to understand why so many die and so few survive to reach adulthood.

Additional reading

Begon, M., Mortimer, M. & Thompson, D.J. (1996) *Population Ecology: A Unified Study of Animals and Plants*, 3rd edn. Blackwell Science, Oxford.

Buckley, L.J., Calderone, E. & Ong, T.-L. (1999) RNA–DNA ratio and other nucleic acid-based indicators for growth and condition of marine fishes. *Hydrobiologia* **401**, 265–277.

Campana, S.E. & Jones, C.M. (1992) Analysis of otolith microstructure data. In: *Otolith Microstructure Examination and Analysis*. (Ed. by D.K. Stevenson & S.E. Campana), pp. 73–100. Canadian Special Publication of Fisheries and Aquatic Sciences 117, Ottawa, Canada.

Campana, S.E. & Neilson, J.D. (1985) Microstructure of fish otoliths. *Canadian Journal Fisheries Aquatic Sciences* **42**, 1014–1032.

Conover, D.O. (1990) The relation between capacity for growth and length of growing season: evidence for and implications of countergradient variation. *Transactions of the American Fisheries Society* **119**, 416–430.

Ferron, A. & Leggett, W.C. (1994) An appraisal of condition measures for marine fish larvae. *Advances in Marine Biology* **30**, 217–303.

Gotelli, N.J. (1998) *A Primer of Ecology*, 2nd edn. Sinauer Associates, Inc., Sunderland, Massachusetts.

Jones, C.M. (1986) Determining age of larval fish with the otolith increment technique. *Fishery Bulletin* **84**, 91–103.

Kamler, E. (1992) Early exogenous feeding period. In: *Early Life History of Fish: An Energetics Approach*, Fish and Fisheries Series 4, Chapter 6. Chapman & Hall, London.

Laird, A.K., Tyler, S.A. & Barton, A.D. (1965) Dynamics of normal growth. *Growth* **29**, 233–248.

Methot Jr., R.D. (1983) Seasonal variation in survival of larval northern anchovy (*Engraulis mordax*), estimated from the age distribution of juveniles. *Fishery Bulletin* **81**, 741–750.

Miller, T.J., Crowder, L.B., Rice, J.A. & Marschall, E.A. (1988) Larval size and recruitment mechanisms in fishes: toward a conceptual framework. *Canadian Journal Fisheries Aquatic Sciences* **45**, 1657–1670.

Quinn II, T.J. & Deriso, R.B. (1999) *Quantitative Fish Dynamics*. Oxford University Press, New York.

Ricker, W.E. (1969) Effects of size-selective mortality and sampling bias on estimates of growth, mortality, production and yield. *Journal of the Fisheries Research Board of Canada* **26**, 479–541.

Schultz, E.T., Reynolds, K.E. & Conover, D.O. (1996) Countergradient variation in growth among newly hatched *Fundulus heteroclitus*: geographic differences revealed by common-environment experiments. *Functional Ecology* **10**, 366–374.

Smith, T.D. (1994) *Scaling Fisheries*. Cambridge University Press, Cambridge.

Thorrold, S.R., Jones, C.M. & Campana, S.E. (1997) Response of otolith microchemistry to environmental variations experienced by larval and juvenile Atlantic croaker (*Micropogonias undulatus*). *Limnology and Oceanography* **42**, 102–111.

Von Bertalanffy, L. (1960) In: *Fundamental Aspects of Normal and Malignant Growth*, Chapter 2. (Ed. by W.W. Nowinski), pp. 137–259. Elsevier, Amsterdam.

Wang, S.B., Cowan Jr., J.H. & Houde, E.D. (1997) Individual-based modelling of recruitment variability and biomass production of bay anchovy in mid-Chesapeake Bay. *Journal of Fish Biology* **51** (suppl. A), 101–120.

Wootton, R.J. (1990) Growth. In: *Ecology of Teleost Fishes*, Fish and Fisheries Series 1, Chapter 6. Chapman & Hall, London.

Chapter 3
Mortality
Edward D. Houde

3.1 Introduction

Teleost fishes have high fecundities and so they must experience high mortality rates to prevent explosive increases in numbers. In many populations, the number of eggs spawned annually or the number of newly hatched larvae may exceed 1×10^{12}. Declines in numbers during egg and larval stages therefore must be precipitous, if juvenile and adult populations are to be maintained at levels that can be supported by an ecosystem. The summed losses, or mortalities, experienced by a population are considered relative to age (age-specific) or ontogenetic state (stage-specific). Natural mortality rates are highest in early life, gradually declining during larval and juvenile stages, and often becoming more or less stable after fishes reach maturity. Fishes may die from many causes, but predation is usually the principal agent of death. Other causal factors are poor nutrition, disease, and unfavorable environmental conditions. Human-related causes of mortality, for example, fishing, contaminants, altered water quality, deteriorating or lost habitats, and power-plant impingement or entrainment may be important sources of mortality in many freshwater and marine fishes.

The range in natural mortality rates of juvenile and adult fishes is variable and strongly related to the life span of a species. Small, short-lived fishes, such as many anchovies and atherinids, may sustain annual mortality rates after recruitment that range from 50% to 90%, making them essentially annual fishes because few individuals live more than 1 year. On the other hand, large and long-lived fishes such as sharks, sturgeons, and swordfish, may experience natural mortality rates of 10% per year or less. There are many medium-sized, slow growing fishes that are long-lived, such as the Pacific rockfishes (*Sebastes* spp.) or striped bass (*Morone saxatilis*) that also have low natural mortality rates. Knowing the natural mortality rate of a population is important, especially for populations that also are subjected to significant mortality from fishing. Despite this need, the natural mortality rates of most fishes are poorly estimated or unknown. It is a difficult and sometimes costly task to estimate natural mortality rates, which vary with age or size, in addition to being inversely correlated with the life span of a species. For heavily exploited species such as Atlantic cod (*Gadus morhua*) and Atlantic herring (*Clupea harengus*), or freshwater percids, fishing mortality rates may be as much as five times higher than the natural mortality rates. Such high mortality rates can destabilize a population and jeopardize the sustainability of a fishery or cause failures in recruitment followed by collapses of stocks.

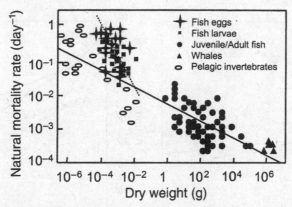

Figure 3.1 Relationship between mortality rate (day^{-1}) and dry weight for marine organisms. Dotted line represents the regression line and slope for fish eggs and larvae, which is significantly higher than the slope of the regression for marine organisms in general. The overall slope is -0.25, while the slope for fish eggs and larvae is -0.85 (reproduced from McGurk 1986 with permission of Inter-Research).

It has been demonstrated that the natural mortality rates of marine organisms in pelagic ecosystems, ranging from the smallest invertebrate larvae to whales, are strongly size-dependent and decline approximately as $M = 0.0053W^{-0.25}$ (McGurk 1986), where W is individual weight. The power relationship expressed here is believed to be the consequence of predation in size-structured aquatic ecosystems. Natural mortality rates of juvenile and adult fishes fit this picture reasonably well, but for eggs and larvae, the exponent tends to be even less than -0.25, indicating a higher than expected mortality rate during these stages and a rapid decline in M with ontogeny (Figure 3.1). The high and variable mortality rates observed for the smallest stages of fishes (Table 3.1) suggest that predation, a primary cause of death, is also high and variable, and potentially a major contributor to fluctuations in levels of recruitment that characterize fish year classes (see Chapter 4).

3.1.1 Survivorship curves

Mortality of a population is usually expressed as, and estimated from, the decline in numbers in relation to age. This relationship, in which numbers of survivors are plotted

Table 3.1 The average relationship between M and W for five species of fishes during the larval stage.

Species	Relationship
American shad	$M = 1.724W^{-0.392}$
Northern anchovy	$M = 1.073W^{-0.353}$
Bay anchovy	$M = 2.284W^{-0.318}$
Walleye pollock	$M = 3.874W^{-0.622}$
Striped bass	$M = 4.875W^{-0.424}$

From Houde (1997).

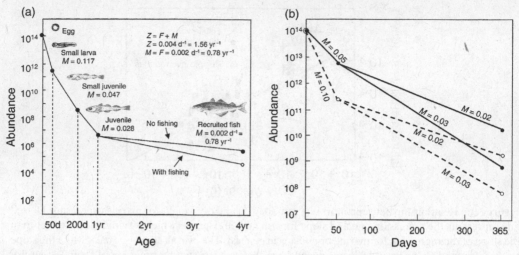

Figure 3.2 Conceptual illustrations of survivorship in a fish cohort from the egg stage through recruited stages. (a) Stage-specific mortality rates for four designated life stages. Effects of fishing are illustrated for the recruited stage. (b) Effects on survival of variability in mortality rates during the larval stage (0–50 days) and in the juvenile stage (50–365 days). Order-of-magnitude variability in survival results from modest changes in mortality rates (M = instantaneous natural mortality, F = instantaneous fishing mortality, Z = instantaneous total mortality).

against age, is a survivorship curve. Fishery scientists often derive this relationship as a "catch curve" in which the slope of the $\log_e$ abundance of surviving individuals plotted against age (often in years) is the age-specific (or instantaneous) mortality rate. In early life stages, the initial slope (steepness) of the survivorship curve and subsequent changes in the slope during early life illustrate the effects of coarse controls that can determine the fate of a cohort (Figure 3.2a). Even small changes in the slope of the survivorship curve (that is, mortality rate) early in life, when numbers are high, can generate very substantial differences in abundances by the end of the larval stage* (Figure 3.2b). It is easy to demonstrate by simple simulations that very high loss rates early in life, from whatever cause, can coarsely control the fate of a cohort's abundance later in life, resulting in a failed or poor recruitment. The converse, however, is not true; low mortality rates during the egg and larval stages do not guarantee high survivorship at recruitment, if mortality rates during the juvenile stage are elevated even modestly above average levels during this life stage, which may be hundreds of days in duration (Figure 3.2b).

3.1.2   *Estimating mortality*

Survivorship curves generally provide the basis for mortality estimation in fishes. This is true whether we are estimating annual mortality rates of recruited fishes or the losses

---

*Here, "stage" is used in a broad sense to represent major phases of a fish's life history (such as, embryonic, larval, and juvenile), rather than as a more instantaneous depiction of ontogenetic state as in Chapter 1.

during egg, larval, and juvenile stages. The catch curves that are fitted to abundance-at-age data provide estimates of losses and the rates of mortality. It is important to accurately estimate mortality rates in any stage to understand how factors such as fishing and natural mortality rates act to control population abundances. It is particularly important, but also especially difficult, to estimate mortality in larval stages. A major impediment to obtaining good estimates of larval mortality is our poor ability to obtain representative samples of larvae that are patchily distributed in both freshwater and marine habitats. The open nature of many aquatic environments, in which complex hydrography and larval behavior interact to disperse, transport, and distribute larvae over broad regions, challenges the ability of fishery scientists to estimate larval mortality.

3.1.3 The components of mortality

The losses in a population, which are proportional to the mortality rate, are expressed as

$$-dN = M \cdot N_t \, dt \tag{3.1}$$

where N is the number of individuals in a population, M is the natural mortality rate, and t refers to age. Solving

$$-M \, dt = \frac{1}{N_t} \cdot dN \tag{3.2}$$

and

$$N_t = N_0 \cdot e^{-M \cdot t} \tag{3.3}$$

Mortality rates can be partitioned into component causes. For post-juvenile and adult stages, natural mortality M and fishing mortality F (for species that are fished) are the two components that contribute to total mortality Z $(= F + M)$. For unfished species and for life stages on which no fishing takes place, $F = 0$ and $Z = M$. These additive "instantaneous" rates (per unit time) are related to more conventional, conditional loss rates (fraction dying per unit time) by the following relationships

$$A = 1 - e^{-Z} \tag{3.4}$$

$$m = 1 - e^{-F} \tag{3.5}$$

$$n = 1 - e^{-M} \tag{3.6}$$

where A is the total mortality, that is, the proportion dying during a specified period, m is the proportion expected to die from fishing if no other causes of mortality were operating, and n is the proportion expected to die from natural mortality in the absence of fishing. For species or early life stages with $F = 0$, the conditional total mortality simplifies to $A = n = 1 - e^{-M}$. In some cases, the natural mortality rate, M, can be partitioned into its constituent components, for example, predation (M_1), and other causes (M_2), although in most cases, the fractional mortality attributable to specific natural causes is unknown.

Natural mortality of organisms can arise from many sources, but predation is usually the primary cause in aquatic ecosystems. Additional sources of mortality can include starvation, disease, or effects of poor water quality and other unfavorable environmental conditions.

For early life stages of fishes, the role of nutrition and its effect on growth rate, and therefore size-at-age, can be crucial since predation may be size-specific and size-selective in size-structured aquatic ecosystems. Interactions between growth rate and mortality rates experienced by young fishes are believed to be among the more important variables that determine recruitment success in marine fishes. Fishing may become an important source of mortality to juvenile and adult fishes, but larval stages are seldom exposed to fishing mortality. In a few notable exceptions, fisheries are directed at larval stages of anchovies and sardines or newly transformed eels in the Mediterranean Sea and western Pacific Ocean.

3.1.4   Concepts of compensation and density dependence

Natural mortality rates may be rather constant with respect to age and size in fishes that survive early life and a single value for M is often assigned to older life stages after the age at recruitment. But during the larval and early juvenile stages, mortality rates are highly variable and responsive to environmental variability, shifts in predation pressure, or prey availability. To compensate for high variability in mortality and its propensity to destabilize populations, fish stocks have a compensatory reserve that can be expressed through density-dependent mortality acting during early life. This regulatory mechanism results in proportional increases of mortality rates in relation to abundance of the young stages themselves or abundance of adult spawners. If density dependence is operating, mortality rates increase in proportion to abundance, a result of competition for resources or an increase in specific predation rates on the population. While direct evidence for density-dependent, compensatory mortality of young fishes in the sea remains rather uncommon, models imply that even small amounts of density dependence can be very important in stabilizing recruitments (see Chapter 4) and regulating abundances of fish stocks. Without such a mechanism, these stocks would be expected to fluctuate wildly as mortality rates in early life shifted up or down in response to annual variability in environmental conditions.

3.2   Larval mortality: concepts and relationships

Mortality rates in fishes decline with size and age during early life. Initially, rates may be very high, exceeding 50% day^{-1} in some species and commonly exceeding 10% day^{-1}. For example, in a review of mortality estimates (Houde & Zastrow 1993), the mean mortality rate for marine fish larvae was $M = 0.24$, a rate equivalent to a loss of 21.3% day^{-1}. The mean mortality rate for freshwater fish larvae is lower than that for marine fish larvae, but freshwater larvae still die at a high rate, on average at $M = 0.16$, equal to 14.8% day^{-1}. The higher mortality rate for marine larvae probably is a consequence of their much smaller average size and higher vulnerability to a more diverse assemblage of predators. Although significantly different for marine and freshwater larvae, such high mortality rates in both groups lead to rapid losses from cohorts in early life. If mortality rates did not decline during ontogeny, few individuals would remain alive at the end of early life stages to contribute to recruitment. For marine species, only 180 individuals are expected to survive the larval stage (>99.9% mortality) from an initial cohort of one million larvae under conditions of average mortality rates ($M = 0.24$) and larval stage durations (36 days). In the case of an "average" freshwater species, lower mortality ($M = 0.16$) is estimated over a typical 20.7 day larval stage, but 96.4% are expected to die. It seems clear that mortality rates must decline during ontogeny and growth to maintain recruitment levels and insure stabilization of stock abundance (see Chapter 4).

Mortality and growth rates of marine fish larvae are strongly coupled during early life. Species with high mortality rates also have high growth rates, and both of these rates are strongly and positively correlated with temperature. Consequently, species from temperate and high latitudes die and grow at slower rates than species from tropical habitats or species that spawn under summer conditions. The larval stage is protracted for most species from cold environments, a consequence of slow growth rates, thus spreading the stage-specific mortality over a longer time frame. Despite the differences in daily rates for species from warm or cold environments, the stage-specific mortalities are similar because of the strong concordance between mortality and growth rates. For example, gadid species such as cod and haddock (*Melanogrammus aeglefinus*; Figure 1.3p, n), which live in high latitude seas and spawn at temperatures less than 10°C, spend approximately 100 days in the larval stage while tropical gobies (Gobiidae) and damselfishes (Pomacentridae) that spawn at temperatures greater than 25°C spend only 25 days in the larval stage. Yet, average survival rates at the end of their respective larval stages are similar because mortality and growth rates are strongly correlated, increasing at the same average rate with respect to temperature (approximately 0.01 per °C) (Figure 3.3).

Where data are available, it can be demonstrated that mortality rates decline regularly as development and growth of larvae occur, although the rate of decline with respect to body size or age is highly variable among species. The relationships describing the declines in M with respect to weight (W) for five species (including anadromous, estuarine, and marine species) during the larval stage (Table 3.1) ranged from $W^{-0.318}$ to $W^{-0.622}$. In all of

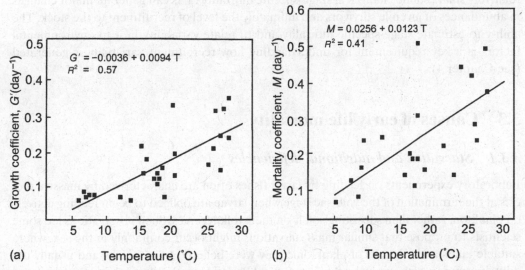

Figure 3.3 Relationships between G', the weight-specific growth rate, and M, the instantaneous mortality rate, with respect to temperature for several species of marine and freshwater fish larvae (derived from Houde 1989).

Table 3.2   The relationships between mortality rate (M) and dry weight (W, in µg) for 6 year classes of American shad and 7 year classes of walleye pollock larvae.

Species	Year	Fitted model
American shad	1979	$M = 4.48W^{-0.564}$
	1980	$M = 0.97W^{-0.319}$
	1981	$M = 1.13W^{-0.339}$
	1982	$M = 1.92W^{-0.381}$
	1983	$M = 0.78W^{-0.292}$
	1984	$M = 33.29W^{-0.889}$
Walleye pollock	1985	$M = 1.72W^{-0.522}$
	1986	$M = 1.68W^{-0.457}$
	1987	$M = 2.43W^{-0.515}$
	1988	$M = 68.59W^{-1.207}$
	1989	$M = 4.38W^{-0.661}$
	1990	$M = 1.31W^{-0.456}$
	1991	$M = 13.52W^{-0.820}$

From Houde (1997).

the cases, declines in M were more rapid than the $W^{-0.25}$ predicted from size-spectrum theory. As one example of declines in mortality, the averaged ontogenetic declines in mortality rates for observations made over 7 years on walleye pollock (*Theragra chalcogramma*, Figure 1.3o) were 5 days old, $M = 0.21$; 12 days old, $M = 0.11$; 37 days old, $M = 0.04$. The mortality rates themselves and the ontogenetic declines in the rates during the larval stage can differ annually in response to variable environmental conditions and predation that larvae encounter from year to year (Table 3.2). Annual variability is important because even relatively small, subtle changes in stage-specific mortality rates can generate major changes in abundances of juvenile survivors and ultimately the level of recruitment to the stock. The ability to estimate stage-specific mortality and to relate variability in it to environmental factors is a key requirement for understanding how recruitment variability is generated (see Chapter 4).

3.3   Causes of early life mortality

3.3.1   *Starvation and nutritional deficiencies*

Laboratory experiments on feeding by larval fishes often are characterized by massive die-offs at the termination of the yolk-sac stage when larvae are obliged to begin feeding actively on plankton, rather than subsisting solely on their yolk reserves. Such observations led some scientists to propose that similar mass starvations might occur commonly in the sea, where suitable kinds and amounts of planktonic prey were believed to be patchy and usually too dilute to support high survival rates. The Critical Period Hypothesis of Johan Hjort emerged from such observations (Figure 3.4). Massive and sudden mortalities of larvae at the first-feeding stage, taken to be evidence supporting the hypothesis, have been

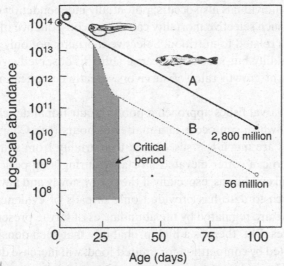

Figure 3.4 Survivorship curves showing effect of a "critical period" on abundance of survivors at 100 days post-hatching. Both populations have an initial mortality rate of $M = 0.10$ day^{-1}. The population that experiences high losses in the "critical period," although recovering to experience the initial $M = 0.10$ day^{-1} rate after the "critical period," is 50 times less abundant at 100 days post-hatching.

observed in some cases. Temporal and spatial matches between larval hatching and the proliferation (blooms) of planktonic prey (the Match/Mismatch Hypothesis of David Cushing [1990]) promote high survival of some larval populations, such as Norwegian cod. On average, the concentration of prey suitable as larval food seems too low to readily support survival of fish larvae, leading to the hypothesis that only under conditions where aggregations of prey occur, for example, in depth-stratified layers, can larvae feed successfully and avoid starvation (the Stable Ocean Hypothesis of Reuben Lasker). On fine- and microscales, survival potential of larvae is elevated in the presence of microturbulence that increases encounter rates of larvae with prey (as proposed by Brian Rothschild and Thomas Osborn). All of these hypotheses to explain recruitment variability are discussed in the next chapter.

How important is starvation as a major cause of mortality to fish larvae? The question is not answered easily because larvae that are poorly nourished can be weakened in a matter of hours, becoming highly vulnerable to predation. Consequently, such larvae may disappear from populations at an accelerated rate and may not occur in collections to allow evaluation of their poor nutritional condition. Nevertheless, there is evidence from many studies that starvation-related mortality of larvae (in Atlantic herring and Atlantic cod, for example [Figure 1.3m, p]), is important. While measures of larval nutritional condition (such as, histological measures, RNA/DNA analysis, or lipid indices) may accurately portray nutritional status of sampled survivors, the indices may not provide a reliable assessment of the overall condition of a larval population or the probability that starvation and mortality from nutritional deficiencies had played a major role. In field collections, we can presume that a larger fraction of larvae than is actually observed was susceptible to starvation mortality. Starvation and nutrition-related mortalities will selectively remove poorly nourished larvae, which are more vulnerable to predation. Selective predation on slower growing, but healthy individuals also occurs, potentially independently of larval nutritional condition. However, such selective mortality can be the consequence of relatively low food intake and, as such, is related to nutrition. Selective mortality not only obscures starvation as the source of mortality but it also may cause shifts in observed size distributions, age structure, and apparent growth rates of survivors, greatly complicating interpretation of dynamics in early life.

During starvation, larval fishes approach a point of nutritional deficit referred to as the "point of no return," which can occur in a matter of hours (see Chapter 1) and leads to death. Juvenile fishes are much less susceptible to mortality from starvation than larvae. Nevertheless, they too can suffer elevated mortality during long periods of starvation or under poor nutritional conditions, especially if they grow slowly and have high vulnerability to predators. Research to date has provided only threads of evidence that survival and growth of larval stages are regulated by the abundances of larvae present that compete for limiting prey resources. But, there is a high probability that such density-dependent mortality, which is regulated by competition for limited food, will increase during the late larval and juvenile stages (Figure 3.5).

In temperate freshwater and marine environments, overwinter mortality of young-of-the-year individuals may occur. Such mortality is usually size dependent, with smaller and poorly conditioned individuals being more susceptible to mortality from energetic deficiencies or predation as the winter proceeds. Size-selective overwinter mortality has been reported for many species, including smallmouth bass (*Micropterus dolomieu*), white perch (*Morone americana*), striped bass, and Atlantic silverside (*Menidia menidia*). Such mortality of

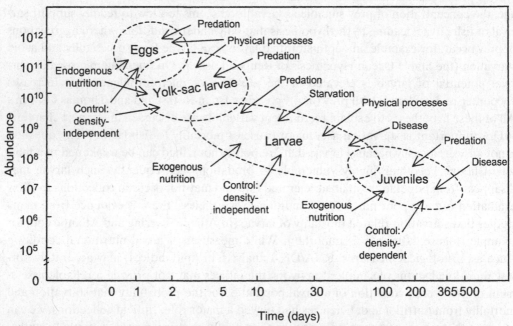

Figure 3.5 Survivorship curve conceptualizing the recruitment process in fishes, including factors that affect mortality and growth. Hypothesized mechanisms of control are indicated (reproduced from Houde 1987 with permission of the American Fisheries Society).

overwintering young-of-the-year fish can selectively alter the size distribution of survivors that will contribute to recruitment as well as the abundance of recruits. In some cases, extreme winter mortalities of young-of-the-year fishes can occur near the range boundaries of tropical and subtropical fishes. For example, annually variable overwinter mortality of Atlantic croaker (*Micropogonias undulatus*) in mid-Atlantic estuaries of North America is believed to be a strong controller of its year-class size.

3.3.2   Predation

Predators inflict a heavy toll on the young stages of fishes and probably are the biggest single cause of mortality. As noted above, predation losses may be linked to nutritional deficiencies that increase vulnerability of young fishes to predators. Predation on fish eggs is size-specific and predation on larvae may be both size-specific and growth-rate dependent. As larvae grow and develop, swimming ability and predator-detection capabilities improve. These developmental factors generally lead to lower predation mortality as ontogeny proceeds (see Chapter 1).

Many organisms, both vertebrates and invertebrates, prey on young fishes. Important predators include juvenile and adult fishes, jellyfishes (ctenophores and medusae), chaetognaths, euphausiids, and insects. The diversity of predators in marine ecosystems apparently is broader than that in freshwater ecosystems. For example, cnidarians and ctenophores, represented by many taxa in marine ecosystems and demonstrated to be major consumers of fish eggs and larvae, are hardly represented in freshwaters. Insects, on the other hand, are essentially confined to freshwater ecosystems and tidal freshwaters of estuaries. Some organisms, not usually considered as predators on early life stages, may be important predators under some circumstances. For example, copepods, amphipods, hydroids, birds, and amphibians all consume young stages of fishes. Some of these predators, such as birds, are significant agents of mortality on small juvenile fishes in both fresh waters and marine waters.

There is a general lack of information on the community of predators that consumes fish eggs and larvae, especially in quantifying effects of specific predators, which makes it difficult to evaluate taxa and their roles as predators. Stated simply, in any region and moment of time, there not only is an array of predator species and sizes that can consume eggs and larvae, but also variable amounts of alternative prey which must be considered in evaluating predation on early life stages and its consequences to a population. Research to date, including modeling approaches, rarely has been able to partition mortality of young stages of fishes among the array of predator taxa and sizes that are present. We are quite confident that predation is a major source of mortality to young fishes, but we are lacking sufficient detail to understand how the process operates in a complex community of potential predators and prey.

Most predators on young fishes feed either by active pursuit or by ambush. Cruising raptorial predators such as small pelagic fishes actively pursue larval fish prey. Cruising, filter-feeding fishes, for example, sardines (*Sardinops* spp.) and some anchovies in marine waters, feed by filtering pelagic eggs or small larvae from the water. Drifting jellyfishes may ambush eggs or larvae by entangling them in tentacles or may employ cruising, filtering behavior to encounter eggs and larvae (for example, lobate ctenophores). Fishes and invertebrates in both freshwater and marine vegetation may be ambush predators or may actively pursue

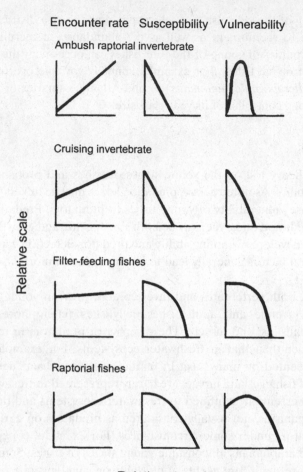

Figure 3.6 Relative size of larva

Figure 3.6 Conceptual models showing encounter rates, susceptibility, and vulnerability of larval fishes to predation. The dome-shaped vulnerability curves illustrate the hypothesized trends in mortality relative to larval size for predators of specific size and predatory behavior (reproduced from Bailey & Houde 1989 with permission of Academic Press).

young fishes. Demersal predators, primarily fishes, may ambush newly settling young that descend to reef or hard-substrate habitats at the end of the pelagic stage.

The relative vulnerability of larval fishes to many kinds of predators can be represented and illustrated principally by dome-shaped curves in relation to larval size (Figure 3.6). The susceptibility of larvae to attack and capture by a particular type and size of predator usually declines as larvae grow, a function of increases in swimming speeds, improved ability to detect predators, and better avoidance and escape capability (see Chapter 1). Vulnerability, which represents the net effect of ontogenetic changes in encounter probability and susceptibility, may increase for intermediate size larvae (Figure 3.6), at least with respect to a particular predator of specific size and capability. Eventually, larger larvae become increasingly more adept at detecting predators and avoiding attacks, thus reducing their vulnerability despite the possibility that encounter rates may continue to increase.

Laboratory experiments have confirmed the high vulnerability of fish eggs and larvae to a variety of predators. The size-specific nature of predation is also illustrated by laboratory experiments which indicate that predators, independent of taxon, tend to consume larval prey that is preferentially about 10% of their body size (Paradis *et al.* 1996). Modeling studies have demonstrated that variability in predation mortality can control or regulate the recruitment process and temporal variability in recruitment success of fishes. Despite the wealth of laboratory research and the proliferation of models, predation remains difficult to detect or evaluate in the sea. Eggs and especially larvae are soft-bodied and are either destroyed upon consumption, making them unrecognizable, or are digested quickly. As a consequence, eggs and larvae may go undetected or be under-represented as prey unless sophisticated immunoassays are conducted on gut contents of suspected predators.

Mortality from cannibalism represents a special kind of predation. Cannibalism on eggs is common in some clupeoid fishes, either by incidental filter-feeding or by selective consumption. A significant fraction of egg and yolk-sac-larva mortality (>20%) in sardines (*Sardinops* spp.) and anchovies (*Engraulis* spp.) in upwelling ecosystems can be accounted for by this type of cannibalism. Cannibalism also is common among fishes that practice parental care (for example, Cichlidae, Gasterosteidae) and apparently becomes significant when parents judge the reproductive value of a clutch to be low, opting to cannibalize it and promote their own nutrition, which potentially results in more successful spawning in the future. Cannibalism also may occur in many species where metamorphosing larvae settle onto a substrate already occupied by older and larger conspecifics. Sibling cannibalism, in which larvae prey upon sibling larvae is reported in many taxa, including freshwater esocids (Figure 1.3e) and characids and marine scombrids (Figure 1.3s). It is beyond the scope of this chapter to critically evaluate the role of cannibalism, except to note that such mortality is density-dependent and thus can serve as a mechanism to regulate survival and recruitment level (see Chapter 4).

3.3.3 *Physics: transport, retention, and dispersal*

Water currents that disperse or transport young stages of fishes into environments unfavorable for survival can be a significant source of mortality. Variability in physical properties of aquatic ecosystems, operating on fine- to basin-wide scales (meters to 1000 km) is strongly dependent on wind and weather patterns that vary on time scales from days to seasons. Fine-scale variability in water-column properties, especially stratification and its effect on the stability of vertical distributions of fish eggs and larvae, their predators, and prey plays a critical role in controlling conditions that determine survival of early life stages. For some species, especially clupeoid fishes in upwelling regions of marine ecosystems, the relaxation of upwelling under low wind conditions is associated with enhanced larval survival and growth. In this circumstance, the stability of the water column under calm conditions allows formation of layers or patches of plankton prey at concentrations suitable for larval feeding, enhancing the probability that larvae will encounter sufficient prey to survive. The breakdown of water-column stability under stormy conditions can disperse larval prey and produce poor feeding conditions, leading to mortality of anchovy and sardine larvae (Figure 3.7).

Hypotheses dealing with recruitment processes historically have focused on transport and retention mechanisms as the critical factors affecting survival and growth of larval

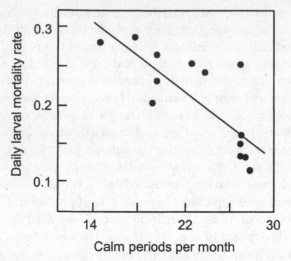

Figure 3.7   Daily mortality rate (day^{-1}) of northern anchovy larvae in relation to calm periods per month in the California Current over a 13-year period. Larval mortality rate declines substantially under calm conditions when water-column stratification occurs that presumably promotes larval feeding success (reprinted from Peterman, R.M. & Bradford, M.J 1987. Wind speed and mortality rate of a marine fish, the northern anchovy (*Engraulis mordax*). *Science* **235**, 354–356. Copyright 1987 with permission from the American Association for the Advancement of Science).

fishes during the pre-settlement stages. Aberrant transport or poor retention will lead to high mortality. Retention mechanisms in estuaries or frontal systems of dynamic marine ecosystems are particularly important to retain larvae in regions where they have a reasonable chance to survive. In recent years, major national and international research programs have attempted to unravel the physical mechanisms that promote survival of early life stages (for example, FOCI for walleye pollock, GLOBEC and OPEN for Atlantic cod, and SABRE for Atlantic menhaden [*Brevoortia tyrannus*]). The complexities of aquatic ecosystems and physical processes therein have required that dual emphasis, distributed between observational science and coupled biological–physical modeling, be coordinated to understand how physics and biology interact to support survival of young fishes.

While it is quite clear that physics at meso- and broader scales (one to hundreds of kilometers) plays a role in controlling levels of mortality of larval stages, physics at finer scales, for example, on millimeter to meter scales, also can be important. Turbulence at those scales plays a role in controlling contact rates between larval fishes and their prey. Rates of contact, controlled by microscale turbulence, can directly influence the nutritional status of larvae and can directly affect their vulnerability to predation. Modeling studies, laboratory experiments, and some field observations on the role of fine-scale turbulence in promoting larval feeding success, and thus growth and survival, have proliferated in recent years. They have attempted to explain how survival of fish larvae in the sea is possible under conditions where average prey levels are thought to be lower than necessary for larval survival. In the case of Atlantic cod larvae, it is apparent that survival is maximized under moderate wind conditions that generate microturbulence sufficient to enhance encounter rates between larvae and prey but which is not so turbulent that larvae are unable to capture prey that they encounter (MacKenzie & Kiørboe 2000).

3.3.4 *Water quality and nursery habitats*

Egg and early larval stages of fishes are surrounded and enveloped by their fluid environment and unable to escape their immediate surroundings because they are immobile (eggs) or swim poorly. This makes them vulnerable to poor water quality, whether a consequence of human activities or natural phenomena. Mortality from poor water quality is perhaps most threatening to young fishes in spatially restricted freshwater environments and in estuarine tributaries that serve as spawning sites for many anadromous fishes. Release of contaminants and toxic materials, either as chronic or episodic inputs, can be lethal to eggs and larvae of fishes or may prevent successful spawning by adults. Other water-quality factors may act in more subtle ways. For example, increased loadings of nutrients such as nitrogen and phosphorous can lead to eutrophication of many fresh waters, estuaries, and coastal ecosystems, which can deplete dissolved oxygen, leading to hypoxia or anoxia that is lethal to fish eggs and larvae. Shifts in ecosystem structure during eutrophication may transform macrophyte-dominated ecosystems into phytoplankton-based ecosystems that lack aquatic vegetation that is critical to shelter juvenile fishes from predation in freshwater and coastal systems. Another threat to eggs, larvae, and juvenile fishes is the acidification of many fresh waters and tidal tributaries that has developed in recent decades as a consequence of industrial combustion of fossil fuels and acid rains that result. Low pH, in the range of 5.0–6.5, has made many fresh waters unsuitable for survival of eggs and larvae of salmonids, and causes mortality of the eggs and larvae of anadromous shads and river herrings in their nursery streams. Low pH events may be episodic, occurring immediately after heavy rains from runoff into poorly buffered streams or ponds, or may be chronic when related to acidification from mining operations or other sources of persistent acidified runoff.

Natural variability in water quality may be caused by weather, precipitation, and runoff events or can be tied to geology and soils in drainage basins. In some ecosystems, these factors present risks to survival or increase the variability in survival of early life stages of fishes. Temperature, nutrient status, levels of alkalinity, conductivity, salinity in estuaries, pH, dissolved oxygen, and turbidity all can vary substantially both temporally and spatially, and on time scales from hours to decades. Lethal levels of these factors, especially as pulsed events, may occur uncommonly on continental shelves and in the open sea, but they can occur more frequently in small estuarine and freshwater ecosystems where water quality and environmental conditions are subject to rapid change.

Water quality can act chronically or episodically to control survival of early life stages. For example, temperature, a primary controller of physiological activity in fishes, can act chronically by controlling growth rates and stage durations, thus indirectly influencing survival potential. In the case of plaice (*Pleuronectes platessa*) eggs, which require many days to hatch at ambient temperatures in the North Sea, egg mortality rate is directly related to water temperature (Figure 3.8). In years of low temperatures, plaice eggs and young experience high survival that supports high recruitments. In fresh waters and in estuaries, heated effluents (episodic or chronic effects) from power plants can be lethal, either directly or indirectly, in addition to causing mortality by entrainment and mechanical damage. Sudden and episodic

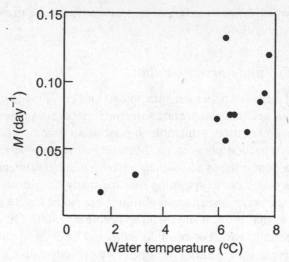

Figure 3.8   Daily mortality rate (day^{-1}) of plaice eggs in the Southern Bight of the North Sea in relation to average water temperature during egg development for an 11-year period (data from Harding *et al.* 1978, figure modified from Van der Veer *et al.* 2000 with permission of Academic Press).

changes in temperatures of nursery habitats that occur naturally are common in temperate regions, especially during transition seasons when many freshwater and anadromous fishes spawn. Such episodic events often are accompanied by other water-quality changes that can be fatal to egg and larval stages or the prey of larvae. For example, striped bass eggs and larvae (Figure 1.3q) in Chesapeake Bay tributaries may experience partial or complete mortality when water temperatures drop below 12°C, which occurs after the passage of weather fronts and cold rains during the April spawning season. Those same weather fronts and rains also can substantially reduce pH, conductivity, and alkalinity levels, thus magnifying the effects of temperature stress and increasing the risk of mortality to eggs and larvae. Chapter 7 provides an extended discussion of spawning and nursery habitat.

3.3.5   Diseases and parasites

Diseases and parasites are agents of mortality to fishes in all life stages. In juvenile and adult fishes, especially in aquaculture and sometimes in natural environments, effects of diseases are recognized, if not quantified, and epizootics sometimes are reported as major controllers of abundance. Except in aquaculture, we seldom consider death of fish eggs and larvae from diseases or parasites as serious sources of mortality. Bacterial and viral diseases certainly must take a toll of young fishes. Parasites of eggs and larvae, including dinoflagellates, protozoans, helminths, and copepods have been reported regularly, although mortality rates and population-level consequences are unevaluated. Under laboratory conditions, external parasites such as copepods of the genus *Caligus* are obvious sources of mortality to cultured larvae. In the wild, diseased, parasitized, and poorly conditioned larvae and eggs disappear rapidly from populations, either through selective predation or decomposition of dead bodies, leading to a poor appreciation of disease or parasites as sources of mortality to early life stages of fishes.

Genetic disorders leading to abnormal development and death during early life are known and reported, primarily from laboratory research and aquaculture experiences. In polluted waters, incidences of morphological abnormalities in eggs and larvae may be elevated, for example, in winter flounder (*Pleuronectes americanus*) in some coastal areas, that are clear expressions of contaminant effects, but the levels of mortality and implications for population-level control are largely unknown.

3.3.6   Interacting factors

It is a simplification to assign conditional probabilities to single causes of mortality without understanding how potentially lethal factors interact. As an example, we are still unable to easily separate the fraction of young stages that die from starvation from those that die from predation. Poorly fed, slowly growing larvae are more vulnerable to predators. In laboratory experiments and in model simulations, such larvae are selectively eaten by predators and have a relatively low probability of survival. Despite such observations and strong intuition, it remains problematic or impossible to quantify the proportional mortalities from starvation or predation in natural ecosystems. There is no denying that the interaction between nutritional state and vulnerability to predation must be important. Low prey levels lead to slow growth, longer stage durations, and greater probability of being consumed during an early life stage. In the case of the bloater (*Coregonus hoyi*, Figure 1.3b), a modeling analysis demonstrated how levels of growth and the variable potential for growth in a larval population modified the effects of predation on survival (Rice *et al.* 1993). High growth rates, and especially variable growth rates, reduced the risk to bloater larvae of size-selective predation and promoted recruitment potential (see Chapter 11).

Physical and biological processes also interact as they affect survival probability. In many cases, failed retention, unfavorable transport, or poor environmental conditions (for example, temperature, pH, hypoxia) may act directly to kill eggs and larvae, but these conditions also will affect the predators and prey of early life stages, creating a complex web of interactions that have important implications for survival of early life stages. Finally, effects of contaminants or poor water quality may alter behavior of larvae, thus impeding feeding and reducing growth rates or making larvae more vulnerable to predation.

Cascading effects can increase the risk of mortality to eggs and larvae in stressed aquatic ecosystems. For example, excess nutrients can lead to eutrophication, which may result in low dissolved oxygen, harmful algal blooms, losses of aquatic vegetation, and probable increases in some larval predators, such as jellyfishes, as the trophic state of the ecosystem shifts. Evaluating and partitioning the effects of such multiple sources of mortality is difficult and seldom accomplished, except in modeling research, which is now contributing valuable insight into the complex and interacting processes that generate mortality in the early life of fishes.

3.4   Estimating larval mortality

Obtaining accurate and precise estimates of mortality rates for any life stage of fishes can be a formidable task. The general pattern of survivorship curves in early life is known

(Figures 3.2 and 3.5), but the levels of abundance and loss rates for a particular life stage can seldom be estimated with certainty. The magnitude of required effort and the cost of the estimation task are greatest in large ecosystems, especially those which are essentially unbounded and subject to significant losses through dispersal and translocation, in addition to losses from mortality. The possibility of success in determining mortality improves in embayments, estuaries, and freshwater habitats, where dispersal losses are minor or can be accounted for by sampling the entire system. Successful estimation of mortality depends upon accurate determination of abundances and dependable assignment of individuals to age classes or stages. Fortunately, most fish larvae and many juveniles can be accurately aged from daily increments in otoliths (see Chapter 2), a great advantage in determining age-specific mortality.

3.4.1   Catch curves and survivorship

As noted earlier, the rate of loss of individuals from a population by death expresses the mortality rate. Age-specific losses often are estimated from a "catch curve" in which abundances-at-age of survivors are plotted against age (Figure 3.2). A log-linear regression equation of $\log_e$ abundances on age then can be fitted to the data to estimate the regression coefficient (slope), which is the instantaneous mortality rate

$$\ln N_t = \ln N_0 - M \cdot t \qquad (3.7)$$

where N_t is abundance-at-age t (often expressed in days for early life stages, but in years for older fishes), N_0 is an estimate of abundance at the beginning of the stage, and the slope M estimates the instantaneous mortality rate, usually expressed as day^{-1} for fish eggs or larvae but $year^{-1}$ for recruited fishes. The cumulative mortality over a period of t days is $M \cdot t$, the survival rate is

$$S = e^{-M \cdot t} \qquad (3.8)$$

and the proportion of a population dying in a time period is

$$A = 1 - e^{-M \cdot t} \qquad (3.9)$$

The hypothetical survivorship curve (Figure 3.2a) and tabulated summary (Table 3.3) illustrate a survivorship analysis based on catch curves for a typical marine fish species. Initially, mortality rates of a newly spawned cohort are high, frequently greater than 10% day^{-1} and sometimes more than 50% day^{-1}. Rates generally decline during early life stages. Cumulative mortalities ($M \cdot t$) usually are highest during the egg–larval or juvenile stages when more than 99.5% of individuals may perish. In the illustrated example, the natural mortality rate of cohorts after recruitment is 52% annually ($M = 0.002$ day^{-1}), a high rate but 58 times lower than the daily rate during the egg–larval stage. Imposition of fishing mortality certainly has a substantial effect on abundances of recruited age classes (Figure 3.2a), but it is early life mortality rates and their variability that are the dominant factors affecting the dynamics, variability, and fates of recruiting cohorts (see Chapter 4).

An example of the application of a catch-curve approach, based upon intensive sampling of the entire Chesapeake Bay ecosystem and its population of bay anchovy (*Anchoa mitchilli*), a species that is essentially confined in the bay during its egg, larval, and juvenile stages, serves to illustrate the approach. Two bay-wide surveys provided data for a

Table 3.3 Instantaneous mortality rates (M, F, Z), cumulative mortalities (cum. Z), and percent mortalities within each larval stage during ontogeny of a typical marine fish $(Z = F + M)$. See Figure 3.2a.

Stage/age	M (day^{-1})	F (day^{-1})	Z (day^{-1})	Cum. Z	Deaths (%)
Egg/larvae (0–50 days)	0.1170	–	0.1170	5.85	99.71
Early juvenile (50–200 days)	0.0470	–	0.0470	7.05	99.91
Late juvenile (200–365 days)	0.0280	–	0.0280	4.62	99.01
Recruited stage (no fishing) (1–4 year)	0.0020	–	0.0020	2.19	88.81
Recruited stage (fishing) (1–4 year)	0.0020	0.0020	0.0040	4.38	98.75

catch-curve analysis on bay anchovy larvae during June and July 1993. In this example, larval mortality was significantly higher in June than in July (Figure 3.9). The higher mortality in June ($M = 0.41$, 33.6% day^{-1}), compared to the lower mortality in July ($M = 0.23$, 20.5% day^{-1}) was attributed to high predation by abundant jellyfish predators in June. The differences in mortality rates indicated that the vast majority of recruited bay anchovy in 1993 was derived from cohorts that originated from July spawning.

3.4.2 Models and data

The catch-curve method described above assumes that an equal proportion of a population dies at each age included in the analysis. If the proportional mortality rate is changing

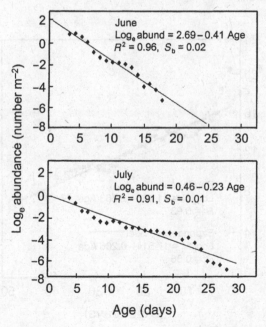

Figure 3.9 Catch curves of log$_e$ abundance vs. age (in days) for larval bay anchovy in Chesapeake Bay during June 1993 and July 1993. The regression coefficients (slopes) are estimates of M in each case (from Rilling & Houde 1999).

during a stage, this assumption may not hold and the simple log-linear model may be inadequate to estimate age-specific mortality rates in the life stage. Other models, such as the Pareto model, may describe the decline in numbers more accurately (Lo 1986), as in larval northern anchovy (*Engraulis mordax*) off the California coast. An example and comparison of the Pareto and log-linear models fitted to data for larval striped bass in the Nanticoke River, Maryland, demonstrates the utility and improvement in statistical fit (higher R^2) provided by the Pareto model (Figure 3.10). The Pareto model, which can be fitted to data by an iterative least squares method, assumes that mortality is a power function of age (or size). It is

$$\ln N_t = \ln N_0 + \beta \cdot t^\alpha \tag{3.10}$$

where β estimates the overall rate of decline and α describes the particular shape of the survival curve, depicting how mortality rate changes with age or size. Survival rates for any period (or size range) can be calculated by estimating the respective abundances N_t and N_{t-i}, and solving for $S = N_t/N_{t-i}$, where i represents the time interval in which mortality rate is estimated. In cases where the mortality rate declines substantially in relation to age (or size), especially at the youngest ages, as in the Nanticoke River striped bass example (Figure 3.10), the Pareto model will be a better choice to estimate age-specific (or size-specific) mortality rates.

In many instances, estimated mortality rates of pelagic fish eggs or larvae will be confounded by unaccounted losses due to dispersal out of the sampling area. In this circumstance, estimated mortality rates will be too high and must be adjusted for the non-mortality losses. Models that include terms to account for diffusive and advective losses in open, dynamic ecosystems, such as that of Helbig & Pepin (1998), may be necessary to separate dispersive losses from deaths.

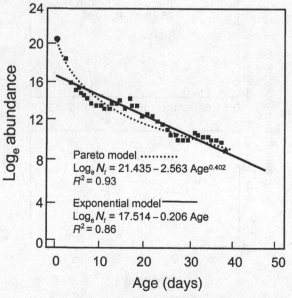

Figure 3.10 Catch curve for larvae of striped bass from the Nanticoke River, 1992. A linear log$_e$-exponential model and a non-linear Pareto model are fitted to the data for comparison. In this case, the Pareto model provides a better fit (modified from Kellogg *et al.* 1996).

Stage-specific mortality

In many cases, our objective is to estimate the mortality that occurs within a life stage or between size groups to evaluate stage-specific dynamics during early life. Here, the linked dynamics of growth and mortality must be considered and can be expressed by evaluating the ratio M/G', where G' is the instantaneous growth rate in weight. This ratio, sometimes referred to as the "physiological mortality rate," can serve as an index of cohort dynamics and productivity that integrates mortality and growth processes (Beyer 1989, Houde 1996). It also can be used to evaluate the potential of cohorts to generate biomass during early life. Stage-specific survival is

$$S = \frac{N_s}{N_{s-1}} = \left(\frac{W_s}{W_{s-1}}\right)^{-M/G'} \tag{3.11}$$

where W_s and W_{s-1} are weights of fish at respective stages (see also Equation 2.9). Relative cohort biomasses with respect to life stages also can be expressed in relation to M/G'

$$\frac{B_s}{B_{s-1}} = \left(\frac{W_s}{W_{s-1}}\right)^{(1-[M/G'])} \tag{3.12}$$

Only when $M/G' < 1.0$ does cohort biomass increase between successive stages $s - 1$ and s. It is observed that the index M/G' and mortality rates (M) generally decline during the early life of fishes. For most marine species and many freshwater species as well, M/G' is >1.0 during the smallest larval stages, signifying that cohort biomasses are declining. As ontogeny proceeds, M/G' eventually declines to <1.0 and a cohort's biomass will begin to increase (see Section 2.5.6).

Variability in the M/G' index and its rate of decline with ontogeny can be used to compare survival potential of larval cohorts among years or to compare different species during early life (Figure 3.11). The size at which $M/G' = 1.0$, and its variability among years for an individual species or among species in comparative analyses, describes the stage-specific survival pattern of a cohort with respect to ontogeny. Such patterns and variability can help to interpret how ontogenetic variability in stage-specific survival and growth relates to recruitment success or failure.

3.4.3 *Projecting and predicting*

The patterns, levels, and variability of cohort mortality in young fishes translate into successful or failed recruitment. Stage-specific mortality rates and cumulative mortality rates during the egg, larval, and early juvenile stages have been used to forecast cohort survival and recruitment potentials (see for example, Bailey *et al.* 1996, Houde 1997). It is important to remember that mortality rates during the earliest life stages are, at best, uncertain predictors of recruitment. Estimates of abundance and mortality rates in natural bodies of water often are inaccurate or have low precision. On the other hand, even relatively crude estimates of larval mortality may serve to predict potential recruitment, if there is considerable annual variability in the mortality rates and if mortality estimates are available for a series of years or for several cohorts

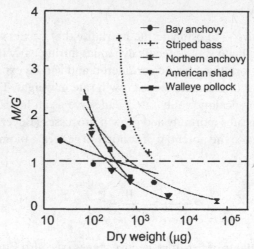

Figure 3.11 Relationships between M/G' and $\log_{10}$ body weights for five species of marine and estuarine fish larvae. Data points are averages for cohorts and year classes that had been studied (reproduced from Houde 1997 with permission of Academic Press).

during a single year. Higher than average mortality rates (or cumulative larval mortalities) generally can successfully predict low recruitment potential for a cohort, but lower than average larval mortality rates cannot reliably predict high recruitments because variable mortality in the relatively long juvenile stage can still generate high cumulative mortality (for example, see Table 3.3). Consequently, only in years when larval stage survival is low can one confidently use mortality in the larval stage to predict levels of recruitment.

Although it is risky to predict recruitment from extrapolated estimates of larval mortality rates and abundances, such estimates can provide important insights into the dynamic processes and their variability within the larval stage. Evaluation of mortality rates in relation to environmental variability and oceanographic factors in studies of larval stage dynamics (see Chapter 5) also may provide valuable insights into processes that control recruitment. An excellent example is provided by the FOCI Program, in which larval and early juvenile mortality rates of walleye pollock were estimated for many years in the Gulf of Alaska. During the 1980s, larval mortality rates (M) and cumulative larval stage mortalities ($M \cdot t$) were inversely correlated with levels of walleye pollock recruitment. In years after 1989, the relationship between larval mortality and recruitment failed when the ocean environment underwent a major change in response to climate variability. In the 1990s, the cumulative mortality ($M \cdot t$) of walleye pollock during the early juvenile stage increased progressively and juvenile stage mortality then became a better predictor of recruitment (Bailey 2000). More information on the FOCI Program is provided in Section 4.8.2.

3.5   Issues and implications

Are survivors of the egg and larval stage favored by some selected qualities that elevated their chances of survival or were they simply lucky? This question has intrigued fishery scientists for nearly three decades. For most typical marine teleosts, mortality exceeds 99%

during the egg and larval stages. For freshwater species, mortality is lower, but still usually exceeds 95% before the juvenile stage. In some research, it is clear that survivors had grown at exceptional rates or that they had experienced better than average nutritional condition (Campana 1996, Meekan & Fortier 1996). Such observations do not necessarily answer the question of whether survivors were predisposed to survive or just lucky. There is increasing evidence, however, that selection for survival favors larval genotypes or phenotypes that grow fast or behave appropriately to reduce their probability of being preyed upon. Alternatively and additionally, the selection process could operate on adults rather than early life stages, favoring survival of eggs and larvae of adults that spawned at favorable sites and times (the Match/Mismatch Hypothesis). These alternative possibilities remind us that there are important dependencies on physical properties of aquatic environments, close linkages among life stages, and feedback mechanisms that act together to regulate or control survival during early life.

Obviously, there are many sources of early-life mortality. Some may act essentially independently of each other, for example, episodic weather events, massive contaminant releases, or harmful algal blooms. But, it is likely that most agents of mortality interact in complex ways to affect mortality and growth rates through the relatively small (and difficult to measure) effects on dispersal patterns, predation mortality, or nutritional deficiencies. It is easy to imagine how greater or less than expected mortality might result from interactions of factors; for example, unfavorable dispersal could deliver larvae to poor feeding areas, making them more susceptible to predation. Under the best of conditions, it is a challenge for fishery scientists to estimate mortality rates with accuracy and precision. Detecting small, but decisive changes or variability in mortality rates, which is essential to understand cohort fates, is difficult under any circumstances and usually cannot now be accomplished with confidence in large or complex aquatic ecosystems. Partitioning the total mortality rate into its constituent components, for example, predation, starvation, and disease, represents yet another challenge that compounds the problem. Simulation modeling convincingly assures us that small changes or variability in mortality and growth rates during early life can control a cohort's fate, but actually estimating rates and slight changes in them is a formidable task.

It is not the magnitude of early life mortality rates so much as variability in the rates that is critical in generating recruitment successes or failures. Small variability in larval mortality rates, for example, 15% above or below average daily rates, when projected over a 100-day period can generate order-of-magnitude differences in recruitment levels. Only the most comprehensive and well designed surveys of eggs and larvae can hope to estimate abundances with sufficient precision to detect such differences in daily mortality. Given the large numbers of eggs spawned by teleost fishes and the high mortality rates that are common in early life, we should not be surprised that order-of-magnitude variability in recruitment occurs. Rather, the fact that even higher variability is seldom seen is perhaps more surprising.

Compensation, exercised through density-dependent mortality, must regulate mortality during early life, thus contributing to stabilization of stock abundance. Unfortunately, we rarely identify or quantify density-dependent mortality in studies on egg and larval stages. Survival of those stages seems to be controlled primarily by coarse, density-independent variability in mortality. Even a small amount of density-dependent mortality in early life stages could provide sufficient compensation to dampen variability and regulate recruitment levels. In the late larval and juvenile stages, density-dependent mortality is recognized more frequently and is demonstrated to be an effective regulator of survival during the relatively long juvenile, pre-recruit stage (Van der Veer *et al.* 2000).

Will new technologies, analytical procedures, or models improve our ability to estimate age- and stage-specific abundances, and thus mortality rates of young fishes? Improvements in sampling technology certainly took place during the 1970s and 1980s. Depth-specific sampling with large, multiple-opening-closing nets equipped with environmental monitoring systems became common and helped greatly to estimate abundances, the key to estimating mortality and its relationship to oceanographic factors. Analytical procedures did not advance so rapidly in the same period, although more sophisticated statistical testing and estimation of variances became common, often confirming that our estimates were imprecise. The biggest advances in exploring how environmental factors relate to mortality were made in modeling the dynamics of early life. Coupled biophysical models became more sophisticated in recent years. Individual-based models, emphasizing biological responses and individual variability, proliferated and are being linked to spatially explicit, hydrodynamic models to characterize and identify mortality factors that determine the fate of cohorts. Consequently, we are able to evaluate and estimate mortality, and the processes that generate it, better than we were able two decades ago. This is particularly true in small, enclosed ecosystems (lakes and estuaries), but also in the coastal ocean where the oceanography is now reasonably well understood.

3.6   Conclusions

In the absence of fishing, most temporal variability in the abundances and recruitments of fish stocks results from variability in cohort survival during early life. Coarse controls on survivorship during egg and larval stages, and then fine-tuning in the juvenile, pre-recruit stage, are the processes that control and regulate recruitment. The interplay of episodic and subtle factors, the relative intensities of density-independent control and density-dependent regulation, the relative roles of predation and nutrition, and the presence of other potentially important causes of mortality all must be considered to fully appreciate how variability in mortality during early life shapes recruitment success. Advances in sampling technology, analytical capability, and modeling applications are improving our ability to relate early life processes to recruitment and to the overall population dynamics of fish stocks. It is often stated that the average fish larva is a dead larva, implying that we should emphasize analysis of unique characteristics of relatively rare survivors. To be sure, much can be learned from that approach, but identifying causes of mortality and estimating mortality rates in early life are the essential steps that we must employ to evaluate the processes and variability in them that determine levels of recruitment to adult stocks.

Additional reading

Bailey, K.M. & Houde, E.D. (1989) Predation on eggs and larvae of marine fishes and the recruitment problem. *Advances in Marine Biology* 25, 1–83.
Beyer, J.E. (1989) Recruitment stability and survival – size-specific theory with examples from the early life dynamics of marine fish. *Dana* 7, 45–147.

Blaxter, J.H.S. (1986) Development of sense organs and behavior of teleost larvae with special reference to feeding and predator avoidance. *Transactions of the American Fisheries Society* **115**, 98–114.

Chambers, R.C. & Trippel, E.A. (Eds) (1997) *Early Life History and Recruitment in Fish Populations.* Chapman & Hall, London, 596 pp.

Heath, M.R. (1992) Field investigations of the early life stages of marine fish. *Advances in Marine Biology* **28**, 1–173.

Houde, E.D. (1987) Fish early life dynamics and recruitment variability. *American Fisheries Society, Symposium* **2**, 17–29.

Houde, E.D. (1989) Subtleties and episodes in the early life of fishes. *Journal of Fish Biology* **35**(suppl. A), 29–38.

Houde, E.D. & Zastrow, C.E. (1993) Ecosystem- and taxon-specific dynamic and energetics properties of larval fish assemblages. *Bulletin of Marine Science* **53**, 290–335.

Lasker, R. (Ed.) (1981) *Marine Fish Larvae. Morphology, Ecology and Relation to Fisheries.* University of Washington Press, Seattle, 131 pp.

Leggett, W.C. & Deblois E. (1994) Recruitment in marine fishes: is it regulated by starvation and predation in the egg and larval stages? *Netherlands Journal of Sea Research* **32**, 119–134.

Pepin, P. (1991) Effect of temperature and size on development, mortality, and survival rates of the pelagic early life history stages of marine fish. *Canadian Journal of Fisheries and Aquatic Sciences* **48**, 503–518.

Sogard, S.M. (1997) Size-selective mortality in the juvenile stage of teleost fishes: a review. *Bulletin of Marine Science* **60**, 1129–1157.

Chapter 4
Recruitment

James H. Cowan, Jr. and Richard F. Shaw

4.1 The recruitment problem: why do fish populations vary?

It may be hard for today's student of fishery science to believe, but consideration of the population dynamics of the early life stages of fishes in studies of recruitment variability has not always been the norm. Indeed, for its first 100 years fishery science focused on the population dynamics and effects of fishing on adults to advance our understanding of population regulation. During these early years, important concepts began to emerge in the fishery literature that recognized the existence of distinct populations within a species' range, the role of density dependence in long term population stability, and how environmental variability influences short term changes in abundance. As our ability to estimate population size improved, however, so did our appreciation for the degree to which year-class success in fishes varies from year to year. This was quickly followed by recognition of the importance of this variability to fishery management as inconsistent year classes recruited to fishable age. Today, we recognize that order-of-magnitude variation in year-class success is the norm.

It was apparent to early fishery scientists that prior understanding of stock dynamics and vital rates of adult fishes (growth rates, death rates, and egg production) were unable to fully explain observed variability in stock size. Thus in 1914 the Norwegian fishery scientist Johan Hjort, while working on his own studies of North Sea herring (*Clupea harengus*), Atlantic cod (*Gadus morhua*) and haddock (*Melanogrammus aeglefinus*) (Figure 1.3m, p, n), and synthesizing work by colleagues of his day,* proposed that variable "year-class success" (he actually coined this term) was most likely determined during early life in marine fishes. He arrived at this conclusion by observing that fish eggs and larvae experienced extremely high mortality rates *in situ*, suggesting that year classes could be reduced dramatically if mortality was nearly total. Thus the notion was born that variability in recruitment – defined here as the survival of an annual cohort to the end of the first year of life – is determined by variability in vital rates of newly hatched fishes, a premise that has guided much research in the intervening years. In the remainder of this chapter, we develop a brief history of the major recruitment hypotheses that followed from Hjort's classic work, emphasizing those that focus on the contribution of early life stages. We end with a discussion of the progress of, and prognosis for, recruitment prediction.

---

* More details of the historical context for Hjort's work are discussed in Section 9.2.

4.2   Predicting recruitment: need and implications

Since Hjort's time our understanding of population dynamics and the consequences of density dependence and environmental variability (even on the global scale) as they affect early life stages and subsequent recruitment variability, has improved dramatically. For example, Canadian fishery scientist Ransom Myers and colleagues have focused on environmental factors and the wide scale correlative nature of recruitment variability in many stocks, as well as the role of density-dependent vs. density-independent controls of recruitment dynamics. Moreover, the long-recognized and powerful stock–recruitment relationship is becoming better understood for many stocks as long time-series data become increasingly available.

Still, our ability to predict recruitment success in any year remains poor, while the need for prediction remains high given the dismal state of many fisheries. Knowledge of processes that strongly affect larval survival would provide early evidence of the abundance of the emerging year class. Despite our inability to accurately forecast recruitment, the focus of recruitment studies on early life stages of fishes is steadfast, although the life stage of emphasis has changed through the years. Combined empirical evidence for when year-class success is determined in marine fishes can be visualized by showing the relative correlation between estimates of successive life-stage-specific abundances (Figure 4.1). By the time the juvenile stage is reached, population or cohort-specific mortality rates generally are lower than for earlier life stages (Figure 3.2), and correlations between abundances of successively older stages are high, supporting the notion that year-class success is determined prior to the juvenile stage.

More precisely, by comparing growth (G') and mortality (M) rates as well as energetics properties of marine and freshwater fish larvae, fishery scientist Edward Houde predicted that recruitment levels and variability of marine fishes will be determined more by larval stage dynamics than juvenile stage dynamics, but that the reverse will be true for freshwater species. This suggests a high potential for environmental variability to drive recruitment variability during the larval stage of marine fishes relative to freshwater fishes. Nevertheless,

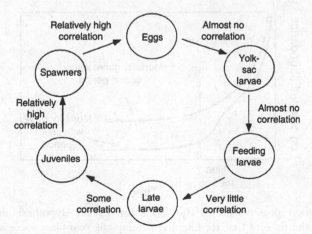

Figure 4.1   The relative correlations between estimates of successive life-stage abundances.

despite some differences in stage-specific characteristics and dynamics (for example, size, growth rate, and starvation potential), the fundamental processes that determine survival of fish eggs and larvae are similar in both lakes and oceans and occur over similar spatial scales. The similarity in processes affecting survival is remarkable given the fundamental differences in the degree of spatial connectivity of marine and freshwater ecosystems. Clearly, the depth of our understanding of the causes of recruitment variability has grown, and it will continue to broaden.

4.3   Hjort's hypotheses: in the beginning

Hjort's discussion of possible causes of inter-annual differences in mortality during early life offered two clearly stated hypotheses. The first suggests that differential mortality between years is a result of variation in food availability at a critical stage during larval fish development. The second hypothesis stresses the influence of transport of eggs and larvae away from appropriate nursery areas due to inter-annual differences in ocean circulation. As will be seen in the following sections, work initially focused on Hjort's first hypothesis but eventually both hypotheses have been addressed in early life-history studies.

4.4   Hjort's Critical Period Hypothesis

Hjort was the first to explicitly link feeding, larval survival, and subsequent recruitment to food abundance during the transition of larvae from endogenous (yolk) to exogenous (plankton) feeding. He proposed what became known as the Critical Period Hypothesis, which stated that when food was limiting during this critical transition, many larvae would die from starvation, but when food was high, survival would be high (Figure 4.2; see also Figure 3.4). He further proposed that these variations in survival owing to starvation could generate recruitment variability and hence explain the variability in year-class strength that he observed in

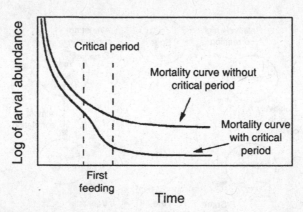

Figure 4.2 Schematic representation of Hjort's Critical Period Hypothesis illustrating the link between survival at the time of first feeding and subsequent year-class success (from Leggett & Deblois 1994 with permission from Elsevier Science).

North Sea fishes. The case study of Japanese sardine described in Chapter 11 includes empirical tests of the Critical Period Hypothesis.

Following Hjort's hypothesis many researchers set out to determine what and how much fish larvae eat, and whether sufficient food was indeed available in the sea. Since the 1960s, an extensive literature based upon laboratory studies has developed on the relationship between food type, concentrations, size, and quality at the time of first feeding, and the subsequent survival of larvae following varying intervals of starvation. In general, laboratory studies have demonstrated that larvae at first feeding and soon thereafter are indeed vulnerable to starvation due to very high weight-specific prey-consumption rates, but that food concentrations in the sea can, at times, be sufficient to support good larval growth and survival. More importantly, strong evidence of a relationship between food abundance and/or quality at the time of first feeding and either larval survival or year-class success has been elusive, as has strong evidence for the existence of high numbers of starving larvae in the wild.

Thus, a feeding paradox appears to exist. Despite very high *in situ* feeding rates, perhaps even higher than predicted in laboratory feeding studies, fish larvae in the sea apparently are able to feed at or near satiation levels, largely independent of food concentrations, even when food concentrations appear to be too low. Several factors may contribute to this apparent paradox, but they can be concisely collapsed into two main issues, both related to difficulties of sampling organisms the size of larval fish prey in the wild. Firstly, traditional estimates of prey abundance, when integrated over large spatial scales, are generally believed to underestimate the effective concentrations of prey available to fish larvae, if larvae have the ability to detect and exploit food patches. Secondly, prey abundance estimates may underestimate prey availability if they neglect the production of new prey biomass, which can be high in coastal ecosystems for prey in the size range generally required by larval fishes. Nevertheless, the feeding paradox exists because prey concentrations, on average, seem low when compared to demand by fish larvae. Several of the recruitment hypotheses that have followed Hjort's contributions have attempted to address this feeding paradox in one way or another.

4.5   Hjort's first hypothesis extended

4.5.1   *Cushing's Match/Mismatch Hypothesis*

Almost 60 years following Hjort's publication, the English fishery scientist David Cushing extended our thinking about the role of early life stages in recruitment variability in his Match/Mismatch Hypothesis. In essence, Cushing collapsed Hjort's original two hypotheses into a single hypothesis which suggested that fish spawn in relation to the particular timing of spring and autumn plankton blooms in the geographic area of inferred larval drift from spawning grounds to nursery areas. He hypothesized that a fixed time of spawning coupled with a variable time of plankton blooms generates variable larval fish survival and, hence, variable recruitment. While the mechanism of larval mortality is again starvation, what Cushing added was the consequences of critical depth (or compensation level in freshwater systems). Critical depth is defined as the depth at which total photosynthesis is balanced by total plant respiration, which must be deeper than the mixed layer (usually above the thermocline)

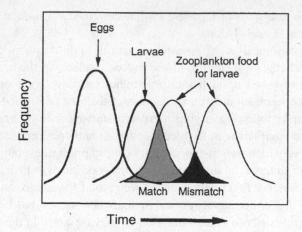

Figure 4.3   Schematic representation of Cushing's Match/Mismatch Hypothesis illustrating variability in the degree of overlap between the timing of a seasonal peak in the production of planktonic food for larvae and the co-occurrence of fish eggs and larvae. The stippled area is a match representing high overlap and the darkened area is a mismatch representing low overlap (redrawn from Leggett & Deblois 1994).

if the plankton are to stay up in the well lit water column. Cushing's modification highlights the dependence of secondary production cycles (production of prey for larval fishes) on variation in the development of water-column stratification/destratification and primary production in the spring and autumn. This generates variability in the magnitude of temporal overlap between the production of larvae (spawning) and the production of larval fish food through the development of appropriate food webs (Figure 4.3). The Match/Mismatch Hypothesis removed the restriction that food-mediated mortality leading to recruitment variability is limited to a particular critical developmental period. Rather, food limitation during any part of the larval period could be a major contributor to recruitment variability, and abiotic factors that regulate water-column destratification and the timing and intensity of seasonal production cycles may be involved.

In the years since Cushing published his ideas, many studies have attempted to test the Match/Mismatch Hypothesis in marine, estuarine, and freshwater environments. Much of the available evidence is broadly consistent with the existence of the hypothesized relationship between overlap in the development of seasonal food production and larval abundance and survival, although the link to recruitment is less well established.

4.5.2   Lasker's Stable Ocean Hypothesis

Reuben Lasker and his colleagues, working with northern anchovy (*Engraulis mordax*, similar to Figure 1.3l) off the California coast, developed an extension of Hjort's hypotheses. Lasker accepted the initial premise that food for first feeding larvae may be limited, but suggested that there are times and places in the sea where food aggregations occur, upon which larval fish survival depends. Based upon laboratory and field studies, he established the existence of minimum thresholds of prey concentrations for larval feeding and that first feeding larvae could detect good feeding areas in the sea. In a series of cruises in the

California Current in 1974 and 1975, he further showed that maintenance of "threshold-for-feeding" concentrations of food were associated with chlorophyll maximum layers in a stratified water column formed during calm and stable ocean conditions, hence the Stable Ocean Hypothesis. Thus, he hypothesized that poor recruitment would occur in stormy or windy years when too few calm periods (later called "Lasker events") occurred. Lasker suggested that measures of wind effects on water such as the cube-of-the-wind-speed, upwelling, and wind-curl indices could be correlated with larval mortality, larval food distributions, and subsequent year-class success. Indeed, relatively stormy conditions in 1975 off California produced one of the worst northern anchovy year classes previously recorded.

In the years since Lasker formulated the Stable Ocean Hypothesis, other researchers working off California in highly productive upwelling environments have shown good correlations between year-class success in Pacific mackerel (*Scomber japonicus*) and upwelling and wind-curl indices. Similarly, others have related mortality rate in larval northern anchovy to wind activity, presumably through its effect on food concentration. Strong correlations between larval mortality rates and subsequent recruitment to the juvenile stage, however, have been more elusive. It may be that stable ocean conditions are more important to larva feeding success in upwelling areas than in other locations in the sea.

4.5.3 Rothschild and Osborn's Plankton Contact Hypothesis

In a novel attempt to explain the aforementioned feeding paradox (see Section 4.4), the American fishery scientist Brian Rothschild and his colleague Thomas Osborn hypothesized that fish larvae can survive in the sea at lower prey densities than expected, if encounter rates between larval fishes and their prey increase as a function of small scale, wind-driven turbulence. This small scale turbulence should not be confused with Lasker's large scale vertical and horizontal mixing events. In the absence of turbulence, encounter rates between larvae and their prey result primarily from the combined swimming speeds of both. Turbulence at scales relevant to predator–prey interactions (centimeters to meters) increases effective swimming speeds, and thus relative motion, thereby increasing the potential for contact between larval fishes and their prey (hence the Plankton Contact Hypothesis). This is especially true when prey density is low. In periods or years when wind speeds are relatively low, it follows that larval fish feeding rates would decrease, reducing larval survival and subsequent recruitment.

Others have used Rothschild and Osborn's theoretical constructs to develop turbulence–encounter rate models to predict encounter rates that could be expected under natural levels of both wind-driven turbulence and plankton and larval fish densities. Failure to consider the effects of small scale turbulence when prey densities are low could result in greater than a 10-fold under-estimation of encounter rates under natural conditions typical of continental shelf waters in summer. In one study, it was demonstrated that under-estimates of up to 112% are possible for larvae at a depth of 20 m during turbulence that is generated by wind velocities of 5 m s^{-1} (Figure 4.4). The magnitude of under-estimation increases as prey density declines and when prey are distributed in patches. It also has been suggested that turbulence may slow the rate of starvation in larvae because weakened, slowly swimming individuals are more likely to benefit from the effect of turbulence than healthy larvae.

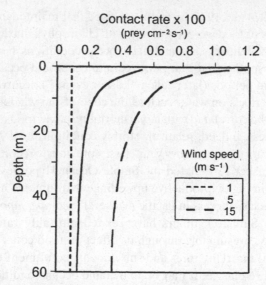

Figure 4.4   Effect of turbulence on encounter rates between larval fishes and their zooplankton prey. The relationship is for 6-mm larvae feeding at a prey density of 5 nauplii dynamic m^{-3} (reproduced from Mackenzie & Leggett 1991 with permission of Inter-Research).

This is not to say that increased turbulence always has a positive effect on larval feeding rates. Recent research has shown that when turbulent velocities become too high, the ability of larvae to capture prey declines. Perhaps this also contributes to low recruitment in stormy years, as first described by Lasker. It is probable that wind speeds greatly in excess of $10\,\mathrm{m\,s^{-1}}$ are no longer beneficial to fish larvae feeding in surface waters. In 1990 Norwegian fishery scientists showed that gut fullness for Arcto-Norwegian cod in the field increased in relation to wind speed, reaching a maximum when wind speed exceeded $6\,\mathrm{m\,s^{-1}}$. No direct tests of the Plankton Contact Hypothesis as it relates to recruitment variability have been made to date, however.

Upwelling systems such as those studied by Lasker and colleagues also generate small scale turbulence that is consistent with the velocities described above. While wind-generated upwelling generally is believed to enhance biomass production via persistent nutrient enrichment to surface waters, a dome-shaped relationship between upwelling and recruitment of pelagic fishes has been observed. In populations of such fishes as Peruvian anchoveta (*Engraulis ringens*) and Pacific sardine (*Sardinops sagax*), maximum recruitment occurs in periods of intermediate upwelling velocities of $5-6\,\mathrm{m\,s^{-1}}$ (Figure 4.5). It is not clear, however, whether the increased recruitment reported for these fishes is related to enhanced feeding success due to more food production, to turbulence-enhanced feeding associated with upwelling, or to the occurrence of more Lasker events (wind speeds $<10\,\mathrm{m\,s^{-1}}$ for four or more days) in years of moderate upwelling-producing winds.

4.5.4   Freshwater discharge and riverine plumes: prey concentrations increase

In addition to the cases already discussed, there are places in the sea where concentrations of prey-sized organisms for larval fishes are consistently high, thereby diminishing the

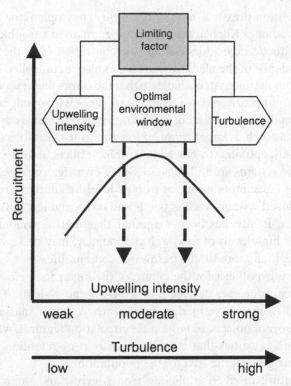

Figure 4.5   Theoretical relationship between recruitment and environmental factors in upwelling areas (redrawn from Cury & Roy 1989).

feeding paradox. It is generally recognized that hydrographic density fronts or discontinuities and riverine/estuarine plumes are important sites for energy transfer and intense biological activity, potentially supporting large phytoplankton and zooplankton standing stocks. Larval fish aggregations also are common in areas of concentrated nutrients and chlorophyll found in oceanic fronts and plumes, such as the Columbia and Mississippi Rivers, the Frisian Front in the North Sea, and the Amazon River plume. These small scale elevated gradients or patches of microzooplankton food are relevant to larval fish searching behavior and may enhance survival. Recruitment may be enhanced at such sites, where physical processes generally insure that biological conditions are favorable for survival and are stable over time.

Simple linear food chains and trophic cascades

Correlations between fishery production and nutrient input from river runoff often have suggested a "bottom-up" linear food chain. Simply stated, freshwater discharge increases the nutrient pool available to primary producers, which increases primary production, which can stimulate zooplankton production, resulting in increased ichthyoplankton survival and fishery production. Estuaries may work this way, as we will discuss later. In some freshwater systems, the strength of trophic linkages apparently is great, whereby one trophic level (usually a consumer), can control the biomass of its prey (zooplankton), which in turn releases predation pressure on the prey's prey (phytoplankton). Fishery scientists Steven Carpenter and James Kitchell labeled this phenomenon a trophic cascade, such that recruitment and production dynamics can also be controlled from the "top down," especially when the abundance of the planktivorous fishes also is controlled by predation. Most field observations from marine systems indicate that the direct linear food chain argument – that is, either a bottom-up or top-down trophic cascade – works only up through primary producers and the smallest zooplankton; cascading effects do not always appear to extend to higher trophic levels such as larger zooplankton and fishes. Reasons for apparent differences in degree of susceptibility to cascading trophic effects among riverine, lacustrine, estuarine, and marine systems are still debated. Some investigators have suggested that the cumulative effects of consumption on one trophic level by another are likely to be more profound in small, closed systems, such as ponds and lakes, and less obvious in larger, more open marine systems. It also has been suggested that marine communities, with many omnivorous taxa and high levels of food-web redundancy, may be less susceptible to cascading trophic effects than more simple freshwater communities.

Another alternative hypothesis for the failure of the linear food chain is that for higher trophic levels, physical processes dominate in open marine systems. These processes are exemplified by local frontal convergences (places where water masses come together), which accumulate buoyant objects, or large scale circulation features, which can determine current directions and velocities that influence transport or retention of eggs and larvae within the appropriate geographic area for the population.

An interesting feature of riverine plumes, fronts, gyres, and their associated convergences, however, is that they are often ephemeral in space and/or time. They frequently meander, migrate, form, strengthen, relax, and dissipate on daily or seasonal time scales, being influenced by tides, winds, and seasonal changes in hydrology (river discharge) and meteorology. The point along this ephemeral development–decay continuum that a larval fish encounters a convergence zone influences what it may encounter, since it affects how long the convergence may have been actively accumulating buoyant particles such as prey or potential predators. In addition, predators have been known to actively seek out discontinuities or gradients in salinity, turbidity, temperature, and food, and to actively maintain contact with such gradients once encountered. Thus, any feeding advantage for larval fishes gained at frontal convergences may be counteracted by increased potential for predation mortality resulting from co-occurrence with increased concentrations of active predators, including cannibals and cruising or ambush predators such as other fishes and jellyfishes.

The Mississippi River discharge plume

One specific example of the type of environment described above is the Mississippi River, which has a great influence on fisheries in the Gulf of Mexico. This large riverine plume is characterized by a buoyant, shallow lens of low salinity, nutrient-rich, and turbid surface water that flows out of the delta's distributaries and expands over the high salinity (more dense) oligotrophic (nutrient limited) Gulf of Mexico shelf waters (see Box 10.4 for more information). A third water mass is formed between the plume and shelf waters called the frontal zone, which consists of a broad zone, 6–8 km wide, containing one or more turbidity fronts and having intermediate temperature and salinity signatures. This frontal zone is characterized by high primary productivity resulting from phytoplankton in the nutrient-rich waters being no longer light limited by the river plume's high sediment load. The frontal zone is believed to be the accumulated effect or sum ("memory") of repeated formations and dissipations of individual convergence zones generated by the horizontal density gradients between the plume and shelf water.

Definitive evidence of the role of the Mississippi River plume on recruitment in the northern Gulf of Mexico has been elusive. In some studies, larval fishes associated with the plume and frontal waters have exhibited a clear feeding and growth-rate advantage over those that were outside of the plume, while in other studies, differences are less clear. Similar results have been reported for mortality rates. In no study has a direct relationship between plume/front dynamics and year-class success been demonstrated, despite the aforementioned general correlation between recruitment and freshwater discharge.

Nevertheless, the magnitude of large continental shelf fronts and gyres associated with oceanic circulation systems is sufficient to influence the survivorship of a larger number of recruits, thereby influencing the adult stock or population. Other fronts may have spatial and temporal scales that only influence the local dynamics of small portions of the population. Yet, at the level of the individual larvae they may still be important.

4.6 Hjort's second hypothesis extended

4.6.1 *Larval transport, larval retention, and recruitment*

Nursery areas of fishes are often distant from oceanic or coastal spawning grounds. In order to reach nursery areas at the proper time, size, and condition, fish larvae require appropriate currents and sufficient and suitable food during transit (see Chapter 7 for more details). Reproductive activity, however, generally occurs at selected sites or only within a relatively small portion of a species' total range. Spawning aggregations for migratory species are often concentrated in geographic areas that historically provide conditions for reproductive success, that is, areas of relatively long term hydrographic stability but not necessarily consistent year-to-year predictability. Fishes with pelagic eggs often reproduce in gyres and fronts, thus making areas of upwelling and boundary currents among the most productive sites of fisheries in the world. Environmental variation, however, is always a factor, necessitating adaptable life-history characteristics. Some coastal species have increased their reproductive potential with multiple spawning sites, extended spawning seasons, or high fecundities to counteract mortalities associated with unpredictable environments. Yet, other species appear to simply maximize larval survival by taking advantage of normal oceanographic conditions. There are relatively few kinds of current systems, and these influence transport of young fishes in predictable ways. Therefore, as Hjort first hypothesized, displacement of spawning grounds or spawning products can have an adverse effect on recruitment. A one-time transport loss involving a fraction of the cohort and occurring early in its existence may not be as detrimental, however, as the cumulative and long term exposure to high daily mortality rates. If these two sources of mortality act in concert, the prognosis for cohort survival may be poor. Therefore, anomalous physical and biological conditions can either enhance or inhibit survival and, ultimately, recruitment success.

4.6.2   Migration Triangle Hypotheses

In the late 1960s and early 1970s, after the flurry of research activity related to the Critical Period Hypothesis, some recruitment hypotheses specifically advocated the importance of the migratory life-history circuit. This concept recognized that spawning grounds often are located within a residual circulation system that transports eggs and larvae to favorable downstream nursery grounds, with the pre-adults migrating back to adult areas (for example, the Amazon catfish described in Section 2.3.3). The circuit is usually completed by eventual upstream migration to spawning grounds. While it has been difficult to test Migration Triangle Hypotheses in the field, they can best be exemplified in concept by the following two cases.

Townsend's coastal conveyor belt

Biologically productive areas can occur where thermally stratified and tidally mixed coastal waters meet and are mixed by strong tides such as those found in the Gulf of Maine, where tidal ranges are from 2 to 6 m, because deeper nutrient-rich waters are brought toward the surface and stimulate productivity. This is the case off Grand Manan Island in northeastern Maine, a known spawning area for Atlantic herring. Herring usually spawn in tidally well mixed waters from late summer through the autumn. They have an extended larval stage and complete metamorphosis when 5–8 months old.

A portion of the larvae spawned at the mouth of the tidally energetic Bay of Fundy are transported to the southwest within the eastern Maine coastal current. As the herring larvae drift further away from the high energy, vertically well mixed area, the nutrient-rich coastal water column begins to stratify. This leads first to increased phytoplankton production and biomass, soon followed by zooplankton production. Hence, the larvae find themselves in an environment of elevated food as they begin exogenous feeding after yolk absorption. This residency in the downstream coastal current can provide a more favorable feeding environment than if the larvae were retained in their spawning area, which has lower production because the phytoplankton are light-limited due to tidal mixing (that is, the phytoplankton are often mixed below the critical depth). David Townsend coined the phrase "coastal conveyer belt" to describe this phenomenon where larvae develop while drifting along with their developing food supply. He also pointed out that this conveyer belt had the advantage of taking the larvae past numerous estuarine and coastal nursery areas before turning offshore to become part of the Gulf of Maine's Jordan Basin cyclonic gyre (Figure 4.6). Entry into the basin gyre may also serve as a retention mechanism for keeping herring larvae within the inner portion of the Gulf of Maine, which may have ramifications for helping to maintain stock integrity.

Larval Retention or Member/Vagrant Hypothesis

As mentioned above, spawning often consistently occurs at fixed locations for many species. This consistency not only fixes the initial position of larval drift but also determines the position of the nursery grounds when the current regime is relatively predictable. A given stock, therefore, may be determined by the constancy of, or be contained within, a migratory

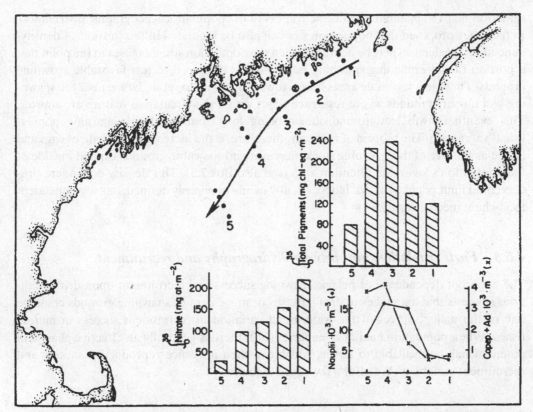

Figure 4.6 Summary plot of the changes in nitrate, chlorophyll, and naupliar and copepodid stages of copepods along the eastern Maine coastal current/plume system for July 1985, used here to illustrate the coastal conveyor belt. The nitrate and chlorophyll histograms are the averages of the vertically integrated (to 35 m depth) values at the stations shown for each of the five transects. The arrow is a streamline of the geostrophic current (reproduced from Townsend 1992 with permission of Oxford University Press).

circuit. The resultant larval drift pattern, when combined with the return migration of adults to the upstream spawning grounds, would then form the geographic base for the stock. In the North Atlantic, the number of genetically distinct herring stocks may be a direct result of the number of oceanographic, larval retention areas. Overwintering larval distributions indicate little stock mixing during the larval phase. Such distinct larval distribution patterns are associated with hydrographic features that show a remarkable degree of year-to-year consistency within the same geographic areas. The physical retention mechanisms for segregating the young of individual stocks appear to be counter-clockwise gyres, which are tidally induced.

Because of the existence of retention areas, fishery scientists Michael Sinclair and T. Derrick Iles further extended Hjort's second hypothesis in an idea that has been referred to as the Larval Retention or Member/Vagrant Hypothesis. It states that through the years species tend to spawn at specific times and places within predictable and distinct circulation features. This selection process enhances the retention or residence time of its "members" within these features by limiting dispersal of passive eggs and weakly swimming larvae until they are able to develop the ability to control their own distribution. In this way, retention helps to maintain the distinctiveness of the population's geographic distribution. In fact, they proposed that population size can also be regulated in this fashion. A density-dependent mechanism may be activated when the population size increases to the point that a portion of the expanding spawning population is displaced to less favorable spawning grounds. These less favorable areas have a lower capability to retain larvae near the spawning and nursery grounds where residence times within the circulation feature are shorter. This eventually will lead to individuals being lost from the local spawning population ("vagrants"). The process of vagrancy, therefore, is the increase in loss rate of eggs and larvae as the size of the spawning stock increases and spawning grounds expand outside of the population's favorable retention area (see also Box 7.3). This density-dependent process could limit population size independently of any previously defined, density-dependent food-chain mechanisms.

4.6.3   Further relationships between hydrography and recruitment

The apparent dependence of pelagic spawning success and recruitment upon drift conditions suggests that anomalies in drift patterns or movement of spawning grounds could be one of the major sources of the widespread variation in recruitment success in marine fishes. When population habitat boundaries shift, there is generally an effective change in potential habitat available to larvae, which can then influence reproductive success and recruitment variability. We offer a few more examples.

Upwelling environments

In the northern hemisphere when wind blows over the surface of water, the resultant current flows to the right of the wind direction (or to the left in the southern hemisphere). This is known as Ekman transport. Where such wind-induced transport is offshore (Ekman divergence), displaced inshore surface waters are replaced by cold, nutrient-rich waters upwelled from below. Persistent upwelling generally occurs along western continental margins and is associated with prevailing winds blowing toward the equator. Well known examples include the Peruvian and the North African Mauritanian upwelling systems. Upwelling also takes place to a lesser degree along the eastern coast of continents, where it is induced by winds blowing toward the poles. Continental shelf or oceanic fish larvae will benefit from increased productivity associated with upwelling, while larvae needing estuarine or coastal nursery areas will benefit from onshore transport in the deeper waters. If the upwelling system decays or disappears, both groups of larvae may experience increased mortality. Recall the role of upwelling environments in the Stable Ocean Hypothesis (Section 4.5.2).

Estuaries

Estuaries are important because many fish species are considered to be estuarine dependent, whereby they must utilize the estuary during some portion of their life. This generally occurs during early life, as estuaries are believed to be enriched in nutrients from freshwater runoff and hence richer in food for larvae and juveniles, thus providing important nursery habitat for young fishes. Estuarine dependency implies that recruitment success may be determined for some species by the degree to which fish eggs and larvae are transported to, and retained within, the estuary. There is a tremendous body of literature on this subject, but indications of direct relationships between estuarine residency and recruitment variability, again, are difficult to find.

Estuarine circulation is influenced by tide, river flow, local wind, topographically induced circulation, and non-tidal forcing from the coastal ocean. Estuarine circulation can also be altered by short term meteorological events and by seasonal or annual river discharge variability. Wind-induced transport toward coastlines (often eastern continental margins) – known as Ekman convergence, coastal setup, or downwelling – can produce significant non-tidal variation in sea level, which can directly affect estuarine currents because the estuary tends to fill with more water. Several of the world's coastlines have such seasonal, non-tidal changes, which can result in seasonal or short term net coastal flows into estuaries. In addition Ekman divergence or convergence can change the density structure of the coastal water column, thereby altering the gravitational circulation and exchange between estuary and ocean.

There are fundamentally two types of estuaries: positive estuaries, in which river discharge and local precipitation exceed evaporation, and negative estuaries, where evaporation dominates (for example, lagoons). The normal exchange pattern for a positive estuary is a residual downstream movement of fresh water at the surface, a level of no net motion at mid-depth, and a net, non-tidal upstream counter current of denser, saltier water at the bottom. This exchange pattern facilitates the passive upstream transport of fish eggs and larvae (or any other neutrally or negatively buoyant planktonic organism) in the estuarine bottom layer with varying degrees of success due to turbulent-mixing-induced downstream losses. The vertical position of an organism in the water column over time can determine the direction of its transport.

Space-limited environments

There are species of marine and freshwater fishes for which recruitment may be limited by the availability of space when pre-recruits settle from the three-dimensional planktonic environment to an essentially two-dimensional benthic habitat. Most notable are the coral reef fishes whose larvae are pelagic, but whose juveniles and adults are associated with a reef. We mention this here, however, because there continues to be debate among fishery scientists as to whether recruitment by these species is indeed limited by post-settlement processes or is limited by the supply of larvae transported to, and retained upon, the reef. Post-settlement processes include reductions in growth rate due to intraspecific competition for food and increased mortality due to a lack of shelter, among others. Similar arguments have been made for the flatfishes, another group whose larvae (Figure 1.3t) are pelagic before settling to the bottom to assume a benthic existence. Like coral reef fishes, post-settlement mortality in flatfishes and other demersal species may be mediated by predation, if the supply of larval pre-recruits is not limiting. Pre-settlement processes include all of those related to egg and larval stage dynamics as discussed elsewhere in this chapter.

Space also can be limiting for freshwater species that spawn in streams, as adults compete for spawning sites and young of the year must compete for a finite number of optimal locations for growth and survival within the complex habitats of streams. In this scenario, currents again play a role because optimal locations often are defined by the balance between flow rates that are not too energetically demanding for young fishes struggling to maintain their position in the stream and flow rates that are high enough to deliver enough drifting food.

4.6.4   Relationships between hydrography and recruitment: examples

Haddock

The Gulf of Maine was an early testing ground for relating plankton distributions to circulation patterns. The normal cyclonic circulation pattern over Georges Bank can cause the loss of haddock eggs and larvae (Figure 1.3n) to the deep waters, since near-surface currents on the bank are generally southerly and offshore, except for short periods in the summer. Transport off the bank can result in mortality due to removal of eggs and larvae from feeding areas and to transport to depths too great for settlement of larvae.

Walleye pollock

Walleye pollock spawn in the Gulf of Alaska mostly along the Shelikof Straits, where eggs and larvae (Figure 1.3o) drift within a southwesterly current along the Alaska Peninsula and enter fjords where their first summer is spent. This current system has abundant food resources, which favor larval survival during transit. Several oceanographic fronts in the Bering Sea separate characteristic water masses that support distinct zooplankton populations with varying food quality. Recruitment of walleye pollock, therefore, is likely to be affected by these oceanographic systems, which can influence the distribution of concentrations of their eggs, larvae, and food supply. We will discuss this example further in Section 4.8.2.

Capelin

The survival of larvae can also be influenced by winds, since winds can alter local conditions by exchanging or replacing water masses. For example, onshore winds along coastal Newfoundland can exchange the local water mass, which normally has a high concentration of predators, for one that has fewer predators and higher densities of prey for larval fishes. The characteristics of this new water mass trigger the emergence of larval capelin from demersal eggs buried within the sand of the intertidal zone. Capelin larvae, therefore, may initiate their drift and first feeding in a "safe-site" that is mediated by the wind. If there is a survival advantage to such a water-mass exchange and transport event, then any synchrony with the time of hatching could be positively reinforced. Other examples of safe sites may exist, but it remains to be seen whether the oceanographic features or the physical processes producing them, and their subsequent impact on larval survival and recruitment success, can be sufficiently documented over time.

4.6.5   Some large scale examples

In recent years, fishery scientists have begun to discover even larger scale links between recruitment variability, oceanography, and climatology. For example, the periodic phenomenon known as the El Niño–Southern Oscillation (ENSO) has been shown to dramatically alter sea surface temperatures in the eastern Pacific, rainfall patterns and sea surface temperatures in the Atlantic, and the frequency of Atlantic tropical storms. All of these changes have been implicated as factors that affect recruitment dynamics to some degree, especially in the upwelling environments of the eastern Pacific. Also in the Pacific basin, large scale and persistent changes in patterns of atmospheric high and low pressure systems result in oceanographic regime shifts in the North Pacific Ocean that have been implicated in the recruitment variability of many species there, especially Pacific salmons.

A somewhat similar atmospheric–oceanic coupled system, albeit much less studied, exists within the North Atlantic (the North Atlantic Oscillation, NAO). That climatic system, however, involves a north–south dipole (low-frequency atmospheric pressure anomalies between the Icelandic/Greenland low pressure and the Azores high pressure systems) rather than an equatorial, east–west dipole as in the Pacific. The NAO, however, may also prove to have some degree of coherence with the tropical North Atlantic Ocean by way of the Mauritanian upwelling system off northwest Africa (such as, correlations with its coastal upwelling index) and perhaps the Caribbean Sea.

Also in the North Atlantic basin, a change in the circulation pattern of the North Sea resulted in a dramatic increase in the abundance of cods and haddocks and declines in Atlantic herring, which persisted for many years but has since reversed. Herring stocks are now increasing. Cod stocks in the North Atlantic Ocean exhibit positive correlations in year-class strength across broad spatial scales (that is, patterns of strong and weak year classes are correlated among stocks), implying a link to climate.

Finally, in oceans worldwide, anchovy and sardine* stocks synchronously alternate in abundance, one replacing the other at a location for long periods of time. Fishery scientists suspect that this periodic alternation is driven by large scale climatic and oceanographic variability, although the mechanisms remain unknown. As our ability to measure synoptic ocean and atmospheric conditions improves as a result of technological advances such as high altitude and satellite remote sensing, coupled with recent dramatic improvements in weather forecasting and ocean circulation computer models, our ability to unravel these large scale links will undoubtedly improve. Such advances will become increasingly important to fishery science in the face of global warming. We will discuss other technological advances in a later section.

4.7   Predation mortality and the paradigm shift

Recently, a dramatic shift in thinking about the causes of recruitment variability has emerged, potentially making the work of fishery scientists more difficult. As described above, most of the previous recruitment hypotheses elaborated on one or the other of Hjort's original two

---

*A detailed analysis of population fluctuations in Japanese sardine, resulting from variable recruitment, is summarized in Chapter 11.

hypotheses and described mechanisms that generally resulted in episodic year-class successes or failures – sort of an all or nothing proposition. Since the mid- to late-1970s, however, it has become increasingly clear that predation is a (perhaps *the*) major source of egg and larval mortality in fishes (see Chapter 3). Although largely forgotten in today's literature on the subject, David Cushing was perhaps the first to recognize the potential importance of predation as a recruitment regulator in his 1975 Single Process Concept. He suggested that as the length of time that eggs or larvae remain vulnerable to predators increases, so will cumulative mortality rate increase. More specifically, larvae that experience favorable feeding conditions and grow more quickly will complete metamorphosis at earlier ages and thereby experience lower cumulative predation mortality during the larval stage, the time when mortality rates are known to be high. More importantly, it suggests a mechanism by which episodic processes are not the only means by which a year class can succeed or fail.

4.7.1   Houde's subtleties and episodes: Cushing's Single Process rediscovered

In several important papers in the late 1980s, Edward Houde revived Cushing's Single Process concept by showing that relatively subtle changes in larval growth and mortality rates potentially could result in strong or weak year classes. In these papers, Houde argued that rate changes, the magnitude of which would be difficult to detect in the field, were sufficient in exponential decline models to generate more than 100-fold variability in the numbers of larvae surviving to the juvenile stage (Table 4.1). Other researchers later called this argument the Stage Duration Hypothesis. Houde furthered argued that predation was the likely source of most egg and larva mortality.

4.7.2   The Stage Duration Hypothesis

Unlike the other recruitment hypotheses that we have discussed, the Stage Duration Hypothesis is important because it represents a different way of thinking about the recruitment problem. And if correct, it may be more difficult to predict recruitment success because of the difficulties in understanding all of the environmental factors that can contribute to subtle changes in larval growth and mortality rates. This may best be understood by viewing Figure 3.5, which shows the magnitude of decline in numbers typical of many marine and freshwater fish species, and the many biotic and abiotic factors that can

Table 4.1 Hypothetical recruitment of young fish under one "good" and three possible "bad" conditions, the latter represented by 25% changes in mortality or growth rates (as age at metamorphosis). Recruitment is defined here as the number of survivors at the end of the larval stage (data from Houde 1987).

Condition	Initial number in cohort	Instantaneous mortality (Z, day^{-1})	Age at metamorphosis (day)	Number of recruits
Good	1×10^6	0.100	45.0	11 109
Bad-1	1×10^6	0.125	45.0	3607
Bad-2	1×10^6	0.100	56.2	3625
Bad-3	1×10^6	0.125	56.2	889

contribute to recruitment success or failure, including density-dependent (or compensatory) feedbacks that arise when numbers in a population become too high or too low (see Section 3.1.4). The extremely high mortality rates implicit in Figure 3.5 also imply that the average fish egg or larva in the sea is destined to perish. Such high mortality has led many investigators to search for characteristics among the relatively few survivors that distinguish them from their departed siblings, especially with respect to their ability to avoid predation. Interestingly, some marine and many freshwater species, those employing the equilibrium or *K* reproductive strategy discussed in Chapter 1, have greatly reduced recruitment variability associated with egg and larval stage dynamics because they invest more energy in parental care, mainly by building and subsequently guarding nests. In these species, many of which are found in resource-limited environments, high fecundity is traded for large eggs and large, well developed larvae upon hatching. The result is the production of a more constant, but low number of new recruits. Space, again, can limit recruitment for these species because adults compete for the best territories.

4.7.3   The fate of the average fish larva

Because the average fish larva dies soon after hatching, and predation is now believed to be the major source of larval mortality, it follows that survivors from cohorts exposed to predators may be exceptional individuals with respect to characteristics that shape predation vulnerability. This notion led many fisheries scientists in the late 1980s and early 1990s to search for simple conceptual models that could offer a better understanding of how predation works. These conceptual models ultimately focused on the empirically derived relationship between increased larva size (or in the degree of ontogenetic development) and decreased probability of being captured by any of a host of different types of predators (Figure 4.7) as a means to simplify our thinking. Some in the fishery literature later referred

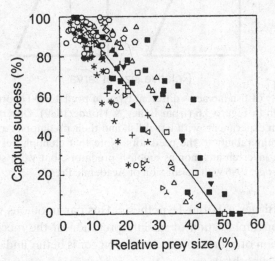

Figure 4.7   Capture success as a function of prey to predator length ratio for different combinations of fish species as predators (12 species) and prey (nine species). Each symbol identifies a different combination of predator species, prey species, and temperature (reproduced from Fuiman 1994 with permission of Academic Press).

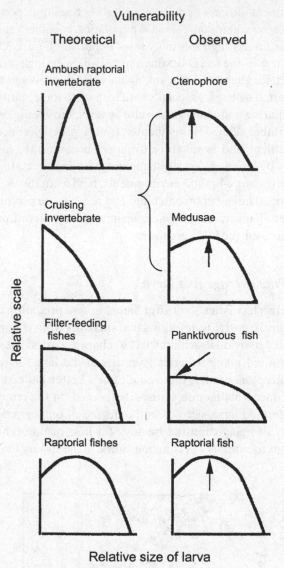

Figure 4.8   Vulnerability of fish larvae to different types of predators. Theoretical relationships are the same as those shown in Figure 3.6 (from Bailey & Houde 1989). Observed relationships were derived from mesoscosm experiments with larval fish and their predators. Arrows indicate the relative size of maximum vulnerability to the predators. Note that ctenophores and medusae did not behave as true cruising invertebrates nor as ambush predators, but were somewhere in between (modified from Cowan *et al.* 1996 with permission of Academic Press).

to this notion as the Bigger-is-Better Hypothesis. This parsimonious view, in our opinion, initially led to an oversimplification or misinterpretation of the mechanics of predation, resulting in the elevation of connate ideas such as bigger is better and stage duration to the level of contradictory hypotheses.

We now know that these mechanics are far too complex to be easily described by simple models that relate vulnerability to any one larval characteristic. As we learned in Chapter 3, empirically derived vulnerability curves are often dome-shaped as predicted by theory (Figure 4.8), but these curves can be generated by very different properties inherent to specific predator–prey interactions. Moreover, because it is likely that members of most cohorts of larvae *in situ* are exposed to a continuously changing gauntlet of predators during early life, where the mix of predator types and sizes is highly variable, it seems unlikely that any simple conceptual model could apply, nor does it appear likely that survivors are indeed inherently exceptional individuals. In fact, it has been suggested that characteristics of individual larval survivors may be more influenced by attributes of the predators to which they were exposed in early life than by their initial sizes within the cohort and their potential individual growth rate.

What does seem clear, based on results of numerous experimental and modeling studies, is that it is likely that survivors are derived from faster growing cohorts of larvae because they spend less time in the stage vulnerable to predators, as Cushing stated so well in his definition of the Single Process. It is less likely, however, that large size or other single attributes of an individual within a cohort will afford it a clear advantage for survival over others. More importantly, faster growing cohorts appear to experience considerably lower cumulative mortality rates regardless of the types of predators to which they are exposed, suggesting the importance of relationships between cohort mean growth rate (not body size *per se*), predation, and recruitment success. Thus, while is has been difficult to directly test the Stage Duration Hypothesis *in situ* by virtue of the mechanisms proposed in the hypothesis itself (that is, subtle changes), its logic and implications are clear.

4.7.4   Peril of the unfit or the unfortunate?

Even while considering the arguments made in the previous section, we believe that the jury is still out regarding the attributes that may make individual fish larvae more or less vulnerable to predation because larval size and growth rate are not the only attributes that need to be considered. Toward this end, fishery scientists Lee Fuiman and James Cowan are currently (as of 2002) asking the following question in a combined laboratory, modeling, and field study. Is there a subset of individuals in a cohort that have a suite of physiological or behavioral characteristics that confer greater fitness in predator–prey situations, leaving less fit individuals to succumb to predators? Preliminary results indicate that larval "athletes" – individuals that consistently see and hear better, swim faster, and respond more quickly and more vigorously when exposed to a threat – appear to exist. These larval athletes are less frequently consumed by predators in the laboratory. It remains to be determined whether these differences are important in a field setting.

4.8   Predicting recruitment: progress and prognosis

It should be clear that larval fish distribution, abundance, and survival is controlled by both biotic and abiotic factors. Biotic factors include: adult spawning condition, behavior, and abundance; environmental optima and tolerances; *in situ* concentrations of suitable food and potential predators; and larval behavior. Physical factors include patterns or cycles in the climatology, hydrography, and oceanography of the area (that is, water temperature and salinity; vertical and horizontal gradients in water density; turbidity; and water current speeds, directions, and anomalies). Moreover, it should be apparent that the processes underlying the recruitment hypotheses we have discussed may represent endpoints in a continuum of processes that shape the recruitment function. In other words, processes emphasized in specific hypotheses may work simultaneously or sequentially to determine year-class success, both through episodic losses of eggs and larvae in some situations and through subtle but cumulative changes in vital rates in others. In this light, the ultimate goal of recruitment prediction seems daunting. Yet, progress is being made through a series of technological innovations fashioned in the past decade that have provided new insights on, and increased detectability of, the recruitment variability problem. Some of these include: otolith aging and chemical analysis of skeletal material (see Chapters 2 and 6 and Section 12.6); new methods of mass marking of eggs and larvae for use in experiments (see Section 12.9); molecular genetic approaches to stock identification (see Chapter 6); new acoustic and video abundance measurement methods (see Section 12.2); new statistical and time-series analysis methods; incorporation of stochasticity (variability, random and otherwise) into population models; assembly of large data bases; and individual-based modeling (see Section 12.8). We will briefly discuss some of these in more detail, ending with an example fishery to which many of these innovations have been applied.

4.8.1   Technological innovations

The following information about technological advances was taken largely from a review paper on compensation by fishery scientist Kenneth Rose and colleagues. Because many of these innovations also have bearing on the future of recruitment prediction, some of the information is repeated here. This list is not meant to be comprehensive, but rather to illustrate that we are at a point where diverse technical advances are providing an opportunity for a major leap in our understanding of recruitment. The growing emphasis on synthetic and comparative analyses, coupled with advances in measurement, statistical, and population modeling methods, is encouraging and absolutely critical for progress in understanding recruitment dynamics and in the effective management of fish populations.

Mass marking of eggs and larvae

The traditional mark–recapture approach to estimating mortality and movement has been greatly expanded by the use of chemicals for mass marking of eggs and larvae, tags that permit information on individual fish to be retained and stored, and by the use of ultrasonic telemetry that allows continuous tracking of marked individuals.

Acoustic and video measurements of abundance

Video and acoustic technologies are two examples of new abundance measurement methods that have recently become available. Video and acoustic methods have been used to augment and calibrate traditional sampling gear, and to permit sampling in situations where traditional gear cannot. They have also been used to simultaneously monitor biological and environmental variables on the scale of meters or less, as well as over large areas, and for recording detailed behavioral interactions between individuals.

New statistical data analysis

New statistical methods are available that are well suited for analyzing fish population-dynamics data. Statistical analysis in ecology in general has been moving from hypothesis testing and linear models to multiple hypothesis evaluation (such as, maximum likelihood) and non-linear models. New methods, such as generalized additive models, non-linear time series, neural networks, fuzzy mathematics, classification and regressions trees, geostatistical methods, and methods that explicitly account for sampling and measurement error, are being applied to fishery-related data. Recent advances also allow for much greater flexibility in time-series model formulation. A promising trend is the focus on the interaction between environmental stochasticity and density dependence, and how they combine to control the long term dynamics of populations. Bayesian approaches using maximum likelihood methods are being used to robustly estimate the many unknown parameters in population dynamics and stock-assessment models. Synthetic analyses involving diverse studies can now be rigorously analyzed statistically using meta-analysis methods.

Explicit treatment of stochasticity in population models has increased realism. Population modeling has moved from deterministic models of simple equilibrium to non-equilibrium approaches that explicitly include stochasticity and uncertainty. The definition of a regulated population has been expanded from simple statements about equilibrium densities to more encompassing definitions appropriate for highly stochastic populations, such as a bounded variance of population densities and a long term stationary probability distribution of population densities. Embracing the stochasticity that is characteristic of almost all fish populations, rather than using models that attempt to average the variability away and produce precise but inaccurate predictions, will increase model credibility.

Individual-based models

Individual-based modeling offers a promising approach for understanding population and community dynamics, and has features that should help in quantifying recruitment variability of fish populations. Representing local interactions in space, size-based interactions, episodic effects, movement, and stochasticity, all of which are important to realistically simulating fish population dynamics and recruitment, is relatively easy in individual-based models. Intuitively, if one can realistically represent how individuals grow, survive, reproduce, and move, then realistic estimates of population-level phenomena such as recruitment can be obtained by simply summing over all of the individuals in the model.

4.8.2   FOCI and walleye pollack: the state of the art

The Bering Sea supports some of the most productive fisheries in the North Pacific Ocean, and the walleye pollock is the most abundant of the species harvested there, accounting for more than 65% of the total groundfish biomass. During the 1980s total pollock biomass in the Bering Sea was estimated to exceed 20 million tonnes and was harvested in the exclusive economic zones of both Russia and the United States. The annual United States catch of pollock in recent years has averaged 1.3 million tonnes, with an ex-vessel value of $210 million. Besides their economic importance, pollock are important to the Bering Sea ecosystem, providing most of the food for the extensive marine mammal and bird populations found there.

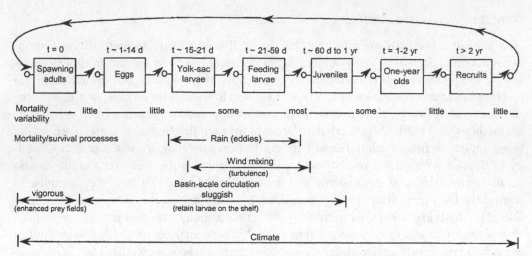

Figure 4.9 Conceptual model of Gulf of Alaska walleye pollock survival at different life stages. Relative mortality, important environmental processes, and the life stages they affect are indicated. Note the similarity to Figure 4.1 under Mortality Variability (redrawn from Megrey *et al.* 1996).

As for most marine fish species, Bering Sea walleye pollock experience large inter-annual variations in recruitment and these determine population size and thus fishery quotas and harvest levels. Recruitment is thought to be largely set during its first year of life. The need for early prediction of year-class success for this ecologically and economically important species is self evident. Thus, to develop an understanding of stock structure and recruitment variation in Bering Sea walleye pollock, the United States National Oceanic and Atmospheric Administration established in 1992 the Bering Sea Fisheries Oceanography Coordinated Investigations (FOCI) program, the most ambitious program to date devoted to the subject of recruitment. This program of study included a major emphasis on the dominant physical oceanographic features that could directly or indirectly influence survival of pollock larvae, through modulation of larval food production. As a result of this effort, more than 400 papers have appeared in the published fisheries literature and most of the technological innovations that we listed earlier have been employed and further developed. More importantly, this effort has led to the ability to predict, although coarsely, walleye pollock recruitment. FOCI scientists now can review a series of biological and physical inputs and provide fishery managers predictions of recruitment that can distinguish between strong and weak year classes. Prediction is based upon model-generated relative mortalities of early life stages, including eggs, yolk-sac larvae, feeding larvae, juveniles, 1-year olds, and recruits, and the environmental processes that drive their population dynamics (Figure 4.9). While this level of prediction may seem crude, it represents the state of the art in fishery science and fishery managers are extremely happy to get it.

4.8.3   *Epilogue*

The following paragraphs are taken almost verbatim from a paper describing the history of recruitment fishery science in the United States written by Arthur Kendall and Gary Duker. It eloquently summarizes where we are today.

Although much information has been gathered and analyzed, and numerous publications completed, understanding of the mechanisms that drive recruitment remains an elusive goal. Some have suggested, as Hjort himself did to the newly formed International Council for the Exploration of the Sea, that understanding recruitment processes is not worth the effort; managers merely need relative estimates of recruitment strength. We disagree with this admonition and further suggest that just correlating year-class strength with environmental variables is not enough; a true understanding of the processes involved in variations in survival of young stages, as Spencer Fullerton Baird (the first Commissioner of the United States Bureau of Fisheries) advocated, is required.

"Since the 1920's, correlations of the strength of year classes with environmental factors... began to take a certain melancholy consistency. Initial data might suggest a high correlation... but eventually the correlation would fail."

Could the founders of recruitment fishery science have anticipated the complexity of the recruitment process when they first advocated an ecological approach to the study of fluctuations in fish populations? Even with increased awareness of the importance of fish recruitment to management and recent technological and conceptual advances, many of the questions and hypotheses remain unanswered and untested.

Additional reading

Carpenter, S.R. & Kitchell J.F. (Eds) (1993) *The Trophic Cascade in Lakes*. Cambridge University Press, UK.

Cushing, D.H. (1975) *Marine Ecology and Fisheries*. Cambridge University Press, London, 278 pp.

Cushing, D.H. (1995) *Population Production and Regulation in the Sea. A Fisheries Perspective*. Cambridge University Press, London, 354 pp.

Cushing, D.H. (1996) *Towards a Science of Recruitment in Fish Populations*. Ecology Institute, D-21385 Oldendorf/Luhe, Germany, 175 pp.

Grahame, J. (1987) *Plankton and Fisheries*. Edward Arnold Press, London, 140 pp.

Hjort, J. (1914) Fluctuations in the great fisheries of northern Europe viewed in the light of biological research. *Rapports et Procès-verbaux des Réunions, Conseil International pour l'Exploration de la Mer* **20**, 1–228.

Kendall, A.W., Jr. & Duker, G.G. (1998). The development of recruitment fisheries oceanography in the United States. *Fisheries Oceanography* **7**, 69–88.

Lasker, R. (Ed.) (1981) *Marine Fish Larvae. Morphology, Ecology, and Relation to Fisheries*. University of Washington Press, Seattle, 131 pp.

MacCall, A.D. (1990) *Dynamic Geography of Marine Fish Populations*. Books in Recruitment Fishery Oceanography, University of Washington Press, Seattle, 153 pp.

Royce, W.F. (1996) *Introduction to the Practice of Fishery Science*, revised edn. Academic Press, San Diego, CA, 448 pp.

Sinclair, M. (1988) *Marine Populations. An Essay on Population Regulation and Speciation*. University of Washington Press, Seattle, 252 pp.

Chapter 5
Population Analysis

Pierre Pepin

5.1 Introduction

To manage wild fish populations conservatively we must have information on the rate at which a stock of reproductively mature animals can renew itself. The underlying stock–recruitment relationship can take a variety of forms, depending on the life history of a species as well as the nature of the environment(s) in which it occurs. Identifying the form of this relationship has not been easy but two key elements necessary to achieve this goal are a measure of the reproductive potential of the population and the number of young that will reach maturity. Fishing can then be targeted to exploit the excess production. Although this may appear relatively simple, the task is in fact rather difficult. To obtain an accurate measure of the abundance of reproducing adults requires surveys that provide quantitative estimates of catch per unit of sampling effort. Often, reproductive potential is taken as the biomass of adults beyond a threshold age or size, without taking into account variations in fecundity. Recruits can sometimes be sampled just before they enter a commercial fishery because they are approaching a size that can be effectively captured by standard fishing gears. However, if managers could obtain an accurate estimate of the true reproductive output of a stock, or some earlier insight about the upcoming fluctuations in recruitment, exploitation strategies might be altered to take advantage of coming booms or to prevent collapse because of persistent low egg production or poor recruitment.

Since the end of the 19th century, scientists realized that the abundance of planktonic early life stages could be used to measure the reproductive output of a fish population. In a classic treatise of 1914, Johan Hjort proposed that the abundance of young surviving through a critical stage in early life could serve as an indicator of the strength of coming year classes. Even if one cannot estimate the number of spawning fish because they can avoid nets or, for whatever reason, cannot be scientifically surveyed, it may be simpler to sample their planktonic offspring. If there is a predictable or measurable relationship for the number of eggs or larvae produced by the average adult, then the information from plankton surveys can be used to approximate the number of adults reproducing, or at least obtain a relative index of their numbers. A good part of this chapter will be dedicated to the concepts surrounding this approach. In addition, the development and use of pre-recruit indices will also be discussed. Identifying a critical stage at which year-class strength is established has proven to be difficult, whether in general terms or within a species or stock, because of the many factors that affect variations in growth and mortality (see Chapters 2 and 3). Despite this, surveys of young fishes are becoming increasingly important in forecasting recruitment, particularly in heavily exploited populations, and they are providing increasing insight about how individuals are not all equally likely to survive.

5.2   Measuring stock size from egg or larval abundance

5.2.1   *The basic concept*

For fishery scientists engaged in estimating the size of fish populations, one of the most difficult tasks involves obtaining a measure of abundance whereby all the necessary parameters used in their assessment can be measured and none have to be assumed. The data may come from a commercial fishery, where a measure of effort can be derived, or from research surveys, where a consistent approach to sampling can be used to derive an index of the density of fish in a population. Standard fishery science texts discuss the inherent problems in using commercial catch information, the least of which is our lack of understanding of how fishers alter their tactics. In the case of many populations, such as free-swimming fishes (for example, herring, anchovy, sardines, mackerel, tuna) or in cases where bottom trawls are ineffective at capturing adults, quantitative samples of the density of adult fish derived from scientific surveys are difficult to obtain. It is under such circumstances that techniques based on measures of egg or larval abundance were devised to provide an assessment of population abundance or biomass.

In their simplest form, surveys of planktonic fish eggs and larvae can be used to derive an *index* of stock abundance. I use the plural when dealing with the number of surveys because in most instances, it is essential that an index of ichthyoplankton abundance provides a representative sample throughout the spawning period, which for many species extends over several weeks or months. One may try to schedule sampling during the peak reproductive period but any changes in the timing of spawning, whether caused by the environment or the fish, may lead to inaccurate measures of stock abundance. In addition to using several surveys to describe the spawning cycle, the spatial extent of the sampling should go beyond the spawning range of a population because in many species the range of a population expands and contracts as their numbers increase or decrease, as described for Japanese sardine (*Sardinops melanostictus*) in Section 11.3. If an unknown portion of the spawners is outside the survey area, the index is likely to be biased and thus will not provide a reliable measure of the state of the population. Finally, the index must be derived from the abundance of a specific stage or the abundance of a specific size range of larvae because mortality during early life is generally high, which necessitates the comparison of year classes using animals at similar states of development. The concept is simple:

(1)   sample the plankton in a consistent manner during the reproductive period of the species of interest;
(2)   ensure that you obtain samples over the entire spawning range;
(3)   design the survey so that you can weight the contribution of each station to the index; and
(4)   focus on a stage of development that you can effectively and consistently sample.

A high index would be indicative of a large spawning stock and a low index should indicate the converse, or at least that something is happening in the population that you should pay attention to. It is important to remember, however, that an index is not an absolute measure of spawner abundance but it does serve as a starting point to deal with the concept of stock estimation from egg or larval surveys.

The relationship between the spawning biomass of a fish stock and the production of offspring of age t is easily derived. Simply stated, the production of offspring (P_0) must be equal to the female biomass that produced them multiplied by the number of eggs produced per unit weight of female (E)

$$P_0 = (B \cdot R) \cdot E \tag{5.1}$$

Female biomass is represented as the biomass of the entire stock, B, both males and females, and R, the portion of the entire stock that is producing offspring (that is, reproductively mature females). Note here that spawning is measured over a specified period of time, normally, the entire spawning season. The number of offspring of age t produced by the population (P_t) is related to offspring production by taking into account losses that occur between spawning and time t

$$P_t = P_0 \cdot e^{-Z \cdot t} \tag{5.2}$$

where P_0 is the number of offspring produced by the population (Equation 5.1), and Z is the daily mortality rate (Chapter 3). By summing P_t over all surveys, the cumulated value provides an index of spawner abundance.

In deriving a simple index of abundance based on catches of a specific stage of development (that is, individuals at some age t), one assumes that development (or growth), mortality, and egg production per female remain relatively constant. Only wishful thinking would make this true. All these elements are known to vary among cohorts and year classes and there is ample evidence that regional differences can be substantial even within a stock's range. Consider the example of 10% variations around an average mortality rate (Figure 5.1). If it were possible to measure offspring production as the eggs were extruded by the females (that is, $t = 0$), variations in mortality rates will not have sufficient time to cause changes in the relationship between offspring abundance and spawning-stock biomass. As one bases the index on later stages of development, however, the potential error caused by unknown changes in losses from the population will increase exponentially as the age of the index stage increases (Figure 5.1).

A key aspect of Equations (5.1) and (5.2) is that they point to the need for basic biological knowledge about the state of a fish population. If we can measure all the parameters in those equations, then we can derive an estimate of absolute abundance for which we need not make any substantive assumptions. An appealing aspect of moving from an index to a measure of abundance is that fishery scientists are required to make more extensive observations on both the adults and their offspring as well as the relationship between the two stages. There are substantial difficulties and potential sources of error that need to be considered in order to use abundance of eggs and larvae to estimate stock size, but with careful design and planning, the major problems can be overcome.

Various methods have been devised to derive spawning-stock biomass from surveys of early life stages: Annual Egg Production Method (AEPM); Daily Egg Production

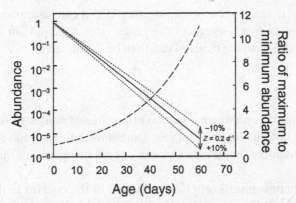

Figure 5.1   Abundance in relation to age for a population subject to a daily mortality rate (Z) of 0.2 (solid line) and the change in numbers caused by variations of ±10% (dotted lines). The dashed line shows the ratio of abundance indices of minimum (−10%) to maximum (+10%) mortality rates.

Method (DEPM); Larval Abundance Index (LAI); Larval Production Method (LPM); and Daily Fecundity Reduction Method (DFRM). A few other methods exist and they will be discussed briefly, but these five represent the most frequently applied approaches. Some of the major developments in the concepts surrounding these various methods are the result of work by scientists from the Coastal Fisheries Resources Division of the Southwest Fisheries Center (California) and scientists from diverse member nations of the International Council for the Exploration of the Seas.

5.2.2   Egg production methods

If we start with Equation (5.1) and assume that $t = 0$ in Equation (5.2) and measurements of production and fecundity are derived for the entire spawning season, we effectively have the basis to estimate AEPM (Figure 5.2). In species where it is possible to obtain an accurate measure of the number of eggs produced during a spawning season by sampling

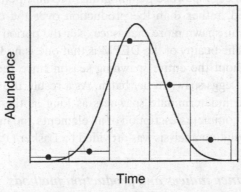

Figure 5.2   The solid line shows the hypothetical offspring production curve for a stock of fish in relation to time, while the circles show the estimated abundance from population surveys, each leg of which is represented by the open bars. Integrating the area under each block and summing among survey legs would provide the estimated offspring production from the population.

female gonads prior to the onset of spawning, the AEPM can be applied because fecundity is considered determinate, whether the eggs are released in small batches or during a single event. Spawning biomass can then be estimated as

$$B = \frac{P_0}{E \cdot R} \tag{5.3}$$

where R is the ratio of reproductive biomass of mature females to total biomass of the population (that is, both males and females, sometimes referred to as the sex ratio). By its name, the AEPM requires that the production of eggs be measured throughout the entire spawning season.

In cases where there is insufficient differentiation of the oocytes in the gonads to assign them to a class that will be spawned during the reproductive season, fecundity is termed indeterminate, and one must turn to the DEPM to estimate spawning biomass from egg production. To achieve this, Equation (5.1) must be modified so that

$$R' = R \cdot f \tag{5.4}$$

where R' is the proportion of females producing biomass, which is composed of the sex ratio, R, and f is the fraction of females spawning during the time interval over which egg abundance is measured (that is, during the surveys). The relationship in Equation (5.4) can be estimated if

(1) females possess a characteristic that indicates when spawning will or has taken place;
(2) the length of time such a characteristic remains detectable can estimated; and
(3) the spawning rate (or frequency) remains constant during the sampling interval over which f is estimated (that is, the number of times a female spawns during the survey period).

Equation (5.3) can then be rewritten as

$$B = \frac{P_0}{E' \cdot R \cdot f} \tag{5.5}$$

where E' is the number of eggs spawned per kilogram of female per batch during the period over which f is estimated, rather than the production over the entire spawning season (fecundity). If a female can spawn more than once over the period of estimation, then the value of f can exceed 1. The beauty of the DEPM is that one is no longer required to sample the plankton throughout the entire spawning season since there is only the need to determine the number of eggs spawned per batch. As a result, the method can be applied to both determinate and indeterminate spawners as long as the three conditions listed above can be satisfied. A complete description of all elements and methodologies for applying this approach to population analysis was provided by Lasker (1985).

5.2.3 *Larval abundance indices and production methods*

For species that produce demersal eggs, it is rarely possible to sample this stage of the life cycle because the habitats over which spawning occurs are poorly known or cannot be sampled easily or effectively. Stocks of herring (for example, in the North Sea) and

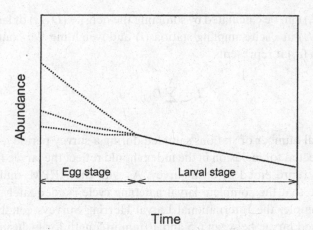

Figure 5.3 Change in abundance observed during the larval period (solid line) in a species for which the egg stage cannot be sampled. The dotted lines show possible paths that could be taken to estimate absolute egg production under scenarios where the environment in which the eggs occur is half, equally, and twice as risky as that of the larvae.

capelin (along the coasts of Newfoundland, Iceland, and Norway) spawn over broad areas of the inter- and sub-tidal zones that are difficult to access or delineate. Many freshwater species build nests or redds, and several guard their young (for example, bass; Figure 1.3i) or brood their young in places inaccessible to easy census (mouth-brooding catfishes and centrarchids). There are viviparous species such as redfish (*Sebastes* spp., Figure 1.1) that produce no eggs but extrude well developed larvae into the plankton. Alternatively, the development time of some eggs may be so short that accurate estimates of their production are not possible. In such cases, the LAI or LPM may be the only suitable approach. These methods, however, can also be applied to species that have planktonic eggs and larvae.

The fundamental principles of the egg and larval production methods are basically the same. In both instances, the objective is to derive a measure of offspring production that closely reflects fluctuations in spawning-stock biomass. In the case of LAI and LPM, however, there is a period separating spawning and hatching during which there can be substantial variations in survival (Figure 5.3). Differences in the nature of the environment may result in differences in mortality rates. Relatively little is known about the magnitude of losses that take place during the incubation of demersal eggs (for example, due to disturbance, predation, or oxygen depletion). For example, between 10% and 90% of the eggs laid on Atlantic herring spawning beds may be lost before emergence of the larvae, and similar values have been reported for egg-to-fry survival in different species of Pacific salmon. Similar values also apply to the survival of pelagic eggs. As a result, any correlation between an index of annual larval abundance and spawning biomass is likely to become weaker with increasing age of the larvae simply due to interannual variations in survival prior to sampling. The objective of sampling programs for LAIs and LPMs should then be to obtain as accurate an estimate of the abundance of hatching larvae as possible. As a result, the application of such methods tends to concentrate on the abundance or production of larvae shortly after hatching. One is still obliged, however, to assume or extrapolate egg mortality from other sources.

A simple LAI (I_t) can be calculated by summing the density ($D_{\Delta L,i}$) of larvae of a selected length interval (ΔL) at each sampling station (i) and weighting the contribution of each station by the area (a_i) it represents

$$I_t = \sum_{i=1}^{X} D_{\Delta L,i} \cdot a_i \tag{5.6}$$

where X is the total number of stations sampled during a survey period t. The limits of the length interval selected for inclusion in the index should reflect the larvae that are produced during the period represented by each survey. As with the AEPM, multiple surveys are required to ensure that the complete larval hatching cycle is adequately described by the sampling. For example, the International Larval Herring Surveys conducted around the British Isles selected larvae between 6.5 and 10 mm in length for inclusion into the index. Earlier field investigations had shown that growth rates ranged from 0.2 to 0.3 mm day^{-1}. Hence the LAI included larvae up to 18 days old. Since each survey was based on a 14-day period, the index would have reflected the production of larvae over the sampling period with little overlap with prior or subsequent surveys. This approach assumes there are no variations in the mortality or growth rate of larvae between regions surveyed, dates of surveys, or different year classes. A more realistic approach is to estimate the production of hatchlings using a method analogous to that used in the egg production methods.

LPMs aim to estimate the number of larvae hatching on successive days throughout the reproductive season. The objective is to back-calculate the abundance of larvae at the hatching length for each length interval that makes up the length–frequency distribution. There are three underlying assumptions to this approach:

(1)   growth in length over time can be accurately described using a known formulation (for example, linear or exponential);
(2)   mortality and emigration rates are constant over time and independent of length (Equation 5.2); and
(3)   there is no immigration into the survey area.

The advantage of this approach over the LAI is that it no longer requires multiple surveys to ensure that the total production cycle is described. As long as sampling is conducted at a time when larval production is complete and when all length classes can be sampled, then it is possible to back-calculate the production cycle. If multiple surveys are available, or required, then independent estimates of larval production can be averaged to increase precision. Michael Heath (1993) provided a simple example of how larvae sampled on different dates can be used to estimate daily production (Figure 5.4). The back-calculated day of production is dependent on knowing the growth rate. The magnitude of the production depends on estimating the mortality rate.

5.2.4   Egg deposition

In species that produce demersal eggs, an assessment based on egg deposition may provide an estimate of spawning-stock abundance that is less subject to error than by using approaches based on measures of larval abundance (LAI or LPM). Some of the life-history

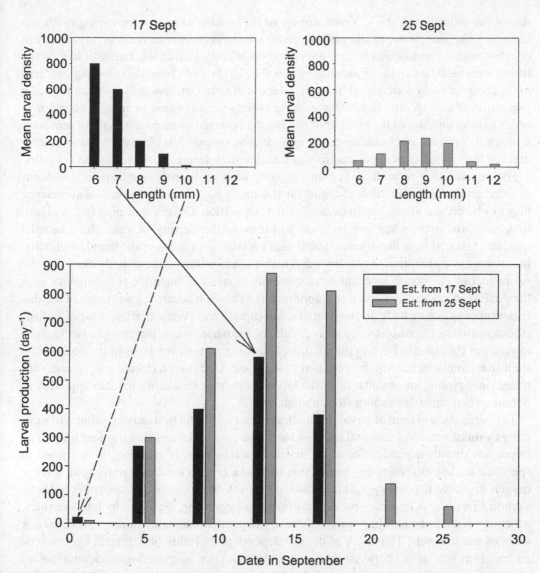

Figure 5.4 Illustration of the back-calculation of larval production based on sampling from the 17th (black bars) and 25th (grey bars) of September. The growth rate (*AGR*) of larvae is assumed to be 0.25 mm day^{-1}, and the length-specific mortality (*Z*/*AGR*) is 0.35 mm^{-1}. Note that as the length of the larvae decreases, the date of production moves to a later time. The earlier sampling period (17th) does not permit the complete description of the production cycle because hatching was not completed. Because larvae are measured to the nearest millimeter, the temporal resolution of the back-calculation is limited to 4-day intervals. Based on Heath (1993).

characteristics of species with demersal eggs limit the use of egg deposition as an absolute measure of spawning-stock biomass. In many species, from both marine and freshwater environments, the production of demersal eggs is often associated with some degree of nesting or parental care (see Chapter 1). When a clutch of eggs represents the investment of a single pair, any removal sampling represents a highly destructive method of estimating the size of the reproductive stock. Visual surveys of the number of nests or spawning redds can serve as a measure of the number of successful spawners in an area. Such surveys are most effective when spawning and nesting are restricted to easily identifiable habitats, such as the riffle zones used by salmon or the shallows used by largemouth bass. As spawning becomes more geographically widespread, the effectiveness of visual surveys is likely to be reduced. Nest surveys also require reasonably good knowledge of spawning or mating behavior in order to have an index of the number of females represented by each nest. As with any other attempt to estimate stock size based on reproductive output, it is essential to have an estimate of size- or age-dependent maturity of females, their fecundity, as well as the sex ratio.

For species that reproduce in large groups, such as herring and capelin (*Mallotus villosus*), sampling that involves the removal of some eggs to obtain a measure of abundance may not be critical to any one individual or the population. Once a spawning bed is identified, standard approaches can be used to measure the density of eggs (for example, quadrats taken along a line transect), and with knowledge of the average female fecundity, the frequency distribution of mature individuals (as a function of age or body size), and the sex ratio, one can obtain an estimate of spawning biomass in much the same manner as in the AEPM. A critical element of this approach is to have an accurate knowledge of the distribution of spawning beds. In the case of some capelin and Pacific herring (*Clupea pallasi*) stocks, spawning can be restricted to inter-tidal or nearshore areas. But in some stocks, such as many of the Atlantic herring populations, spawning occurs over areas of the continental shelf that cannot be accurately identified or surveyed. Under such circumstances, methods based on sampling the abundance of the larval stages may be a more reliable approach to obtain an estimate of spawning-stock abundance.

The separation in time of larval production from egg production may be subject to mortality by unknown environmental sources, but in the case of demersal eggs, there may be an important density-dependent element that comes into play. If spawning is restricted to specific habitats, increasing egg production beyond a certain level may reduce the level of oxygen available to some eggs, thus reducing survival. Competition for spawning redds by salmon or nest sites in other species may result in eggs being disturbed by late spawners, which in turn may be another source of density-dependent mortality during this very early stage of the life cycle. The result of density-dependent mortality of demersal eggs may be an apparent uncoupling between true egg production (that is, spawning-stock abundance) and a subsequent estimate of larval abundance or production. If other methods are being used to estimate stock abundance (for example, hydro-acoustic integration), one might be tempted to discard some information, such as the data from one assessment method, because of inconsistencies between surveys when, in fact, it is pointing to a significant process affecting that stock. Consequently, care should be taken in interpreting differences in measures of abundance between life-history stages.

5.2.5   Estimating development and mortality rates

To this point, I have intentionally restricted any reference to offspring production as though it were a single stage in order to develop the concepts of spawning-stock estimation using methods that rely on sampling of eggs and larvae. However, we never sample eggs or larvae at a single stage of development. Since production is protracted over time and we cannot capture the moment of extrusion or hatching at multiple sites, surveys are not instantaneous and only provide a limited snapshot of conditions during the period over which they are conducted. Most methods of estimating stock size from egg or larval surveys rely on developing a catch curve – an age- or length-structured description of the number of individuals caught by the samplers (but see the discussion of catch curves in Chapter 2). From the moment eggs are fertilized and offspring are released into the environment, they begin to undergo developmental changes and their numbers begin to decrease because of mortality. The sequence of developmental changes and increasing length can be used to determine the age of the eggs or larvae. The production of eggs or hatchlings on successive days throughout the spawning season is estimated by back-calculating the abundance of the youngest age category from the age– or length––frequency distribution averaged over the survey area. The rate of change in numbers of offspring (N) over time is often described using a simple exponential decay equation

$$N_t = N_0 \cdot e^{-Z \cdot t} \tag{5.7}$$

where t is time (in days) and Z is the daily mortality rate (day^{-1}) (this is equivalent to Equations 3.3 and 5.2). Other functional relationships have been used to describe the change in numbers over time, particularly when mortality is not constant with age or size. The formulation used in Equation (5.7), however, is the one most typically applied in population analyses based on early life stages. A number of regression procedures can be used to estimate N_0 and Z based on observations of N_t and t, depending on the design of the survey. Equation (5.7) can be used to back-calculate the production of eggs or larvae from several sampling dates using different age classes of eggs or larvae in each case. In order to apply this principle, we must be able to determine the age of early life stages.

The determination of age for eggs differs from that used for larvae. Developing embryos undergo a series of predictable and identifiable developmental changes (Figure 1.4). Following fertilization, the development of different externally identifiable features during cleavage, gastrulation, and subsequent embryonic development can be used to classify eggs into different stages with the aim of determining the age at the transition from one stage to another or the midpoint of the stage. There are numerous staging schemes used to describe the development of fish eggs. Differences in the level of detail among schemes are based on the developmental characters that can be accurately scored in repeated observations of an egg. As the number of stages increases, it may be possible to classify eggs into narrower age categories, but this may be less reliable if the effectiveness of the scoring is decreased by subtle differences in development. Laboratory observations of the sequence of changes can then be used to derive the age of the egg from fertilization. Because development is independent of external energy sources, development rates in fish eggs, like that of most ectotherms, is determined primarily by environmental temperature. By using a series of controlled temperature baths that span the environmental conditions typical of a species range, it is then possible to derive temperature-dependent relationships that describe the time to the onset of any given stage. Typically, development times of fish eggs follow a negative exponential relationship with temperature whereby the development rate accelerates with increasing temperature (Figure 5.5). Once laboratory data on temperature-dependent development times are obtained, measurements of the temperature in which the eggs are captured can be used to estimate their age. The temperature-dependent response, however, may also place limits on our ability to assign ages to developmental stages in some species. For many warm

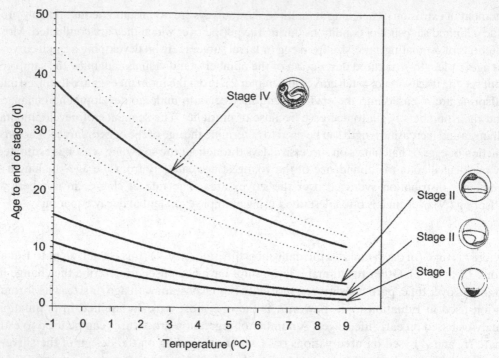

Figure 5.5 Time until the end of four stages (solid lines) used to describe the development of Atlantic cod (*Gadus morhua*) eggs from a study by Pepin *et al.* (1997). The dotted lines show the lower 10th and upper 90th percentiles of the distribution of development times until the end of stage IV. These serve to illustrate that not all individuals pass through a transition point at the same age. Similar distributions could be applied to earlier developmental stages, although the breadth of the distribution would be wider based on the results of that particular study.

water or tropical taxa, incubation times may be so short (<1 day) that sampling programs may not be able to resolve time-dependent changes in development, unless spawning occurs continuously through the day. It is reasonable to assume that the conditions in which each egg is captured reflects the environment in which development has taken place. This is more appropriate than using an overall average for the whole spawning range or period because of spatial and temporal variations in temperature that are likely to occur.

In contrast to ageing eggs, the approach used to age larvae is relatively simple. Once the larvae hatch and absorb their yolk, they begin to feed and allocate most of their surplus energy (that is not used in metabolic processes) to growth and ontogeny. Larvae sampled as part of population surveys are generally classified into 1-mm length intervals, although shorter intervals are used for particularly small species (for example, anchovy; Figure 1.3l). The age of larvae within each interval is generally estimated using a simple age–length relationship. The nature of this relationship is of little importance but the choice must be an accurate reflection of the population's pattern of change in length over time. As seen in Chapter 2, there are many approaches available to estimate the average growth in length of larvae.

It is clear that if we can age larvae using the daily growth increments found in the otoliths, it is possible to obtain an accurate description of the population's average rate of growth in length. Length–frequency analysis for species with relatively short spawning periods can be used as an alternative to the analysis of otolith microstructure but this approach may be more prone to error (see Chapter 2). As in the assessment of adult fish populations, the development of an age–length key would probably provide the most accurate approach for deriving the age–frequency distribution. The level of sampling required to construct an accurate age–length key, however, is often beyond the scope of most surveys of early life stages. Whereas the sampling of adult fishes can be done effectively onboard ship, plankton samples typically require considerable effort before the larvae can be sorted and measured. A similar contrast in the time required to age adult and larval fishes also applies.

It is also possible to describe the change in numbers with increasing development using a length-structured, rather than age-structured, approach. Prior to the common use of otolith microstructure, length–frequency distributions were most often used in population analysis. Equation (5.7) can then be converted to a length-based form

$$N_{L_t} = N_{L_0} \cdot e^{-\left([Z/AGR][L_t - L_0]\right)} \tag{5.8}$$

where L_0 is the average length at hatching, L_t is the length at time t, Z is the daily instantaneous mortality rate (day^{-1}), and AGR is the absolute growth rate (mm day^{-1}; Equation 2.3). The basic assumption in this transformation is no longer that growth can be described using any functional relationship, but instead length increases linearly with age. The length-specific mortality (Z/AGR) then becomes a critical parameter that must remain constant over the length range used in the analysis. The term on the right-hand side, however, can be modified if there are deviations from linear growth and constant mortality rates are not a reasonable assumption (for example, Lo 1985). In fact, there is a considerable body of evidence that indicates that increases in length over time are not necessarily linear and that mortality rates are size-dependent (see Chapters 2 and 3). If the functional forms of the growth and mortality rates in relation to increasing body length cancel each other out so that the length-specific mortality is constant, Equation (5.8) may be valid. However, if they do not and the length range sampled differs among year classes, then it is quite likely that the back-calculation to the youngest and smallest stage will lead to erroneous projections.

A critical assumption of all methods used in the estimation of egg and larval mortality rates is that of equilibrium conditions. In deriving the age– or length–frequency distribution from which mortality rates are calculated, it is essential that egg or larval production be approximately constant over the duration of the survey (or at least that the averages for the largest and smallest size classes be equal) or that it be a reflection of the accumulated production over the entire spawning season. Furthermore, growth and mortality rates must be relatively constant over space and time. If either of these conditions cannot be satisfied, back-calculation of the abundance to the youngest stage will not provide an accurate measure of production because determination of the rate of change in abundance with age will reflect the effects of variations in both production and mortality.

The derivation of mortality rates requires that the spatial and temporal pattern of spawning be considered. If there is more than one major spawning area or there are differences in the schedule of production among parts of the population's range, then the survey information should be stratified to reflect those patterns. Not only is it likely that there are differences in the growth, mortality, and production rates over time and space, but variations in the abundance of the different spawning components are likely to reflect significant changes in the dynamics of a stock. It is common for a stock's range to contract as the biomass decreases, and increase as biomass increases. Although this would be reflected in the distributional data obtained from egg and larval surveys, differences in the realized production among regions may have consequences on the level of exploitation a population may sustain.

5.2.6   *Relating egg or larval production to spawning-stock abundance*

Production methods designed to estimate the number of eggs spawned or the number of larvae hatching are all based on the premise that the rate of change in numbers over time (or length) can be described using a simple functional relationship, such as an exponential decrease in abundance. Mortality rates are estimated for each cohort and the intercept of this relationship – the initial number of eggs or larvae – is then related to spawning-stock abundance or biomass. There are two elements that can contribute to uncertainty in predicting the intercept: increasing age of the youngest stage that can be sampled and environmentally driven variations in the mortality schedule.

As discussed previously, the correlation between spawning-stock abundance and off-spring abundance is likely to be stronger if the population is sampled soon after the off-spring are produced because there will be fewer uncertainties about variations in the unknown losses among cohorts or year classes. This does not apply, however, to instances where development rate changes because of variations in the environmental temperature. If we were to obtain identical abundance estimates from two populations, one of which took twice as long to undergo development, the precision of our estimate of production (that is, $t = 0$) based on simple linear regression analysis would remain the same (Figure 5.6). For these two populations, the confidence intervals at the origin are dependent on the precision of the relationship between abundance and age, which should not be mistaken for uncertainty in events that are not observed by our sampling.

Beyond the simple statistical consideration, there is also a biological uncertainty associated with the state of a population of eggs or larvae. Does the stage-dependent pattern of loss remain the same as environmental conditions change? Production methods assume that vital rates are constant with age or size, or that they change following a consistent schedule from cohort to cohort. As we have seen in earlier chapters, this assumption may not be reasonable. For instance, differences in the type or abundance of predators from year to year may alter the pattern of vulnerability of eggs and larvae, or there may be simple physiological constraints that alter the pattern of mortality. Developmental abnormalities are known to increase in frequency at the extremes of a species range. For example, in a controlled laboratory study of the embryonic development of Atlantic cod (*Gadus morhua*), my colleagues and I found that the pattern of losses over time varied in relation to environmental temperature (Figure 5.7). Greater overall losses occurred early during development as temperature decreased, and the schedule of losses showed substantial variation among treatments. It is easy to see that assuming a constant mortality rate throughout early development could lead to inaccurate extrapolations of the number of eggs fertilized. Physiological abnormalities that result in temperature-dependent variations in mortality in the laboratory may not play an important role in natural populations if other processes, such as predation, are of greater magnitude, but the issue is worth considering when dealing with data from natural populations.

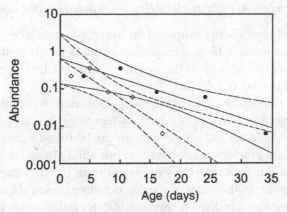

Figure 5.6 Time-dependent catch curve and confidence intervals about the regression of two hypothetical populations of fish eggs. The population represented by the open diamonds and the dashed lines has a development rate that is twice as fast as the one represented by the solid circles and lines. In this example, the catches from both populations for each "stage" are identical.

It might appear that these points raise serious concerns about the applicability of egg or larval production methods simply because some of the processes are not fully understood. My purpose in mentioning them, however, is simply to raise the reader's awareness that such factors might come into play. Departures from constant development or mortality rates with changing environmental conditions may result in changes in the slope of the production curve or in the pattern of residuals, suggesting a change in the functional form of the relationship. My point is to be observant when applying any methodology or fitting a model, and to be aware that some assumptions may not always hold, even if they are valid most of the time.

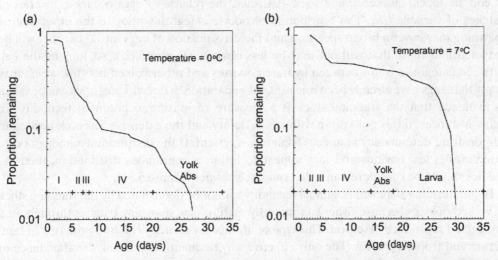

Figure 5.7 Proportion of individual Atlantic cod eggs and yolk-sac larvae surviving in relation to time, through the various developmental stages (dotted line at the bottom of each panel), for two laboratory temperature regimes. Note that losses during the egg stage occur principally during developmental stages I–III. The high losses that occur during the larval stage at 7°C are caused by food deprivation. Based on data from Pepin *et al.* (1997).

5.2.7   *Measuring spawner characteristics – knowledge and bottlenecks*

Knowing the potential reproductive output of the average female is critical if one intends to relate egg or larval abundance to spawning-stock biomass in terms of absolute numbers. Unfortunately, this is probably one of the weakest elements when we try to link early life stages to the production rate of a stock.

There is no doubt that fishes come from fish eggs, at least in the immediate sense, and their numbers are determined to a large degree by fecundity, the number of eggs produced by a female. As mentioned earlier, egg production can be broadly classified into two categories, determinate and indeterminate, based on our ability to assess the number of eggs likely to be produced during a reproductive season from the standing stock of advanced oocytes in the ovary prior to the onset of the reproductive season (Figure 5.8). The determination of oocyte maturity is done by making histological sections from the gonads of mature female fish and categorizing the oocytes into maturity stages. Fecundity is then determined by counting yolked or hydrated oocytes from a subsample of the gonad using gravimetric or volumetric methods. The gravimetric method is based on counts of mature oocytes from a weighed subsample of the total gonad. The volumetric method is based on egg-size frequencies from an aqueous solution of all oocytes in the ovary.

For determinate spawners, there is often a clear separation in the categories of oocytes present in the ovary, providing a demarcation between small, unyolked oocytes and those undergoing relatively synchronous maturation. In a number of temperate and boreal species, from both freshwater and marine environments, maturation of all the oocytes to be spawned in a season is highly synchronous and their release takes place over a relatively short time period, at least for an individual female (Figure 5.8a). In such instances, estimating fecundity, and any relationship it has to a characteristic of the female (for example, length, weight, condition), can be relatively straightforward. In determinate spawners such as cod, haddock, mackerel, and some flatfishes, the release of eggs occurs in a series of batches of variable size. The distribution of oocyte categories prior to the onset of first spawning may cover a broad spectrum and the classification of eggs into those that will be spawned and those that will not may be less obvious (Figure 5.8b). As long as the gap between categories is not between hydrated oocytes and other yolked oocytes, which may imply that eggs have already been released, the separation in oocyte categories can be taken as evidence that the standing stock is a measure of maximum potential fecundity. In many instances, this is not known with any certainty and the extensive research needed to demonstrate determinacy in annual fecundity is absent. If the discontinuity among oocyte categories is less obvious and they appear to follow a continuous distribution, then the species should be considered an indeterminate spawner (Figure 5.8c).

In indeterminate spawners, annual fecundity cannot be determined by the standing stock of yolked oocytes because fishes continuously mature new spawning batches throughout a protracted reproductive season. This type of life-history strategy is found primarily in temperate and tropical species. The only effective way to quantitatively link the abundance of early life stages with that of the adult spawning stock is to determine the batch fecundity and the spawning frequency. Batch fecundity is based on the number of hydrated oocytes, distinguishable by their relatively large size and greater translucence (water content), observed in weighed subsamples of the ovary. The determination of spawning frequency

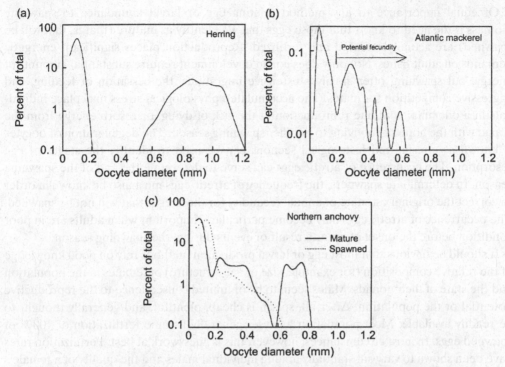

Figure 5.8   Relative size frequency distribution of the oocytes in ovaries of Atlantic herring (Blaxter & Hunter 1982), Atlantic mackerel (Priede & Watson 1993) and northern anchovy (Hunter & Leong 1981). There is a clear hiatus in the distribution of oocyte diameters in herring but there are several peaks in the oocyte size frequency in mackerel which may lead to ambiguity in the determination of annual fecundity. The rightmost peak found in mature northern anchovy (——), reflects the diameter of hydrated eggs, which are no longer present in females after spawning (·········).

relies on knowledge of the cycle of oocyte maturation and vitellogenesis, being able to identify the proportion of a sample of the adult population that shows evidence of being at a particular stage of the oocyte maturation cycle, and knowing the duration of that stage. Laboratory observations of the spawning cycle coupled with the collection of histological samples are necessary to correlate the periodicity in egg production with changes that take place in the developing ovaries. The latter are then used to describe the development of oocytes and post-ovulatory follicles into identifiable categories.

   For indeterminate spawners, the coupling of laboratory and field observations is essential in order to derive an accurate understanding of the short term cycle of egg production. Field observations alone would only be adequate under exceptional situations where most of the population followed a synchronous periodic pattern in oocyte maturation. The DEPM was developed specifically for application to indeterminate spawners. Knowing the characteristics of spawners during the course of a population survey is critical for an accurate estimate of spawning biomass. Although a single survey of eggs and larvae can be used to derive an estimate of the abundance of adult fish, as long as most of the population is in the survey area, it is absolutely essential to sample the adults in order to take into account any seasonal or interannual variations in the potential production of eggs.

Of equal importance to any method relating egg or larval abundance to spawning biomass is the need to know that those eggs that we identify as mature (that is, that will be spawned) are actually released and fertilized. Reproduction places significant energetic demands on adult fishes. Not only does gonad development require substantial amounts of energy, but spawning often involves extensive migrations, the cessation of feeding, and aggressive competition for mates. The accumulated physiological stress may place individuals in a dilemma: continue reproduction at the risk of dying, or resorb energy from the gonad with the hope of surviving to another spawning season. The degeneration of oocytes is known as atresia, and histological sections can be used to identify the stages of egg resorption. The occurrence of atretic eggs can serve to determine the end of the spawning season. In determinate spawners, the frequency of atretic eggs must also be known in order to correct the original estimate of annual fecundity for those eggs that will not be spawned. The occurrence of atretic eggs may become particularly important when adults are in poor condition before the onset of, or as a result of events during, the spawning season.

It should be obvious that most egg or larval production methods rely on good knowledge of the relative composition (for example, size or age structure) of females in the population and the state of their gonads. Males seem to be of limited consequence to the reproductive potential of the population. After all, sperm is cheap, plentiful, and generally thought to be readily available. Most population analyses generally assume fertilization of 100% of spawned eggs. In most circumstances, however, this is guesswork at best. Fertilization rates have been shown to vary substantially among individual males and the quality of a female's egg may also play a considerable role in the success of reproduction. As with other elements of production methods, fertilization rate should be assessed from samples taken in conjunction with egg and larval collections. Any deviation from a fertilization rate of 100% will result in an underestimate of spawning biomass. This is an aspect of population analysis, however, that remains an element of research and has yet to be routinely applied to approaches based on egg or larval production.

It is essential to measure the character (for example, fecundity, egg quality) of spawning fish frequently, not only when trying to relate egg or larval abundance to adult biomass, but also to obtain some indication of the population's health and reproductive potential. One of the most complete examples available at this time deals with Atlantic cod in Icelandic and Norwegian waters. Individual egg production increases with increasing size of female but unless the relationship is isometric with weight, changes in the adult population's size or age structure will alter the average production of eggs per kilogram of spawning fish. Thus any loss of larger adults, say through fishing, will alter the reproductive potential per unit of spawning biomass. In the case of Atlantic cod in the Barents Sea, there is clear evidence that an individual's potential annual egg production is affected by its condition at the onset of the spawning season. In a subsequent analysis, Marshall *et al.* (1999) showed that variation in condition was closely linked to variations in the abundance of capelin, the dominant prey of adult cod. Changes in fecundity will lead to corresponding but opposite changes in the relationship between egg or larval production and the estimated spawning-stock abundance. For example, if fecundity decreases but the number of adults remains unchanged, fewer eggs and larvae will result. If fecundity is not monitored carefully, the decrease in egg and larval abundance would suggest a decrease in adult population abundance when, in fact, none has taken place, even though the reproductive output has decreased.

The effect of spawner characteristics may go beyond this. The eggs produced by first time spawners may be less likely to survive: their eggs are smaller and have lower energy reserves for the developing embryo and larva than the eggs of repeat spawners. Although the production methods outlined above should capture the changes in mortality rates of developing eggs and larvae, the implication of the findings for Atlantic cod is that as one moves toward a population structure made up of smaller and younger spawning fish, the reproductive potential is likely to decrease more rapidly than the spawning biomass. It is also possible that smaller adults may be more susceptible to variations in their environment, thus increasing the potential variation in reproductive output and subsequent recruitment.

5.3   Recruitment indices

If the number of animals that will enter the fishery or the spawning population could be measured just before the onset of the fishery or reproduction, then one would likely obtain an accurate measure of recruitment. In most cases this is not possible and scientists providing advice on the status of a stock are in need of an index of year-class strength. Thus, there is a need to identify a stage of the life cycle when the relative strength of a year class is established.

Following Hjort's hypothesis of a critical early life stage during which year-class strength might be established, considerable research effort was directed towards identifying when this might occur as well as the factor(s) that regulated survival (see Chapter 4). In a number of instances, scientists were able to identify an element of the environment that was closely related to the pattern of survival during the course of a study. One factor that received considerable attention was the availability of food following yolk absorption, since failure to successfully feed would likely result in significant mortality because a larva's energy reserves would quickly be exhausted. A demonstration of the significance of such a mechanism would require the observation that a substantial portion of the larval population was subject to starvation. Although starvation does play a role during early life, there is little evidence that a valuable index of recruitment could be derived from estimates of the abundance of the very early larval stages (Leggett & Deblois 1994, see Figure 4.1).

An alternative approach in forecasting recruitment could involve the measurement of an environmental proxy that provides an accurate reflection of the state of the environment in which reproduction and early development take place. Research would identify important environmental factors that affect survival during the egg or larval period. Retrospective analysis can then provide evidence that the strength or occurrence of one or several factors exhibited a correlation with subsequent recruitment. There is substantial risk in taking such an approach, however. For example, a number of studies found that the outflow from the St. Lawrence River, an index of the influx of nutrient-rich waters into the Gulf of St. Lawrence and onto the Scotian Shelf, as well as other environmental indices were closely correlated with the landings of many species in the region. When those relationships were revisited, Drinkwater & Myers (1987) found that a large proportion of them was no longer useful in forecasting recruitment. The reasons for the failure of such environmental proxies to forecast recruitment beyond their period of development probably have more to do with an oversimplification of the complexity of the ecosystem. Many aspects of marine environments undergo changes over decadal time scales. When many aspects are in phase, certain relationships between the number of recruits and an environmental proxy may arise. On the other hand, as the elements of an ecosystem move out of phase with one another, for whatever reason, proxies may no longer be adequate descriptors of complex interactions. In fact, the use of environmental proxies has probably done more harm than good to the incorporation of knowledge gained from studies of early life stages into the assessment of freshwater and marine fish populations.

The most effective approach to providing an index of recruitment appears to come from indices of abundance immediately before or after the transition away from the planktonic phase of the life cycle, when the animals move into juvenile habitats. As we learned in Chapter 1, transition from the larval to the juvenile period is often associated with important changes in morphology and behavior as well as in niche shifts. For example, the settlement of juvenile plaice (*Pleuronectes platessa*) in the Wadden Sea (Netherlands) is associated with metamorphosis and movement into a benthic habitat where the young fish encounter a new array of prey and predators. One of the key mechanisms determining the strength of a plaice year class appears to be the transport of eggs and larvae from the spawning grounds in the southern North Sea to nursery areas in the Wadden Sea. A reliable estimate of year-class strength (recruitment index) can be obtained from a population survey conducted shortly after larvae settle to the bottom. There can be a variable decline in the number of recruits after settlement because of variations in predation following settlement. There are similar observations for other species that undergo a shift from a pelagic to a benthic or lotic lifestyle. The development of physical and behavioral competence on the part of young fishes appears to be a critical element in reducing their vulnerability to environmental variations and thus making abundance indices of juvenile stages useful tools in population analysis. The case study of Japanese sardine populations in Chapter 11 demonstrates the use of the abundance of age 1 fish entering juvenile feeding areas as a reliable index of recruitment.

5.4   Sample collection

5.4.1   *Sampling systems, net avoidance, and extrusion*

Population analysis based on the sampling of planktonic fish eggs and larvae faces particularly difficult problems because development is rapid and mortality rates are generally high. This results in substantial changes in the nature of the organisms being sampled over a relatively short time period. Rapid changes in body form (weight and length), swimming capability, and sensory development over the course of the first few weeks of life result in substantial changes in the vulnerability of specimens to collecting gear. The reduction in the number of eggs and larvae over time, their dispersal by currents, and their scarcity relative to other planktonic organisms of similar size requires the development of survey designs and methods that will provide an accurate representation of the population and the changes it is undergoing.

Sampling planktonic fish eggs and larvae is done with nets made of fine mesh. Mesh size depends on the dimensions of the animal to be collected. Devices and methods used to collect fish eggs and larvae represent a balance between the need to collect a sufficient number of animals to obtain a representative sample of the population while attempting to maintain capture efficiency across the range of stages (or sizes) used in the analysis.

When animals are abundant and poor at avoiding sampling gear, such as in the case of eggs and yolk-sac larvae of abundant and fecund species, small devices towed vertically through the water column from a standardized depth may provide an adequate method to sample the population. For example, the CalVET net, a small (25 cm diameter) paired net deployed vertically, is often used in population estimation of anchovies and sardines based on the DEPM. The depth to which the net is lowered is chosen to ensure that most of the eggs or young larvae are included in the collection. Because most planktonic fish eggs are positively buoyant, the vertical distribution of the population can be approximated using knowledge of the buoyancy of the egg and the vertical density structure of the water column.

As the abundance of organisms decreases and their ability to avoid nets increases, the volume of water sampled, as well as the capability to capture specimens, must increase. A common method of sampling early life stages in order to increase the volume sampled and reduce the ability of larvae to avoid the net is to tow the net while the ship is underway. Nets are progressively lowered to a desired depth and gradually retrieved, which results in a saw-toothed profile through the water column, commonly referred to as an oblique tow. The choice of gear type and deployment strategy, which includes towing speed and distance as well as depth of tow, is determined by the need to maintain capture efficiency over a significant portion of the target species' early life. Being able to estimate capture efficiency is particularly important in population analyses aimed at back-calculating the abundance of the youngest age classes, either eggs or hatchlings, since such approaches rely on obtaining an absolute measure of abundance.

In order to estimate capture efficiency, we must be able to determine the probability that an organism will be caught or, alternatively, determine the effective volume sampled by a gear type. The ability of a larva to avoid a sampler will depend on the distance at which it can detect it, and its ability to get out of the path of the gear. This will depend on the distance the larva can cover by swimming at maximum speed in the time available before the sampler reaches the larva (Figure 5.9). It is simple to see that as the sensory (both visual and mechano-receptors) and motor skills of a larva improve through development, both the perception distance and the maximum swimming speed will increase. Although on the surface this may appear simple to determine, the collection of the information needed to do so is not. To ensure that the probability of capture, or the effective sampling volume, does not decrease too sharply through development, increasing the speed of the sampler or increasing the expanse of the mouth is an obvious solution. Increasing towing speed, however, may also increase the distance at which the net is detected because the pressure wave at the front is more intense, and increasing sample volume may also result in more difficult sample processing.

A more common approach to estimate capture efficiency of larvae is to use the ratio of night-to-day catches. The rationale is that if larvae are relatively inactive and visual cues are eliminated or reduced to minimal levels because of low light, the probability of capture should be maximal. If larvae respond to non-visual cues, this method for estimating capture efficiency will be inaccurate. Nevertheless, by contrasting the night–day catch ratio across age or length classes, one can determine how capture efficiency decreases in older and larger larvae. It is important to note that a catch ratio of 1 does not necessarily imply a capture probability of 1 because small larvae may still be able to avoid the nets to some degree. One possible diagnostic to determine if the catchability of small larvae is nearly 100% involves the use of the catch curve of both eggs and larvae. If there is an abrupt drop in the

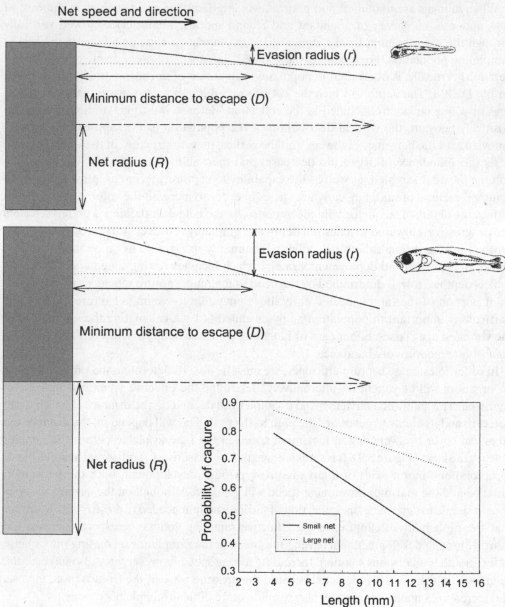

Figure 5.9 Schematic of the elements that determine the vulnerability of larval fishes to plankton nets. The maximum evasion radius is directly related to the maximum swimming speed of a larva, which in turn determines the minimum distance at which a net can be detected. As the length of the larva increases, the maximum evasion radius increases, which results in a decrease in the probability of capture. There are two solutions to reduce the possibility of net avoidance: increasing the net speed or increasing the net radius. The effect of the latter is illustrated in the lower portion of the figure where the lighter shading indicates a doubling of the net radius. Although the maximum evasion radius remains unchanged for either large or small larvae, the likelihood of avoiding the net is decreased because a greater proportion of the net's area is beyond the larva's ability to escape, which results in a greater probability of capture as illustrated in the lower right panel. (Graphic is partly based on Gartz *et al.* 1999.)

number of animals following hatching, this might indicate that catchability is not 100%. Alternatively, there could be an increase in mortality associated with hatching, simply because of physiological demands, or because larvae are being extruded through the mesh of the net. Eggs are more resistant to extrusion than young larvae.

Given the dimensions of planktonic fish eggs and their relatively hard chorion, it is reasonably easy to ensure that they will not pass through the mesh by choosing a spacing that is considerably smaller than the minimum axis. In the case of larvae, extrusion may be important depending on the gear type used. When larvae hatch, their small size may allow them to pass through the mesh. Similar situations may arise for more advanced larvae when sampling with larger plankton nets with wide mesh. The capture efficiency will be dependent on body shape; larvae that have a broad body form for a given length are more likely to be captured than ones that are more elongate.

As sampling moves to later stages of development, such as in the collection of pre-recruits, it becomes increasingly difficult to determine capture efficiency. Although the catchability of eggs by plankton nets can be assumed to be near 1 so that estimates of absolute abundance can be considered as accurate, the same cannot be said for late larvae and early juveniles when estimates of abundance become indices.

5.4.2   Sampling variability and patchiness

When a plankton net is towed through the water, the degree to which one sample will resemble the next depends on the time interval and distance between samples and the local patchiness in the distribution of eggs and larvae. Fish eggs and larvae, like most other plankton, are not uniformly distributed through the water at any one site. Most plankton occur in clumps, or patches, whose cohesiveness is dependent on the balance between forces that tend to maintain aggregations, such as convergent currents or active movement to remain close to conspecifics, and those that tend to disperse them, such as current shear.

There is little information available on the scale of patches *per se* but it is worth considering the factors that influence the scale of separation between individual eggs and larvae. The initial scale is determined by the pattern of spawning by the adults. One could consider the mating pair as the smallest level of aggregation (<1 m), most often we can consider spawning aggregations, whether schools of fish or spawning habitats, to be on a scale of tens to hundreds of meters. Even within this spatial scale, the distribution of eggs and larvae is not uniform. A succession of samples taken at a single site reveals that the statistical distribution of catches is typically skewed to the right (Figure 5.10). The frequency of the occasional large catches is dependent on the level of aggregation of the animals within the area sampled and the likelihood that plankton nets will encounter these small patches. As the volume of water filtered by nets is increased, small patches of eggs or larvae are more likely to be encountered and consequently estimates of abundance from large volume samples will have a greater level of precision than those obtained from smaller volumes (Figure 5.10). The catches of other zooplankton of about the same size as ichthyoplankton do not appear to be affected similarly by the volume of the sample, suggesting that the small scale distribution pattern of fish eggs and larvae differs from that of other organisms that occupy a similar trophic level. Unfortunately, there is insufficient information to determine the cause of this apparent difference.

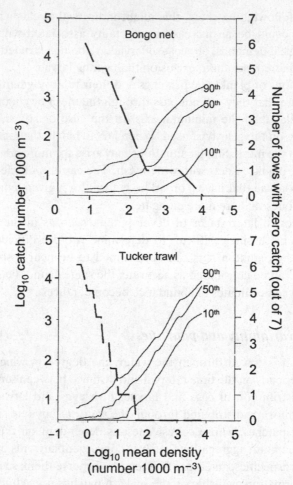

Figure 5.10 Cumulative probability distribution of non-zero catches from two plankton nets, showing the 10th, 50th and 90th percentiles (solid lines), in relation to the mean density of fish eggs and larvae from seven replicate samples at each of four sites (based on data from Pepin & Shears 1997). The bongo net had a 60-cm diameter (area = $0.28\,m^2$), while the Tucker trawl had a 2-m square frame (area = $4\,m^2$). Note that the larger net can detect smaller concentrations of larvae, the likelihood of obtaining a zero catch (dashed line) is substantially higher for the smaller bongo net, and the variability in catches (measured here as the distance between the 10th and 90th percentiles) is lower for the Tucker trawl.

Following their release, either as planktonic eggs or as hatchlings from spawning beds, physical forces will begin to transport and disperse the offspring over scales of thousands of meters (Figure 5.11). At the largest scale, the distribution of eggs and larvae will be determined by the size of the population's spawning area, which is an important consideration for survey design. Transport will continue until the young fish become physically competent to carry out directed swimming for a prolonged period of time. The scale of patches may then be dependent on the dominant physical processes of the system in which they are released, such as those defined by eddies, currents, and fronts. Dispersal, on the other hand, may decrease well before that stage of competence because individuals may be capable of forming small local aggregations because of changes in behavior, such as

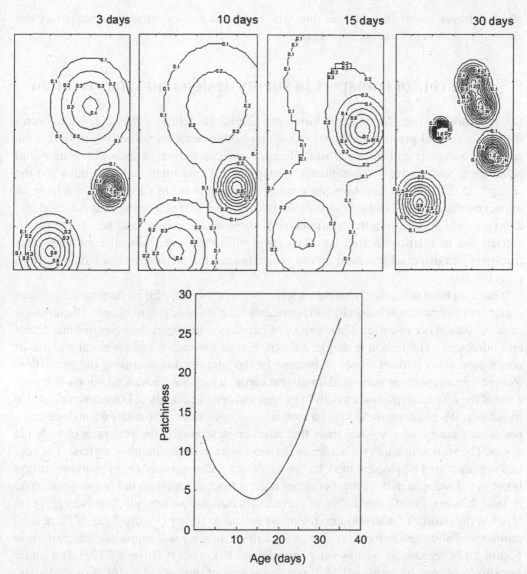

Figure 5.11 Schematic diagram of the changes in distribution and patchiness of planktonic fish eggs and larvae in relation to time since spawning. The contour lines are relative. The pattern shows an increased spread in the distribution associated with a decrease in patchiness. The lower illustration is based on the observations of Matsuura & Hewitt (1995) for the California Current, where the lowest level of aggregation is reached sometime between 8 and 20 days after spawning.

increases in the tendency to school. Biological processes that result in local losses, such as predation by schooling planktivores or the settlement of faster growing individuals, may also begin to counteract physical dispersion forces by reducing the number of eggs or larvae at the scale typical of those processes (100–1000 m). There is no single scale that typifies patch size for early life stages because young fishes are continuously being affected by diverse physical and biological factors. We can safely conclude, however, that the level of aggregation is initially high, and that it will decrease through the early stages because physical forces dominate until the time when internal and external biological processes allow aggregation to increase once again.

5.5   The role of transport in survey design and interpretation

Because most marine fish eggs and larvae are planktonic, whereby they may have control over their vertical position in the water column but are entirely vulnerable to horizontal currents, transport and dispersal will influence our perception of their distribution and abundance. Surveys are not instantaneous snapshots of environmental conditions and the changes in distribution that take place while sampling is being carried out will have an impact on the accuracy and precision of the information. The size of the area surveyed, the degree to which the strength and direction of currents are known, and how the survey is carried out in relation to that knowledge will ultimately determine the influence that transport and dispersal will have on our perception of the distribution and abundance of early life stages.

It has long been recognized that the early life stages of fishes are highly dependent on ocean currents to determine their drift to, and retention within, suitable environments. The inclusion of the effects of ocean currents in the study of population dynamics, however, is fairly recent and infrequent. The reason is simple, a description of the spatial and temporal patterns of ocean currents is difficult to obtain because of the complex forces driving the circulation. Recent developments in numerical simulation methodology have advanced our basic knowledge of the relationship between early life stages and ocean circulation, but this work is still in its infancy. We can nevertheless take a look at some basic rules to consider when designing a population survey, or any other study that relies on determining the vital rates of early life stages. The section that follows is intended to serve as an intuition-building exercise. The routine application of biophysical modeling to the design and interpretation of plankton surveys is not well developed at this time, but rather is the subject of sophisticated ongoing research.

Take the situation of a population of particles (in our case we can call them fish eggs) centered in the middle of a 10 000-km^2 region we intend to survey (Figure 5.12). If there is no motion and abundance does not change over time, then we will reproduce the pattern in Figure 5.12a reasonably well by sampling stations 10 km apart (Figure 5.12b). But under any survey scheme, currents will alter our perception of the overall distribution and abundance of planktonic organisms. If there is a uniform current flowing toward the east at a speed of 10 cm s^{-1} (Figure 5.12c), then a survey of the area along north-to-south transects starting from the northwest corner of the region (assuming it takes 1 h to sample each station; Figure 5.12b) will give the impression that the particles are more broadly distributed because the center of mass is moving in the same direction as our survey (left to right) (Figure 5.12d). If on the other hand, the survey is conducted along east–west transects (Figure 5.12e), the view of the population's spatial distribution will be substantially different. A clear skew in the distribution will appear because eggs and larvae in the southern part of the region will have moved further east (Figure 5.12f). If we now move to a situation where there is a horizontal gradient in the strength of the current (from 10 to 40 cm s^{-1}) from the north of the region to the southern end (Figure 5.12g), that is to say that there is an increase in the variability within the system we sample, the distortions become even

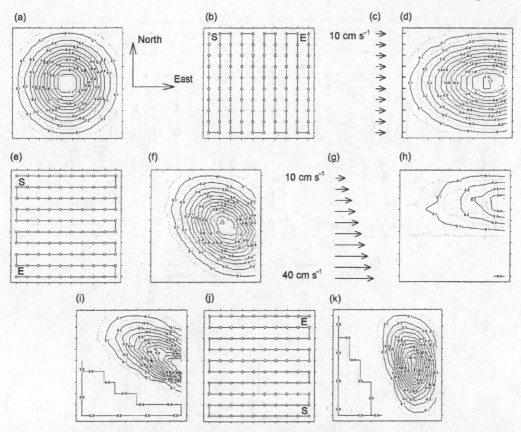

Figure 5.12 Illustration of the effect of currents and the choice of station sequence on our perception of the distribution of passive particles, such as fish eggs, from a "synoptic" survey. The instantaneous production of particles is shown in the top left corner. In this illustration, there is no development or mortality and production of particles occurs only at the start of the simulation. The area under simulation is 10 000 km² (100 km on each side) and station spacing in all surveys (b, e and j) is 10 km. The capital letters S and E represent the start and end points of the surveys. The time required to reach each station and perform the sampling is simulated to take 1 h, for a total survey time of a little more than 4 days. (d) and (f) show the perceived distribution of particles given a uniform current of 10 cm s^{-1} (c), for survey sequences (b) and (e), while (h) and (i) show the result for a spatially varying current (g). (k) shows the perceived distribution of particles for the station sequence shown in (j). In all simulations, diffusion is assumed to be negligible.

more pronounced depending on the survey design we apply. If we use north–south transects starting from the northwest corner (Figure 5.12b), the perceived distribution suggests there are relatively few particles in the southern range, due to more rapid advection of particles to the east along the southern edge (Figure 5.12h), and the center of mass is even further to the east, relative to the case with uniform currents. The distortion in the perceived distribution is substantially different when the transects are in a west-to-east direction, and the survey is again started in the northwest corner of the region (Figure 5.12e). Portions of the area that have low current velocity appear to be sampled with less distortion than the portions with high flow rates (Figure 5.12i). If on the other hand, we had started a similar

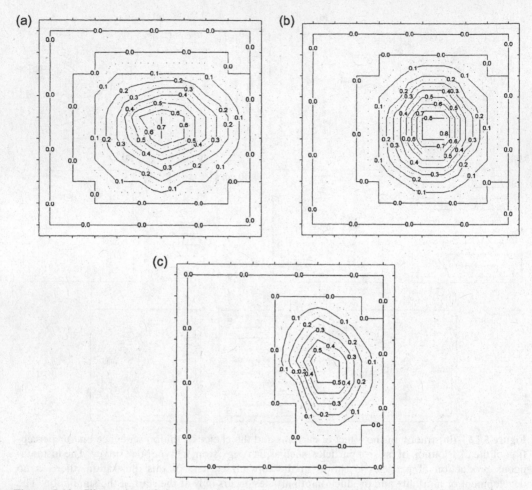

Figure 5.13 Illustration of the perceived distribution patterns of particles for simulations where there is no current (a), a uniform 10-cm s^{-1} current throughout the region (b), and a spatially varying current (c) but with the addition of a spatially uniform mortality rate (Z) of 0.2 day^{-1}. The station sequence in (a) and (b) is the same as shown in Figure 5.12b, while the station sequence in (c) is from Figure 5.12j.

survey in the southeastern corner, using east-to-west transects (Figure 5.12j), then the entire population might have been sampled with less distortion (Figure 5.12k).

The addition of mortality to the population's dynamics causes a distortion in the perceived distribution as areas that are surveyed later suffer greater cumulative loss (Figure 5.13). In the case where there is no drift and a population that is subject to a mortality rate (Z) of 0.2 day^{-1} is surveyed along north-to-south transects as in Figure 5.12b, the center of mass appears to be shifted to the west and the range appears contracted. The interaction of mortality and transport on our perception of the distribution will depend on the magnitude of both factors and the timing of our observations at each site. In essence, a good survey design will cover portions of the area of interest that would likely suffer the greatest losses from the combined factors early in the survey. This example is not based on a true situation but simply serves to illustrate the interaction between survey design and the physical environment on our perception of

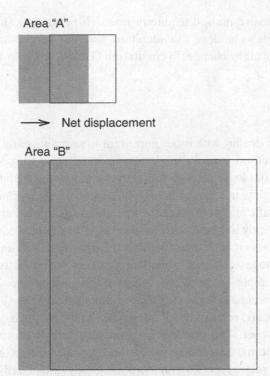

Area "A"

⟶ Net displacement

Area "B"

Figure 5.14 Consider two square survey regions, A and B, in which B has 10 times the area of A. Both are to be surveyed simultaneously at times t_1 and t_2. They are subject to a current of the same intensity, which flows from west to east, and results in a net displacement (transport) toward the east equal to the length of the arrow shown between the two areas. Although the level of detail in population distribution measured will be different in the two surveys, because of the time it takes to sample each station, the relative proportion of the water that will have moved out of each area will be substantially different. The shaded grey areas show the original survey regions and the black lines show the position of the projected grids after 12 h. The white portion shows the fraction of the original survey area that would be displaced during that period. Since surveys are generally conducted over the same grid for a given region, the white portion would represent the amount of water (and corresponding eggs or larvae) that would have "emigrated" from the study sites. In the case of the smaller area, 43% of the water would have been advected from the original area while slightly less than 14% of the water from the larger area would have been transported away.

distribution patterns. In most applications of egg or larval abundance or production measurements, the goal is to ensure that the entire distributional area of a population or spawning component is surveyed and delineated by regions with minimal catches so that the impact of transport, either as emigration or immigration, is rendered negligible.

Whether the effect of variations in current strength and direction has an influence on our ability to estimate the abundance of the youngest stages is unclear. It is simple to understand that in a fixed length of time, the loss due to transport will be proportionately greater if a survey is conducted over a small area than over a large one (Figure 5.14). Under this scenario, as long as the time required to survey a region does not increase in direct proportion to the area, the information from a survey conducted over a large area is less likely to be influenced by unexpected variations in currents. Present approaches to measuring or describing the circulation in a region and coupling this with observations of biological conditions rely on intensive multi-disciplinary research programs. There is still considerable research that needs to be done to understand how our perception of abundance and distribution are influenced by changes in circulation (Helbig & Pepin 1998).

5.6 A final note

Throughout the section dealing with issues important to sampling early life stages, I avoided dealing with specific methodological considerations. In addition to species-specific requirements, there are numerous logistical aspects of surveys, sampling, and laboratory protocols as well as resource availability, in terms of funds, equipment, and personnel that play key roles in determining the specific approach taken for an individual population analysis. One aspect seldom considered explicitly is the necessity for thorough taxonomic knowledge of the species of interest as well as closely related ones. Without the information to unambiguously identify early developmental stages of fishes, abundance indices based on back-calculation may be unreliable. For example, in temperate areas distinguishing the embryonic stages of Atlantic cod (*Gadus morhua*), haddock (*Melanogrammus aeglefinus*) and witch flounder (*Glyptocephalus cynoglossus*) remains nearly impossible. In the future, biochemical approaches may prove useful, however. The problem is made considerably more difficult in tropical ecosystems where taxonomic descriptions of eggs and larvae are often limited because of the large number of closely related species with morphologically similar early life stages.

5.7 Summary

Population analyses that use approaches based on the sampling of early life stages often do so because methods that traditionally rely on surveying the adult population are not possible when trying to determine spawning-stock biomass or an early indication of year-class strength. Consequently, comparing the accuracy of different individual approaches to population assessment is rarely feasible but we can at least contrast their advantages and limitations.

A common method for assessing exploited fish populations involves the use of commercial catch and effort information, which is converted to an estimate of absolute abundance through calibration of catchability using survey indices of adult-stock abundance through sequential population analysis. Calibration can also be done using other indices of abundance, such as those obtained from egg and larval surveys, as is the case for North Sea herring (Heath 1993). Analyses of population abundance based solely on adults depend heavily on accurate knowledge of catchability. In the case of sequential population analysis, abundance estimation is made more complex by the need to have accurate information on commercial catch and effort. Any factor that can lead to inaccurate reports of the actual catch per unit of fishing effort, or changes in fishing behavior caused by changes in the distribution of the target age classes, can lead to errors in the estimate of absolute abundance. In contrast, methods based on sampling early life stages attempt to reduce the number of assumptions by measuring as many of the parameters associated with the assessment as possible. What limits estimates based on early life stages is the level of precision with which each parameter can be estimated. This is because the error in the estimate of population abundance is based on the sum of the error (variance) in each parameter, and the inter-dependence (covariance) of these errors. In addition, each estimate based on egg or larval abundance does not build on previous knowledge of the population's state, whereas sequential population analyses can smooth some of the short term variability in estimates of population abundance because each new year's information is not independent of that obtained previously. The result is that population analysis based only on early life stages may exhibit considerable inter-annual variation that can only be reduced by thorough sampling of both adults and offspring. This may not necessarily be considered a disadvantage, however, because approaches such as sequential population analysis tend to underestimate the level of variability in population abundance.

The precision of population estimates is also an issue when we contrast hydro-acoustic integration and early life-history methods, such as the DEPM, both of which provide estimates of absolute abundance. Hydro-acoustic surveys have the advantage that a large area can be surveyed with reasonably high resolution. This is because sampling is continuous along transects and the number of transects can be increased when the ship is not forced to stop on station as frequently as is required when taking conventional biological samples. Furthermore, there are fewer elements to measure when conducting population analysis based on hydro-acoustic surveys, which can result in a greater degree of precision in the estimate of population abundance. On the other hand, hydro-acoustic integration also depends on the calibration of target strength in relation to abundance, which is critical to obtaining a measure of absolute abundance. In the instances where hydro-acoustic integration and early life sampling have been applied to the same population, however, both appear to predict similar changes in abundance indices over time. The general trends in population abundance appear to follow similar patterns even though the year-to-year changes do not exactly correspond.

One situation where the use of early life stages is directly analogous to methods used on adult fishes involves the application of mark–recapture techniques to population estimation. Whereas adults are generally tagged externally, young fishes are marked by chemical immersion (for example, using oxytetracycline or alizarin complexone) in order to "score" the otolith. Otoliths extracted from individuals captured at a later time are then placed under a fluorescent light to assess the existence of a mark. Other ways of marking early life stages include the use of small pit-tags placed in the snout of young fishes, such as salmon alevins, that can subsequently be identified using electronic sensors, or by immersing young fish in fluorescent dyes to mark bone tissue which can be detected easily upon recapture. Mark–recapture studies require that large numbers of marked larvae or young fish be released into the population in order to obtain a sufficiently large number of recaptures, a matter discussed in Section 2.3.1. Assessment of population size can then be carried out using methods identical to those traditionally applied to juvenile and adult populations. The assumptions concerning limited emigration or immigration into the area of interest as well as the lack of differential mortality between marked and unmarked individuals must be satisfied. Application of mark–recapture techniques has been successful in several instances where small local populations appear to remain fairly cohesive, in locations such as small bays, rivers, and estuaries. The concept may not be readily applicable to large populations distributed over a broad geographic area because of the logistics of releasing sufficient numbers of young fish in order to obtain adequate recaptures (but see the coral reef example mentioned in Section 7.3.7). In most examples where the approach has been successful, the potential for differential mortality could be assessed by sampling soon after the release of marked individuals to determine if substantive losses had taken place. The likelihood of differential mortality appears to be related to the duration of the rearing period prior to release. As the period when young fish are kept in cultured conditions is lengthened, a greater proportion of naïve individuals is produced, possibly because of a lack of appropriate natural stimuli or natural selection.

Approaches based on sampling early life stages are most effective for population assessment when used in conjunction with other sources of information. Their advantages come from the limited number of assumptions they require in contrast to some of the traditional methods that primarily sample adults. Their limitations come from the difficulty in obtaining precise estimates of some elements critical to their application. They are likely to provide a good measure of absolute abundance, however, because they rely on a thorough understanding of the biological characteristics of a species or stock. Therefore, discrepancies with other assessment methods can point to critical gaps in our understanding of the processes affecting a population.

The application of egg or larval production methods has become more widespread because in many instances accurate surveys of adult populations are almost impossible. Population analyses based on early life stages move beyond providing a simple measure of stock size because the basic requirements of such methods rely on observing stock status. We have to consider aspects of both the adults and their offspring in order to appropriately apply the principles on which the approach is based. Population analyses based on early life stages, whether used to estimate spawning-stock abundance or to forecast year-class strength, provide unique information on the biology and processes that affect population dynamics. Estimates of population abundance based on egg or larval production rely on a good understanding of the physiological state of spawning adults and thus provide information on changes in both numbers and reproductive potential of the population. However, these methods can also provide information on the processes that affect year-class strength because the approaches rely on the estimation of development and mortality rates, thereby giving additional insight into the dynamics of the ecosystem in which a stock or species lives.

Additional reading

Drinkwater, K. & Myers, R.A. (1987) Testing predictions of marine fish and shellfish landings from environmental variables. *Canadian Journal of Fisheries and Aquatic Sciences* **44**, 1568–1573.

Heath, M. (1993) An evaluation and review of the ICES herring larval surveys in the North Sea and adjacent waters. *Bulletin of Marine Science* **53**, 795–817.

Helbig, J.A. & Pepin, P. (1998) Partitioning the influence of physical processes on the estimation of ichthyoplankton mortality rates. I. Theory. *Canadian Journal of Fisheries and Aquatic Sciences* **55**, 2189–2205.

Lasker, R. (1985) An egg production method for estimating spawning biomass of pelagic fish: application to the northern anchovy (*Engraulis mordax*). National Oceanic and Atmospheric Administration Technical Report, National Marine Fisheries Service 36, Springfield, VA.

Leggett, W.C. & Deblois, E. (1994) Recruitment in marine fishes: is it regulated by starvation and predation in the egg and larval stages? *Netherlands Journal of Sea Research* **32**, 119–134.

Marshall, C.T., Yaragina, N.A., Lambert, Y. & Kjesbu, O.S. (1999) Total lipid energy as a proxy for total egg production by fish stocks. *Nature (London)* **402**, 288–290.

Chapter 6
Cohort Identification

Karin E. Limburg

6.1 Introduction

6.1.1 What is a cohort?

The term "cohort" is usually associated with people. It is derived from the Latin word *cohors* which referred to each of the 10 divisions within a Roman legion. Since that time the term has come to take on other meanings, such as this one from Merriam-Webster's dictionary:

"A group of individuals having a statistical factor (age or class membership) in common in a demographic study."

In the context of population ecology, we define cohorts similarly: a group of individuals falling into the same age group, or falling into some other classified group (for example, body size). Cohorts are a convenient way to keep track of groups of individuals within a population, for we can follow their fate through time and study how they fare under different environmental or ecological conditions.

6.1.2 Different ways to define a cohort

In the context of this book, we are interested primarily in cohorts of fish larvae. There is no *a priori* "correct" way to define a cohort, but there are several logical ways to do so. For example, in a species of fish that has a protracted reproductive period, we might be interested in keeping track of groups of individuals that were hatched at different times in the spawning season. Alternatively, we might be concerned about size-related risks that larvae might encounter. For example, predators often select the smaller individuals because they are easier to catch and subdue. In this case, we would want to keep track of the numbers of individuals in vulnerable size groups as well as those that have grown into a "size refuge" (a size at which they are no longer vulnerable to the predator). A third way to create a cohort classification might be on the basis of location: for example, if a fish species spawns in a number of different tributaries within a river drainage, or different embayments within a lake or an estuary, we may want to identify the larvae from the different spawning sites, particularly if there are environmental differences such as temperature, pollution, or other habitat characteristics that distinguish each site.

143

Related to definitions based on geographic separation, cohorts may also be different genetically, particularly if the fish populations have high fidelity to specific spawning areas, as do many salmonids. In these circumstances, genetic drift may have proceeded to such an extent that individual subpopulations can be identified by genetic analysis. Thus, we have defined at least four ways to group cohorts for identification and study: time, size, location, and genetics.

6.1.3   The importance of scale in defining cohorts: differences between adults and larvae

Although we emphasize larvae here, a survey of the fishery literature will show that most fishery scientists concern themselves with cohorts defined on a coarser scale. Adult fishes are typically described in year classes or cohorts defined as fish all born in a given year. Managers and scientists often refer to year-class strength within a fish stock and keep track of year-to-year fluctuations in such things as harvests that may be a function of how many fish are in a particular year class.

We need to consider the temporal and spatial scales at which we study larval, juvenile, and adult fishes. Each of these groups usually has its own spatio-temporal scale, which is largely a function of the age and size of the individual in relation to its environment. As an example, consider a fish that is 5 years old: an additional year is 20% of its life, but one more week is only 0.38% ($1 \div 52 \times 20\%$) of its experienced life. In contrast, a week to a 28-day-old larva is 25% of its total life experience. Similarly, a millimeter of growth on a 50-cm adult fish represents a small increment (0.2%) to its length, whereas a millimeter of growth for a 6-mm larva is a large addition (16.7%) to its present size. Thus, the incremental week and incremental millimeter represent enormous gains to fish larvae, in terms of dealing with physical factors such as water viscosity or ecological factors such as predation. Some of these relationships were explored in Chapter 1.

6.2   Processes that create cohorts

The creation of groups that we call cohorts arises from a variety of processes or phenomena. For example, age cohorts of larvae can be formed in species that have spawning periods that are protracted over weeks to months. During some portions of the spawning period reproduction may be very successful and in others very poor. Classifying the age cohorts is often arbitrary, based on the judgment of the investigator, unless the age groups are widely separated in time. For example, bluefish (*Pomatomus saltatrix*) produce an early spring and a summer cohort off Bermuda, and these arrive in northeastern United States estuaries roughly 2 months later. Depending on the temporal scale of interest, the investigator may follow 5-day cohorts (groups defined as having been born in the same 5-day period), or weekly cohorts, or perhaps even daily cohorts.

In spite of the often close relationship between the age and size of fishes, factors such as temperature and food availability can cause variation in growth rates, resulting in size cohorts. There is also evidence to suggest that size cohorts may develop even when temperature, food supply, and other environmental conditions are stable. This variation in growth rates among individuals is often referred to as growth depensation. Keeping track of size cohorts may be important, when there are strong forces that can alter the chances of survival of fishes at certain sizes. For example, fish larvae under a threshold size may be vulnerable to the suite of predators in the area. Conversely, fishes in a particular size range may be especially successful in feeding on the prey items that happen to be available.

Larval cohorts may be spatially structured by spawning area, defined both at local and regional scales. For example, at a local scale, fishes in a lake could swim up different tributaries to spawn. Each stream has its own set of characteristics, such as amount of tree cover that affects water temperature, or the particular communities of fishes and macro-invertebrates present, and these factors, in turn, affect growth and survival of eggs and larvae. At a regional scale, anadromous fishes might enter different estuaries to spawn, or species of coral reef fishes might use reefs arrayed along a large geographic extent.

Even if larvae are spawned in one specific locale, they may be transported by currents, or may actively swim, to other areas where growth occurs. At times, we wish to keep track of their origins, but at other times, it might be of interest to keep track of where they are dispersed as well.

6.3   Methods of larval cohort identification

Until the 1970s, cohorts of larvae were separable based only on size and morphological development. Fish larvae of similar size and developmental stage, collected during the same time period, could be classified into the same cohort. The problem with this method is that it cannot account for differences in growth and developmental rates among individuals, and may therefore lead to erroneous classification (unless size cohorts are the units of interest). This method is used, however, in some aspects of population analysis, where cohorts of embryos (eggs) are identified by their stage of development in order to estimate spawning-stock biomass (Chapter 5). The earlier stage of development and the fact that embryos rely on an internal supply of food ameliorate some of the problems of this method of cohort identification.

Three major developments over the past quarter century have greatly enhanced our ability to group fish into cohorts. The first was the discovery of daily increments in otoliths (Pannella 1971) and the application of this tool to early life stages of fishes (Brothers *et al.* 1976). A second thrust was the more recent application of trace element and isotopic measurements in otoliths and other hard parts of fishes, as a means of determining the "environmental signatures" characteristic of specific environmental conditions or habitats. The third major development was in the area of molecular biology, where resolution of genetic markers has enabled researchers to isolate "genetic fingerprints" of fishes and group them by similarity.

6.3.1   Otolith microstructure

As discussed in Chapter 2, otoliths are calcified structures that form part of the hearing and balance system in teleost fishes. There are three pairs of otoliths that reside in the semicircular canals adjacent to the brain. Movement or vibrations cause the otoliths to move across

Figure 6.1 Photomicrograph of an otolith (radius *ca.* 52 μm) of a larval striped bass (*Morone saxatilis*), showing the core and daily growth rings.

the underlying sensory cells, which transmits a signal to the brain that is interpreted as directional movement or sound. As a fish grows, so do the otoliths. This is accomplished by precipitating calcium carbonate ($CaCO_3$), usually in a crystalline form called aragonite, on a protein matrix. The daily pattern that is formed, from differences in the rate of $CaCO_3$ deposition, can be seen by examination under a compound microscope. These increments appear similar to tree rings (Figure 6.1). By counting the rings, one can determine the age of the fish. One may also estimate growth rate from the relative widths of the increments and relating these to the size of fish. These techniques are discussed thoroughly in Chapter 2.

Because counting daily increments in fishes is time consuming, due to the time it takes to prepare and examine otoliths, many researchers develop age–length keys that relate the length of a fish to its age. This is sometimes done with larvae when large sample sizes are involved. Typically, one determines the age of fish of different lengths and develops a statistical relationship, usually by fitting a line or a curve, between the two variables. Figure 6.2 shows a hypothetical relationship between age and standard length for fish larvae. Note that the data points scatter about the line of best fit because individual variation exists. However, because one can calculate the error (a measure of the degree of variation about the line), it is possible to compute the probability that a fish of a given length falls into different age classes. If the data are normally distributed, or can be transformed to fit the normal distribution, then the probability of a larva of a particular length falling into one age class will be highest and will be lower for adjacent ages.

Once the ages have been determined, larvae can be grouped into cohorts of similar ages. The relevant grouping usually depends on the time scale of interest. For example, in

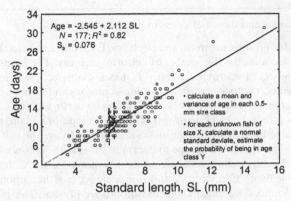

Figure 6.2 Construction of an age–length relationship for fish larvae.

a rapidly changing environment, one might wish to keep track of fish in daily cohorts. On the other hand, over a longer period of observation, it might be of interest to know whether weekly or monthly cohorts survived differently.

6.3.2 Otolith microchemistry

A parallel avenue of research on otoliths has been the study and application of so-called microchemistry techniques. Otoliths are largely composed of the elements calcium, carbon, and oxygen, but they also can contain minor or trace amounts of about 40 other elements. The mechanisms of their incorporation are still a research forefront, but factors such as the presence of the element in the environment or in food, temperature, and growth rate are known to affect the uptake and deposition of some elements in otoliths. In addition to elemental concentrations, scientists can also measure the ratios of stable isotopes of certain elements, such as carbon and oxygen, and derive environmental information from these as well (see Box 6.1).

Otoliths are small, but nonetheless they maintain a faithful record of age and growth, and some aspects of the environmental history of a fish can be interpreted from the elements and stable isotope ratios. One of the more powerful techniques has been the use of strontium in otoliths as an environmental tracer. Strontium, an alkaline earth element, has an ionic radius only slightly larger than that of calcium. It can thus freely substitute into the relatively open crystalline matrix of aragonite (but not as easily into other crystalline forms, such as calcite). Because strontium is generally more abundant in sea water than in most fresh waters, it has been possible to measure strontium in otoliths and use this (normalized as a ratio to calcium) as a proxy for salinity. This has been extremely useful for examining migration between waters of different salinities.

For example, the pattern of Sr : Ca shown in Figure 6.3, from an otolith of a brown trout (*Salmo trutta*) from Sweden, reveals two important ecological facts. First, the Sr : Ca ratio is somewhat elevated in the otolith core. Embryonic development takes a long time in salmonids (larvae are shown in Figure 1.3c, d), and as the embryo develops, so too does its incipient otolith. If the mother develops her eggs at sea, the otolith core is endowed with

Box 6.1   Stable isotopes and their usefulness in the natural sciences.

Atoms are composed of protons, neutrons, and electrons. Each element has a characteristic number of protons and electrons, but the number of neutrons can vary. The forms of elements containing different numbers of neutrons are called isotopes. Consider, for example, the element carbon. The most common isotope of carbon contains six protons and six neutrons and is denoted by the symbol ^{12}C. It comprises about 98.9% of all carbon on earth, and is a stable form. On the other hand, ^{14}C (six protons + eight neutrons) is a radioisotope of carbon, meaning that it is unstable and will tend to lose its excess neutrons via radioactive decay. The half-life (the length of time for half of the initial amount of isotope to decay) of ^{14}C is fairly long, about 5730 years. It is quite rare on Earth. Finally, carbon has a second stable isotopic form, ^{13}C. It is more abundant than ^{14}C but less stable than ^{12}C, comprising approximately 1.1% of the carbon on Earth.

In any given compound, different amounts of isotopes are present. This is because in physical and chemical reactions, the lighter isotopes tend to react first, whereas the heavier ones have a slightly higher "energy hill" to climb to break loose from their chemical bonds. The differential transfer of heavy and light isotopes is called fractionation. Isotopes are typically measured as ratios of the heavier (rarer) isotope to lighter (more abundant) form or forms with a mass spectrometer. The reason we measure ratios rather than absolute values is that ratios are less sensitive to instrument error. When we measure the isotopic ratios in a sample, say a gram of fish muscle tissue, we need to relate these to a standard. We do this by calculating a so-called "del value." For ^{13}C, the equation is

$$\delta^{13}C = \left(\frac{R_{sample}}{R_{standard}} - 1 \right) \cdot 1000 \tag{6.1}$$

where R is the ratio of heavier (^{13}C) to lighter (^{12}C) isotope. The primary standard used for carbon is a Cretaceous marine fossil, *Belemnitella americana*, from the Pee Dee formation in South Carolina (the original Pee Dee Belemnite standard is gone, but there are secondary standards based on the original). Thus, one will often see the notation $\delta^{13}C_{PDB}$. The value is parts per thousand or per mil and is denoted by a "‰" symbol.

Stable isotope analysis is widely used in the natural sciences, from geology and hydrology to biology, ecology, physiology, and nutrition. Stable isotope analysis of different elements provides different kinds of information, depending on the application. For example, in ecology, carbon stable isotopes are studied as a means of tracking energy flow through food webs because once the carbon is photosynthetically fixed by plants, its $\delta^{13}C$ ratio does not fractionate very much when passed along the food chain. Thus, if different kinds of plants (e.g. C_3 vs. C_4 plants) have different $\delta^{13}C$ ratios, these end members create a gradient of sources, from which it is then possible to determine how much of one source vs. another was consumed by a grazer. Equally informative are nitrogen stable isotopes, which are widely used to determine trophic relationships in food webs. Stable isotope analysis is also used in hydrology to trace different sources of water. In water (H_2O), one can analyze the ratio of ^{18}O to ^{16}O ($\delta^{18}O$) or the ratio of ^2H (deuterium) to ^1H (δD); one can also analyze the ratios of stable isotopes of many of the solute constituents (C, N, S, Sr, etc.).

strontium. Thus, one can determine whether the fish's mother was a freshwater resident or a sea-run migrant. The second thing we see is that there is a decline in Sr : Ca away from the central core, followed by a rapid elevation that extends to the outer edges. The low Sr : Ca corresponds to life in the freshwater stream, where the concentration of strontium is

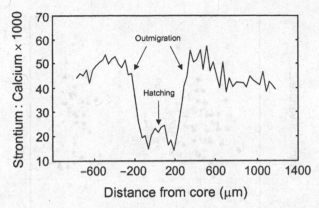

Figure 6.3 Sr:Ca ratios measured along a side-to-side transect of an otolith from a brown trout caught in the Baltic Sea near Sweden. The core, which actually forms in the embryo, is indicated by the "hatching" arrow.

extremely low. The sudden elevation in Sr:Ca indicates movement out of the stream and into the sea. By knowing the physical location of this "jump" on the otolith, one can count the number of daily increments up to this point and determine how old the fish was when it emigrated from its natal stream. In the case of this fish, the emigration occurred at around 3 months of age, rather than at least 1–2 years. Thus, this method provides new insight into a precocious behavior for this species.

Otolith strontium content is sometimes also used as a "natural thermometer" in some species of marine fishes, because the ratio varies inversely with water temperature. However, unlike corals and clams, there is wide variation in the degree to which otolith Sr:Ca reflects water temperatures, and so is not always a useful indicator.

Strontium, calcium, and a number of other elements may be readily measured throughout otoliths using electron probe microanalysis (EPMA). A beam of electrons is focused on a small area on the specimen's surface, the electrons excite the atoms in the target to a higher energy state, and as they decay to their original states X-rays are emitted. The X-ray wavelengths are a function of the element and the particular electron orbitals involved, so elements can be measured quantitatively. Transects can be run from the core of an otolith outward to an edge, with point samples as small as a few micrometers (see Gunn *et al.* 1992 for a thorough discussion of advantages and problems associated with EPMA).

Recent advances in microchemical research have employed high-resolution techniques that can measure elements down to the parts-per-billion level and at spatial scales fine enough to extract temporal information from otolith daily increments. The instruments used for such analyses include inductively-coupled plasma mass spectrometers (ICP-MS), ICP with atomic emission spectrometry (ICP-AES), proton-induced X-ray emission analysis coupled with nuclear microscopy (micro-PIXE). References by Campana (1999) and Thresher (1999) listed under *Additional Reading* provide more details on all the microchemical methods.

Elemental microanalyses, in combination with analyses of stable isotopes, can be analyzed with multivariate statistics to produce classifications of the geographic location from where a fish originates. For example, this method, including analysis with a neural network

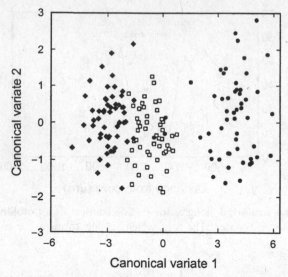

Figure 6.4   Example of a multivariate analysis of trace elements and the stable isotopic ratios of carbon and oxygen, used to distinguish juvenile American shad from different estuaries. Symbols identify fish collected from the Connecticut (□), Hudson (◆), and Delaware (●) Rivers in October 1994 (reproduced from Thorrold *et al.* 1998 with permission of the American Society of Limnology & Oceanography).

(a computer program that can be "trained" to do pattern analyses with complex data sets), was used to classify correctly American shad (*Alosa sapidissima*) born in different river systems (Figure 6.4). This "elemental fingerprinting" is useful to trace the origins of migratory stocks, once the fish leave their birthplaces and mix in the ocean or even in large lakes.

In spite of the great prospects of otolith microchemistry, one should be aware that this science is still in its infancy. We are only beginning to understand the complex mechanisms by which trace elements and isotopes are incorporated into otoliths, and how tightly or weakly bound these are to the carbonate or protein matrices. The instruments used to make the measurements are highly sensitive to a range of errors, and different instruments are better suited than others to particular applications. Nevertheless, improvements in measurement technique, together with careful experimentation to understand underlying mechanisms, continue to push the field ahead.

6.3.3   Genetic markers

Genetic variation within populations arises from mutations and from non-random mating. Therefore, identifying and quantifying such variation allows researchers to understand the forces that structure populations. Population geneticists have a battery of tools for detecting genetic variation. Aside from the ingenious "common garden" experimental method devised by Gregor Mendel in the 19th Century, in which organisms with variable traits are raised under the same conditions and compared, the main tools in genetic research involve isolating biochemical compounds and quantifying their movement in an electrical force field. This general technique, gel electrophoresis, has been in use for over a half century.

Allozyme analysis, or the quantification of genetic variation in enzymes, is a widely used electrophoretic tool. The advantage of allozyme analysis is that it is fairly simple to implement, but it does not provide fine scale information about population structure.

Advances in the field of genetic research have brought about the identification of increasingly finer molecular structures within genomic material. With increased resolution of these molecular compounds, it has been possible to identify more loci (places on genes) that are subject to genetic variation, and thus useful for quantifying population structure. Some genetic material is better suited for studying genetic drift than other material. For example, the DNA in mitochondria (mtDNA) is inherited maternally and is therefore not subject to sexual selection. If populations have been isolated long enough, the mtDNA in subpopulations (demes) will be different, but also it is possible to study the rates of exchange, or genetic drift, between demes.

One of the problems that long hindered genetic research was the difficulty in extracting sufficient quantities of DNA from organisms. The recent and explosive development of a means to amplify DNA, the polymerase chain reaction or PCR, has enabled genetic research to advance exponentially in the past decade. Now it is possible to take a very small amount of DNA and amplify it with PCR. Thus, increasingly finer portions of the genome are amenable to study. For example, variation within so-called microsatellites in nuclear DNA is an active research front. These are short, tandemly repeated sequences of DNA consisting of only one to six base pairs in a repeat sequence. They are highly polymorphic, have discrete loci, and are distributed throughout the entire genome, which makes them particularly useful genetic markers. They serve as key identifying markers in DNA finger-printing, and provide information on paternity and kinship as well. For example, in a recent study of Atlantic salmon (*Salmo salar*) sampled from 29 sites in Europe and North America (King *et al.* 2001), 266 alleles were observed at 12 microsatellite DNA loci. The resolution was great enough to separate populations not only by continent, but by geographic region within continents as well (Figure 6.5). The genetic distance (degree of relatedness) between North American sites was lower than between European sites, and is attributed to a shorter time since separation due to more recent glaciation in North America.

6.4 Examples of cohort identification and analysis

Our interest in identifying and studying different cohorts of larvae often lies in a number of different issues. For example, it may be important to identify the geographically based factors that cause variation in year-class strength. Alternatively, temporally distinct cohorts may be subject to different sources of mortality, such as seasonal changes in water temperatures or the presence or absence of food and predators. Cohorts may also be identified as genetically distinct, and therefore needing special management considerations.

6.4.1 Differential survival of striped bass larvae

Striped bass (*Morone saxatilis*) is a common temperate bass (family Moronidae) along the east coast of North America. It has also been introduced into reservoirs and along the Pacific coast of the United States, and is important in both commercial and sport fisheries.

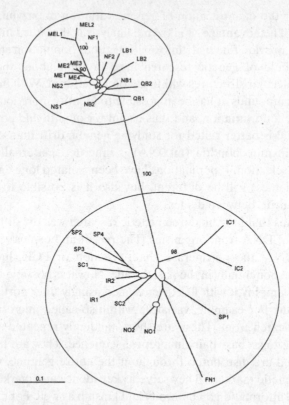

Figure 6.5   Genetic distance phenogram representing the genetic structure of Atlantic salmon. Top cluster represents North American demes, and the lower one is European (modified from King *et al.* 2001 with permission from Blackwell Science).

In the 1980s, the striped bass population in the Chesapeake Bay along the Atlantic coast plummeted, for reasons that were unclear at the time. Among the potential causes of recruitment failure were pollution, overharvesting, and environmentally induced larval mortality. A number of studies of larvae (Figure 1.3q) were therefore undertaken.

Rutherford & Houde (1995) investigated the early survival of striped bass larvae. Larvae were intensively sampled with towed plankton nets in different parts of Chesapeake Bay over a period of several weeks. Fish ages were determined by otolith increment analysis and were classified into 3-day cohorts. Cohort population sizes were tracked over time. The authors were able to demonstrate that some cohorts suffered heavy mortality due to water temperature fluctuations. Specifically, high mortality occurred in cohorts that experienced storm fronts, which resulted in the water temperature dropping below 12°C. Thus, the year class was composed of cohorts spawned later in the season.

In a complementary study, Secor & Houde (1995) examined 6-day cohorts of striped bass larvae in the Patuxent River, a major tributary to Chesapeake Bay. There the researchers also identified a temperature effect, although it was more complex than indicated in the Rutherford & Houde (1995) study. Cohorts that experienced temperatures below 15°C or above 20°C during the first 25 days of life suffered highest mortality

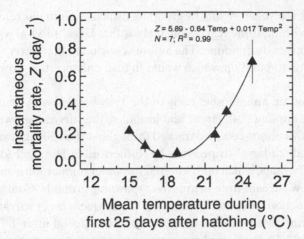

Figure 6.6 Instantaneous mortality rates for 6-day cohorts of striped bass larvae, plotted against the mean water temperature in the Patuxent River, Maryland for the first 25 days post-hatching (reproduced from Secor and Houde 1995 with permission of the Estuarine Research Federation).

(Figure 6.6). Growth rates were also measured and found to correlate neither with zooplankton densities (the primary food of larval striped bass) nor with water temperatures. In this study, Patuxent striped bass hatched during the middle of the spawning period achieved highest survivorship.

As shown in Figure 6.6, abiotic factors such as water temperature can be important in determining year-class strength, but need not be the primary determinants. In a study published in 1999 and carried out in the Hudson River in New York State, my colleagues and I found no relationship between water temperature and cohort-specific survival in 6-day cohorts of larval striped bass. Rather, highest growth rates and lowest survival rates were associated with late-spawned cohorts. In this case, growth appeared to be sustained by high consumption of copepods, and predation on later cohorts was inferred as the primary cause of mortality. The method of estimating mortality of cohorts was two-fold: estimates during the larval period, and retrospective estimates made by sampling juvenile fish later in the summer, determining their hatching dates by counting otolith increments, and analyzing the frequencies of cohorts that had recruited into the juvenile stage. This cross-system comparison (Chesapeake vs. Hudson) for the same species and life stage suggests that differences in whole-system characteristics may affect the relative importance of abiotic vs. biotic influences on larval demography and ultimate recruitment.

6.4.2 Evidence of two-stage recruitment in young-of-year American shad

American shad (*Alosa sapidissima*) is an anadromous herring native to eastern North America and of considerable commercial importance. Spawning typically occurs dozens to hundreds of kilometers from the sea, and emigration of juveniles occurs prior to winter. As with many species, year-class strength was thought to be set in the larval stage. In studies conducted in the Connecticut River, Crecco & Savoy (1985) analyzed otoliths of shad larvae over a number of years. Again, by grouping fish larvae into discrete age cohorts, they were able to show with multiple regression analysis that larval survival was highest in years with low river discharges during June. The authors concluded that lower flows enabled zooplankton populations to build up, which would in turn enhance the growth and survival of young shad.

In 1996, I reported on an intensive study of the 1990 year class of American shad in the Hudson River. By sampling both larvae and juveniles, the survival, growth, and movement (migration) of weekly cohorts could be tracked throughout their first growing season. In that year, as with the study of larval striped bass by Rutherford & Houde, I found that a frontal system brought heavy rains, which both lowered the water temperature and increased flows, dispersing larvae downstream from the up-river spawning grounds. Comparing estimates of egg and larval production from another study to the frequencies of surviving cohorts, more than 85% of the year class recruited from the cohorts hatched after 1 June, even though 94% of all eggs and 82% of yolk-sac larvae were observed prior to that date.

Given luck and persistence, such retrospective analyses can be carried even further. Analysis of Sr : Ca in otoliths of adult American shad returning to spawn in the Hudson River in 1995 provided insight into the age and size of shad when they left the natal river as juveniles for the sea (Limburg 2001). A surprising pattern that occurred across year classes was a bi-modal size- and age-at-emigration. That is, fish appeared to emigrate from the river either as smaller (<60 mm total length) and younger, or larger (>90 mm) and older individuals. The 1990 year class was one of the best represented year classes in the 1995 study, because the samples had been collected by commercial fishermen and the year class had fully recruited to their fishing gear. Thus, it was also possible to make a second retrospective analysis on further changes in cohort survival. Although it was impossible to count daily otolith increments back to hatching in the adult fish, the growth of the larvae's otoliths was strongly and positively correlated both with water temperature and date. By measuring the average widths of the first 10 daily growth rings (part of the larval stage), it was possible to reconstruct hatching dates of the adults. This analysis revealed that a second demographic restructuring occurred, most likely during emigration (Figure 6.7).

6.4.3   Management of northern Atlantic cod

Atlantic cod (*Gadus morhua*) is widespread across the northern North Atlantic Ocean. Until the past four decades, cod was so abundant that it fueled national economies, but with the advent of industrial fishing many stocks were decimated. Of these, the northern cod stocks historically fished around Newfoundland, Canada, were particularly hard hit, collapsing the local economy and necessitating a complete and indefinite closure of the fishery. In order to better understand the stock and manage its recovery, a number of extensive assessments have been and continue to be made. These have consisted of direct tagging as well as the use of genetic markers and otolith "elemental fingerprints."

Previously, the degree to which the populations of northern cod were distinct was unclear. The implications are important for managing the fishery. If the population is unstructured – that is, all individuals are equally well adapted to the various geographic areas that the northern cod use – then one might expect that the recovery of the population could be easily accomplished by allowing the less exploited parts of the stock to migrate in and use the

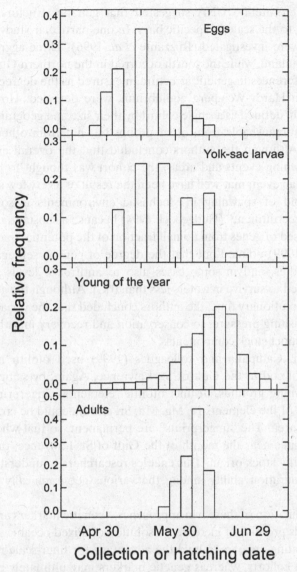

Figure 6.7 Evidence of two-stage differential recruitment of the 1990 year class of American shad in the Hudson River. Top two panels: frequency of occurrence of eggs and yolk-sac larvae during the spawning season. Bottom two panels: cohort frequency distribution of juveniles (YOY) that recruited from larvae, and of adults that survived until capture in 1995 (modified from Limburg 2001 with permission of the Ecological Society of America).

habitat made available by the severely depleted portions. If, on the other hand, the stocks are composed of subpopulations (demes) that are genetically distinct and more closely adapted to one or another part of the region, then the correct management strategy would be to practice conservation of the depleted demes by setting up marine protected areas.

In a review summarizing a series of papers, Ruzzante and colleagues (1999) presented evidence that strongly supports the hypothesis that Atlantic cod is composed of distinct subpopulations. Microsatellite DNA suggested that genetic structure varied from the regional scale down to the scale of specific bays. In one particular study, larvae from four large aggregations were investigated (Ruzzante *et al.* 1996). Three aggregations occurred off eastern Newfoundland, while the fourth occurred in the northern Gulf of St. Lawrence area. Significant differences in genetic structure, measured as the degree of heterozygosity and departure from Hardy-Weinberg equilibrium, were detected. However, one group of similarly-sized fish, defined as a single cohort by their size and geographical distinctness, were genetically distinguishable from other groups but not from other members of the postulated cohort. Although the authors concluded that the overall analysis pointed to several distinct spawning events and areas, the cohort was thought to have formed from a single event. Such an event may well have been the result of a very few spawners. Recruitment from this kind of spawning in stochastic environments is sometimes referred to as "sweepstakes recruitment" (Hedgecock 1994) because a substantial portion of a year class may be composed of genes from a small fraction of the potential spawning population.

Ruzzante *et al.* (1999) pointed out that the degree of genetic segregation corresponds well to tagging studies, and in some cases also to antifreeze levels in the cod blood (antifreeze is required to survive in waters less than 0°C). Although the genetic distinctions may not last over evolutionary time, the authors concluded that the variation has been sensitive to the heavy fishing pressure, so conservation and recovery must be concerned with conserving these distinct stock components.

In a similar vein, Campana and colleagues (1999) used otolith microchemistry to distinguish groups of cod in the Gulf of St. Lawrence. Again, by sampling larvae in the vicinity of the spawning grounds, distinct otolith "elemental fingerprints" based on the relative proportions of the elements Li, Mg, Mn, Sr, and Ba could be created for fish from the same spawning area. The "fingerprints" are permanent, so that when adults disperse into overwintering sites near the mouth of the Gulf of St. Lawrence, the otolith markers can be used to identify stock origin. This enables researchers to understand the degree of both mixing and migration ability among the various geographically distinct spawning stocks.

Clearly, the combination of these two approaches – genetic markers and otolith elemental signatures – holds promise of increased resolution of mixed oceanic stocks. At present, elemental fingerprinting appears to hold the advantage for finer-scale resolution of environmentally distinct cohorts, whereas genetic markers may ultimately elucidate broader-scale population structure. Such knowledge is increasingly essential to manage fisheries where stocks are over-harvested, habitat is depleted, and economic pressures are great.

6.4.4   Gene flow in coral reef fishes

Coral reef fishes, although demersal and reef-resident as adults, begin life as dispersing larvae. Theoretically, the degree of dispersal should depend on the duration of the pelagic larval phase, and the ability of larvae to become entrained in currents that carry them away from their parents' reef. This has implications for the design of conservation plans in coral reef species, in particular, how large an area must be conserved.

The Great Barrier Reef off eastern Australia provides an ideal testing ground for this theory. Peter Doherty and his colleagues (1995) investigated the degree of genetic differentiation in seven species of coral reef fishes from the Great Barrier Reef, representing a gradient of species with different larval durations. One damselfish, *Acanthochromis polyacanthus*, lacks a pelagic larval phase; at the other end of the spectrum, the surgeonfish *Ctenochaetus striatus* has large larvae that spend several months in a planktonic phase. Doherty and colleagues collected larvae over a large geographic range (northern and southern Great Barrier Reef, as well as collection of one species from a distant reef assemblage 1000 km to the east). At least 40 fish from each reef and species combination were analyzed for enzymes and other general proteins. In all, 32 loci were available for analysis. To quantify variation, the investigators tabulated the frequencies of various alleles and genotypes, and used allelic frequencies to calculate average heterozygosity (*H*). They also estimated the degree of population structuring with the fixation index:

$$F_{ST} = \frac{H_T - H_S}{H_T} \tag{6.2}$$

where F_{ST} compares the heterozygosity of the least inclusive groupings (subpopulations; H_s) to the most inclusive group (total population; H_T) as the differences in heterozygosity at the two scales, normalized to the heterozygosity measured in the whole population. Higher values of F_{ST} imply greater genetic differentiation, with a theoretical maximum of 1.0 and a minimum of 0 (panmixis).

Within the seven species investigated, larval stage duration explained 85% of the variation in F_{ST} (Figure 6.8). Larvae with long pelagic durations showed high rates of gene flow; at the opposite extreme, the non-dispersing *Acanthochromis* showed the greatest regional distinctness. Larvae with pelagic durations longer than 1 month were essentially panmictic. Thus, species deserving special attention with respect to conservation areas would be those with shorter larval dispersal periods.

6.4.5   Stable isotope analysis of otoliths and tissue

Both the carbon and oxygen stable isotopic ratios in otolith aragonite ($CaCO_3$) can provide useful information for cohort identification, although the applications are only just beginning to be appreciated. Oxygen isotopic ratios ($\delta^{18}O$) in otoliths reflect the ratios in the ambient water, whereas the carbon $\delta^{13}C$ is a mixture of dissolved inorganic carbon in the water and metabolic CO_2 in the bloodstream.

Although whole otoliths may be ground up to produce a bulk sample, new techniques that involve micromilling of otoliths (Figure 6.9) enable researchers to drill out tiny amounts of material from specific parts of otoliths and enable measurement of changes in isotopic ratios over the lifetime of a fish.

Through the micromilling technique, fossil otoliths have been assayed to reconstruct paleotemperatures. In a modern application in southwestern Australia, carbon and oxygen stable isotopes in pilchard (*Sardinops sagax*) otoliths revealed the presence of three different stocks in catches landed at four different ports (Edmonds & Fletcher 1997). The otolith $\delta^{18}O$ was interpreted to reflect the ambient water temperatures. Indeed, otolith stable oxygen isotopic ratios have generally been shown to be deposited in equilibrium

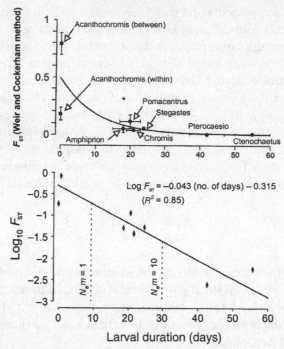

Figure 6.8   Genetic variance (F_{ST}) in seven species of Australian coral reef fishes plotted against the length of their pelagic larval stage. Horizontal lines through points on the upper plot are the range of ages at settlement, the vertical lines are 95% confidence intervals. N_em is the number of migrants exchanged per generation at equilibrium. Populations to the left of the first dashed line ($N_em < 1$; shorter larval dispersal periods) show greater regional distinctness and may deserve special attention with respect to conservation areas. Populations to the right of the second dashed line ($N_em > 10$) should be genetically homogeneous (reproduced from Doherty *et al.* 1995 with permission of the Ecological Society of America).

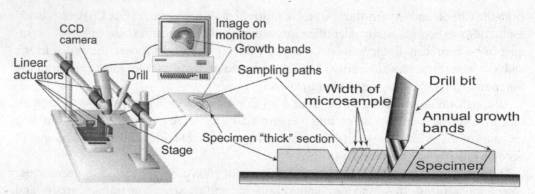

Figure 6.9   Schematic of a micromilling apparatus. Specimen is fixed beneath a dental drill and viewed with a digital camera. Drill motion is controlled by computer through an IEEE-488 interface, permitting stage manipulation in three directions via three micropositioning actuators. Sample is milled along growth banding so that sample paths are significantly narrower than the diameter of the drill (Wurster *et al.* 1999).

with the surrounding water. This probably makes oxygen stable isotope analysis a more reli-able method of inferring water temperatures than otolith strontium.

An unusual application of stable isotope analysis in fish otoliths was reported by Spencer *et al.* (2000). These investigators found that stable isotopes of lead accumulated in otoliths of Hawaiian reef fishes. They were able to establish a gradient of $^{208}Pb/^{204}Pb$ vs. $^{207}Pb/^{204}Pb$ that reflected, at one end of the gradient, ambient oceanic ratios, and at the other end, the "fingerprint" ratio of tetraethyl lead that had been added to gasoline. The latter then indi-cated influence of waters entering the system from an urbanized watershed, and the signal was picked up both by corals and fishes.

Stable isotope analysis has been limited in the past by sample size requirements. How-ever, with such advances as microsampling devices and increased sensitivity of mass spec-trometers, even the otoliths of fish larvae may contain enough carbon and oxygen for analysis. Future refinements may open the way for analyzing the isotopes of nitrogen and sulfur, now widely used in ecological studies, in the otolith protein, although such analytical sensitivity is presently out of reach. On the other hand, somatic tissues of larvae are well suited to stable isotope analysis. Whole fish somatic tissue analysis, combined with otolith microanalysis, can provide information on biotic influences (for example, trophic position in the food web) as well as abiotic ones (such as salinity or temperature).

Stable isotope analysis can be used to quantify rates of tissue turnover as well (Fry & Arnold 1982). With this knowledge, life history events such as settlement by fish larvae, as they shift from a pelagic to a benthic diet, can be determined and cohorts identified. In two recent papers, Sharon Herzka and colleagues determined the turnover time of C and N in larval red drum (*Sciaenops ocellatus*, similar to Figure 1.3r) in the laboratory, and then applied this to estimate the rate and temporal variation of settlement by larval red drum in the wild (Herzka & Holt 2000, Herzka *et al.* 2002). Red drum larvae were found to have only 20% of their original C and N after 10 days on a different diet, and based on growth models combined with isotope mixing models, most drum larvae settled at 4–5 mm standard length, but there was considerable variability in size and date of settlement.

6.5   Summary and conclusions

As we have seen, cohort identification can be carried out in a number of different ways. Age and size identification of larval cohorts enables the tracking of specific groups so that their fates can be followed. The increase in temporal and spatial information that is gained from identifying cohorts helps us understand the ecological (biotic and abiotic) forces that shape their year classes and how these vary from year to year. Fisheries have increasingly become a major ecological factor that structures fish populations. They can do so by harvesting dis-tinct subpopulations when they mix, as in the case of northern cod, or by directly affecting life-history traits, such as size and age at first reproduction, which can in turn affect egg pro-duction and the quality of eggs. Cohort identification has been spurred on by advances in the understanding and analysis of biological characters of fishes, in particular hard parts, such as otoliths, and molecular components, such as DNA and allozymes. Future refine-ments will likely increase the analytic resolution, with concurrent improvements in cohort discrimination.

Additional reading

Campana, S.E. (1999) Chemistry and composition of fish otoliths: pathways, mechanisms and applications. *Marine Ecology Progress Series* **188**, 263–297.

Griffiths, A.J.F. & others (2000) *An Introduction to Genetic Analysis*, 7th edn. W.H. Freeman, New York.

Thresher, R.E. (1999) Elemental composition of otoliths as a stock delineator in fishes. *Fisheries Research* **43**, 165–204.

Chapter 7
Habitat Requirements

Robert G. Werner

7.1 Introduction

Fishery scientists are well aware of the importance of habitat in mediating growth and survival of fish populations. Poor-quality habitat can lead to reductions in individual growth rates and a greatly increased likelihood of mortality. Inadequate habitat can also lead to reduced fecundity, poor reproductive success, and eventually to low rates of population growth. In recognition of this fact, the US Congress passed the Sustainable Fisheries Act, which established requirements for identifying essential fish habitat for important US fish stocks. It required the identification of the sequence of habitats from spawning to maturity, including those critical to early life stages.

Several things distinguish the analysis of habitat of early life stages from that of adults. First, egg and larval stages of fishes are often much more vulnerable to environmental perturbations than adults. Secondly, the time during which egg and larval requirements need to be met is often very short and the requirements change rapidly and continually with development. If the requirement is not met at the right place and right time, mortality is likely to be the result. Thirdly, as we have seen in Chapters 1 and 3, rapid growth moves larvae through windows of vulnerability quickly and thus, reduces mortality. Habitat that provides high levels of prey, and consequently rapid growth, is extremely important to larval and juvenile stages, much more so than it might be for adults. Fourth, the ability of larvae to withstand periods of starvation is very limited. Prey needs to be continually available during the rapid growth phase. Long periods of limited prey availability followed by large surges in prey numbers are not desirable. Fifth, since larvae are very vulnerable to a variety of sources of mortality, habitat that limits mortality is very important, either by limiting the number of predators or providing shelter to larvae during particularly vulnerable stages. The final difference is the fact that eggs and larvae have a very limited ability to position themselves in relation to their habitat needs, whereas adult fishes are much more capable of reaching a critical habitat. The location and sequencing of habitats is particularly important for early life stages.

A definition of habitat appropriate to early life stages of fishes is difficult to develop largely due to the fact that many species of fishes have pelagic eggs and larvae. The pelagic habitat, for the most part, does not possess the structure that characterizes terrestrial or even inshore aquatic habitats, yet the nature of the water mass is often critical to the survival of the egg or larva. Although water masses may not have structure *per se*, they do have habitat characteristics. Water masses are distinguishable on the basis of temperature, salinity, plankton populations, current direction, and other features. Consequently, I prefer a more general definition than that commonly found in ecology textbooks. I will define habitat for fish eggs and larvae as the position in space occupied at any particular point in time characterized by the physical, chemical, and biological features of that space.

Habitat can be recognized on at least two different scales. Habitat on a macro-scale, such as a major estuary or a long stretch of river, creates the general parameters for micro-scale habitat. For example, a tropical estuary will have many characteristics, such as temperature regime, tidal range, and seasonal duration of production that will be relatively similar among other tropical estuaries around the world. Each tropical estuary, however, will possess a multitude of micro-habitats that differ from each other on the basis of water depth, plant species composition, current, and a variety of other site-specific characteristics. The same pattern applies to other habitats: coral reefs, large temperate rivers, lakes, and estuaries, for example. Since macro-scale habitats are not uniform across their entire area, and since the limited mobility of eggs and larvae does not permit them to experience the range of smaller habitats, the most relevant habitat is the contingent habitat – the immediate habitat surrounding the individual fish. The larva responds to the conditions in its immediate environment. What is the prey field in the small volume of water that surrounds the fish at any given time? What are the chemical and physical characteristics? What is the predation risk? How much shelter from predators does the habitat provide? In a discussion of mortality risk in the marine environment, Rothschild (1986) recognized this fact and developed the concept of the "occupancy volume," the volume of water occupied by the larva in space and time. Larvae have their contact with "habitat" within the occupancy volume. Thus, the contingent habitat surrounding the egg or larva and the factors that govern that habitat are the topics of most concern to us in this discussion. And, as I will develop later in this chapter, it is not only the extent and quality of suitable habitat available to early life stages that is critical to growth and survival, but also the pattern and sequencing of these habitats.

To illustrate these points, I will address five questions in this chapter:

(1) What are the basic habitat needs of early life stages of fishes?
(2) How do larvae get to suitable habitat?
(3) Can larvae recognize and select suitable habitat after they arrive?
(4) Do larvae undertake ontogenetic shifts in habitat?
(5) How are larvae able to successfully occupy habitat that maximizes growth and survival?

7.2 Habitat requirements

7.2.1 Physical

A variety of physical habitat requirements can affect the growth and survival rates of early life stages including: temperature, light, protection from mechanical damage, shelter from predators, current, and a host of other factors. Temperature is an extremely important habitat variable for all organisms, but particularly ectotherms such as fishes. Temperature primarily governs rate processes, such as metabolism and growth in eggs and larvae. It also, through control of the above processes, influences size at hatching, duration of developmental stages, feeding rate, rate of digestion, behavior, and swimming speed. Indirectly, temperature also influences larvae through effects on the viscosity and oxygen saturation capacity of water and its productivity. Some of these were discussed in Chapter 1.

Growth and developmental rate determines the length of time a fish spends in a particular habitat. As was noted in Chapter 1 and as we will see later, many species of fishes undergo ontogenetic habitat shifts, which are often keyed to size or stage of development. Thus, sub-optimal temperatures can reduce growth rate, leading to a longer stay in a particular habitat. Inter-annual variation in temperature can influence the length of time a fish spends in any of these habitats, which may, in turn, be important in determining survival.

In addition, there is good evidence that temperature may have a greater effect on early life stages than it does on adults. For example, the metabolic Q_{10} values for embryos and larvae are relatively high, averaging around 3.0, whereas they average around 2.0 for adults. This suggests that the metabolism of early life stages may be considerably more sensitive to changes in temperature than adults, with all of the implications associated with this physiological characteristic.

The temperature ranges tolerated by early life stages are quite broad, ranging from below 0°C for Antarctic fishes to nearly 40°C or more for some tropical and desert fishes. Each species will have its own thermal optima. Spawning is structured to insure that the eggs are not only deposited in an appropriate habitat, but that the timing is such that the temperature regime will be suitable for development. Despite such care, considerable inter-annual variation exists.

Light plays an important role in governing the ability of larval fishes to capture prey and to avoid predators. Larval fishes are visual predators, for the most part, and require adequate light to locate and identify suitable prey. The amount of light available and its duration are thus important habitat variables for larvae. Adaptations to take advantage of light in capturing prey are common in the development of larvae and show the importance of light in the survival of larvae. Blaxter (1986), in a review of the sensory development of fish larvae, pointed out that a pure cone retina is adequate for feeding in fish larvae and that this normally develops at about the time the larva has absorbed its yolk and begun to feed. More recent work has empirically linked light intensity to feeding and has shown that the visual sensitivity of larvae increases rapidly from first feeding to the end of the larval period. Increases of up to three orders of magnitude were observed over this period of development (see Figure 1.8).

Turbidity, which reduces light, may limit the duration of the daily feeding period and the depth over which adequate light would be present. A reduction in the light available would then, both spatially and temporally, limit the feeding area available to the larvae. This would, of course, be modified as larval visual sensitivity improves with development. On the other hand, a reduction in light should work to the benefit of the larva by limiting the success of visual predators in locating their larval prey. Some marine larvae will often remain transparent as long as possible to reduce the impact of visual predators. This may require delaying the development of pigmentation and other visible structural characteristics until they have the ability to avoid predation on their own. A reduced light regime might be beneficial to such larvae.

Eggs and larvae are particularly vulnerable to mechanical damage. Even pelagic eggs and larvae can be damaged, if they come into contact with turbulence close to the water's surface. Those species that spawn demersally or on various substrates or in shallow water may subject their young to severe mechanical damage. In such situations, protection by construction of nests or locating spawning sites out of areas where water movements might propel eggs or larvae into hard objects is a mechanism to avoid this source of mortality.

Shelter from predators is another habitat consideration for egg and larval stages. Many species of fishes, such as trout (salmonids, Figure 1.3c, d) and sunfish (centrarchids, Figure 1.3i), are nest builders, providing protection for eggs until hatching and, in many cases, until larval dispersal. At the point where larvae leave the nest and/or parental protection they become susceptible to predation, leading to a need for shelter. Freshwater fishes or those marine fishes living near structure find shelter in vegetation, reefs, or other submerged objects. Studies on several species of north temperate freshwater and marine fishes have shown a strong tendency for larvae and juveniles to seek shelter in the presence of a predator. But, when predators were absent, they distributed themselves much more widely, occupying most of the available habitat. This indicates how important shelter from predators is to larvae and juveniles. For pelagic larvae, locating shelter is not much of an option. For them, an appropriate strategy might be to remain transparent for as long as possible.

Current is another important habitat variable. Riverine fishes often choose their spawning site to take advantage of current, insuring that sufficient oxygen will reach their eggs. Sturgeon, salmon, and redhorse suckers (*Moxostoma*) all spawn where the current is strong. Eggs tend either to be deposited in depressions in the bottom and then covered with gravel or released over gravel beds possessing large interstices into which the eggs can descend out of the strong current flow. In other species, the response to current is left to the larvae. In the naked goby (*Gobiosoma bosc*), for example, the pattern of settlement by the pelagic early larvae is strongly affected by current. Most individual gobies respond actively to current near the bottom, seeking out low flow areas downstream of bottom structure and adjusting their position as the current changes direction.

7.2.2 Chemical

Water quality in the spawning and nursery areas is an important habitat consideration for eggs and larvae. The concentration of dissolved oxygen, pH, salinity, and pollutants all contribute to the likelihood of success for egg and larval stages.

Oxygen is arguably the most important chemical variable determining habitat for early life stages. As we saw in Chapter 1, at least one ecological classification system for fishes is based on the availability of oxygen to early life stages. Embryos and larvae are known to be metabolic regulators, that is, at low oxygen concentrations metabolic rate is dependent on oxygen levels, increasing as oxygen levels increase, but at higher oxygen concentrations metabolic rate is independent of oxygen levels. The point at which the fish shifts from being dependent to independent of oxygen concentrations is referred to as the critical oxygen tension and it is strongly influenced by developmental stage, temperature, and activity. All early life stages of fishes seem to be more sensitive to low levels of dissolved oxygen than adults. Larvae are most sensitive to reduced oxygen levels but eggs are less sensitive than larvae. Due to the great range in their level of activity, larvae are more variable in their metabolic response to lower dissolved oxygen than embryos.

The explanation for some of these ontogenetic differences probably lies in the changing ratio of surface area to mass, the development of gills, and the proliferation of red blood cells. Initially, in the embryonic period, respiration is accomplished by diffusion of oxygen across the chorion, through the perivitelline fluid, and eventually to the embryo. This is a

---

Box 7.1 Outgrowing cutaneous respiration.

Blaxter (1988) outlined the theoretical relationships between larval surface area, weight, and oxygen requirement. Assuming that the larval surface area available for cutaneous respiration is approximately proportional to the square of body length and that body weight is approximately proportional to the cube of body length, then respiratory area per unit weight is inversely proportional to body length. Since the oxygen requirement of a larva is essentially directly proportional to body mass (Figure 1.11) or to the cube of body length, then the mass-specific oxygen requirement is constant (independent of body weight or length). Thus, on a weight-specific basis, respiratory surface area declines rapidly while oxygen demand remains relatively steady. As a result, the period of early development during which cutaneous respiration can effectively provide oxygen for a larva is relatively brief.

---

relatively inefficient system, but the demands are normally low and thus it generally proves to be satisfactory. At hatching, the larva begins to respire cutaneously. Since the ratio of surface area to mass of small larvae is relatively large, often an order of magnitude greater than it would be for juveniles, obtaining sufficient oxygen is a more tractable problem for small larvae than it is for larger individuals. With increased size, the ratio of surface area to mass is reduced and thus requires a higher partial pressure gradient, leading to a need for water with a high dissolved oxygen level. Since there are obvious limits to this, it ultimately requires, in most fishes, a shift in the mechanism for acquiring oxygen (see Box 7.1).

Fortunately, the establishment of an alternative to cutaneous respiration appears fairly early in post-hatching development. Concurrent with the reduced ratio of surface area to mass, red blood cells begin to develop and after several days to weeks gills form and begin to function. But, it is during this period of rapid development of the respiratory system, in conjunction with greatly increased activity, that the sensitivity to low dissolved oxygen levels is likely to be the greatest. At warm temperatures and high levels of activity, the critical oxygen tension for larvae can approach the saturation value. Some interesting adaptations to such conditions have been developed, particularly among tropical freshwater fishes, where the combination of high oxygen demand and high temperature work to limit the amount of oxygen available to the larvae. For example, the larvae of *Monopterus albus*, found in small ponds and marshes in southeast Asia which commonly become hypoxic, deal with the problem of low oxygen by developing a dense capillary network on the yolk sac. In this network, the blood generally flows toward the head from the posterior part of the body, while the water, propelled by the larva's pectoral buds, flows posteriorly across the vascularlized yolk sac. This generates a countercurrent flow similar to that found in the gill lamellae of fishes, which helps maximize the uptake of dissolved oxygen from the water as it does in the gills. This was clearly illustrated by the fact that oxygen uptake was reduced by one half when the water flow across the yolk sac was experimentally reversed.

Oxygen has an obvious effect on mortality through hypoxia, but it can also affect eggs and larvae in more subtle ways. Low dissolved oxygen levels tend to retard embryonic and larval development by slowing metabolism at lower ambient oxygen levels. Retarded development extends stage duration and results in a longer exposure to the perils of egg and larval life. For embryos, low oxygen levels produce a compensatory mechanism, which reduces the time to hatching by a premature release of hatching enzymes (see Chapter 1). Often the newly hatched larvae that appear as a result of this are ill-prepared for life due to their immature development and smaller size.

With the recognition of the acid rain problem in some lakes, a considerable amount of research has gone into evaluating the effect of lowered pH on egg and larval stages of freshwater fishes. Some deleterious effects have been recorded for salmonids at pH levels as high as 6.5–6.7. More commonly, pH's below 5.5 result in reduced reproductive success. One mechanism for this is the inactivation of the hatching enzymes at low pH. Due to the strong buffering capacity of sea water acid precipitation effects are not as important to marine fishes.

Variations in salinity are potentially most important to estuarine fishes, but variations in salinity can affect buoyancy of marine fishes as well. The ability of fishes to osmoregulate probably develops very early in the embryonic stage, near the time of yolk plug closure, due to the development of chloride cells on the yolk sac. At hatching, or at least by the time the gill epithelium develops a short time later, larvae have switched to the adult mode of using chloride cells associated with the gills for osmoregulation.

Another factor affecting the quality of spawning and nursery habitat is the nature and amount of man-made substances added to the water. Pollutants range from nutrients to heavy metals and have a variety of effects too complicated to explore here. It has been shown, however, that there is considerable variation in the sensitivity to pollutants between developmental stages. Generally speaking, the early life stages are most sensitive to pollutants, and across the spectrum of early development, the yolk-sac stage is most susceptible, followed by the early embryonic stages that occur prior to gastrulation. As larvae develop into juveniles, their body eventually becomes covered with scales, reducing the rate of uptake of many toxic compounds. Concomitantly, their internal organs begin to mature and develop an increasing capacity to detoxify compounds. The combination of these changes decreases the sensitivity to pollutants with further development. Although, most pollutants can be lethal at high concentrations, many have sublethal effects at lower concentrations that modify physiology or behavior. In the long run, sublethal modifications of behavior can be just as lethal if they prevent larvae from obtaining adequate prey or they prevent larvae from escaping predation.

Direct contact with pollutants will have obvious effects, but maternal exposure can also transfer pollutants to the egg and larval stages resulting in significant impacts. In particular, lipophilic compounds can concentrate in the yolk of the oocyte, which ultimately provides a release of contaminants to larvae as they absorb the yolk. Studies have shown that parental exposure to pesticides can cause not only mortality, but also modifications of larval behavior. Behavioral changes can be of a nature to reduce prey capture efficiency and predator evasion. For larvae, this almost always leads to mortality.

7.2.3 Biological

One of the most important needs of larval fishes is an adequate food supply. Abundant food leads to rapid growth, and rapid growth moves larvae through windows of vulnerability quickly, reducing potential mortality rates (see Chapter 3). Several factors need to be considered when evaluating a habitat with regard to the availability of prey to larvae:

(1) the density of prey in relation to larval density;
(2) the size of prey with respect to gape and developmental stage of larvae;

(3)   the nutritional quality of prey;
(4)   the timing of the appearance of prey in relation to their larval predator.

A mismatch in any of these parameters may make a habitat unsuitable for larvae. For example, if zooplankton is abundant, but is composed of large-sized prey at the time that larvae begin to feed, the small larvae may be unable to capture and consume such large prey and thus find it difficult to obtain sufficient food for optimal growth. Conversely, if the larva has grown beyond small-size prey and is ready for larger prey, but none are available, growth is likely to be reduced. Thus, the sequence of prey sizes needs to match the increase in the size of the larval gape.

Equally as important as the food supply are the predators available to potentially prey on eggs and larvae. Since eggs and most larvae are small, unprotected by scales or spines, and have modest swimming ability, they are particularly vulnerable to predation. The suite of predators feeding on larvae includes a variety of invertebrates such as cnidarians, predaceous insect larvae, and crustaceans in freshwater; chaetognaths, cnidarians, and ctenophores in the sea; as well as a myriad of piscine predators in both habitats. The predation sequence for most predators involves encounter, attack, and capture. Aspects of habitat that can reduce the predation rate by interfering with any element of the predation sequence would improve survival. For example, the encounter rate could be reduced by an increase in the structure of the nursery habitat such that larvae were able to avoid detection by predators. The reduction in the encounter rate could also be enhanced by cryptic coloration, typically as a result of camouflaging pigment that helps match larvae to their background. Similarly, availability of cover or other aspects of habitat that reduce the attack and capture rates would also reduce the predation rate.

7.3   Getting to suitable habitat

Quite frequently, adult fishes spawn in an area that is some distance from the nursery grounds (see the Migration Triangle Hypothesis in Chapters 4 and 8, Figure 8.3). Since it is imperative that the early life stages get to suitable nursery habitat in order to enhance their chances of survival, various mechanisms have evolved to accomplish this. Even though some larvae have a limited capacity to swim, their swimming ability is often insufficient, by itself, to accomplish the necessary migration to the nursery habitat. Thus, we are left with two primary factors that determine how this movement is accomplished: parental choice of spawning site and the ingenious use of current. Initially, eggs are deposited in a spawning site chosen by the adult fish. Such sites may be as diffuse as a region of the open ocean or as specific as a nest along the shore of a lake. Whatever the case may be, this is where the egg or larva begins its journey. The initial site is often located either within, or upstream of, the nursery area. Although larvae do have a limited ability to swim short distances, they cannot in any real sense migrate from one habitat to another during the early larval period. Many do, however, have the ability to make adjustments in their vertical position in the water column by either swimming vertically or modifying their buoyancy. This can make big differences in the current systems that the fishes occupy and ultimately in where the fishes are transported. Keep in mind that fishes are the most fecund of the vertebrates, with many species producing a superabundance of eggs. Some species can produce more than a

Box 7.2   Vast production, vast losses.

Rothschild (1986) graphically illustrated the magnitude of egg production that one might expect from a large population of marine fish each year. He assumed that the typical egg is about 1 mm in diameter and that a population might produce approximately 10^{22} eggs. If these eggs were placed side by side they would extend for 10^{6} km. Since the circumference of the Earth is only 4×10^{4} km they could encircle the globe at the equator approximately 25 times. Since the adult population producing the eggs is much smaller, one would expect high losses from the egg and larval population. A considerable fraction of those losses may be attributed to failures to reach appropriate nursery habitat and the consequences of that failure.

million eggs per female. When this is multiplied by a large population of spawning females, the output can be extremely large (see Box 7.2).

Consequently, with such a large initial population and with most of the losses occurring early, there is room for a lot of error in locating the nursery grounds. Only a relatively small proportion of the total needs to be successful in reaching suitable habitat to maintain the population, assuming that mortality is greatly reduced after the drift period.

To characterize the initial habitat for egg and larval stages of fishes, it is useful to refer to the classification of reproductive guilds developed by Kryzhanovskii and Balon and mentioned in Chapter 1, which outlines the variety of spawning habitats and spawning behaviors present in fishes. This classification system demonstrates that fishes use a wide variety of habitats for spawning, and that these habitats are the geographic/ecological starting points for the egg and larval periods. The range of initial habitats occupied by fishes is quite broad and the diverse adaptations quite remarkable. For those species classified as bearers, which includes fishes that brood their eggs and sometimes their young on or in their bodies, as well as the viviparous fishes, the initial and subsequent habitat is well defined and controlled by the parent. Release of the young is dependent on development and typically does not occur until the young are in the late larval or even early juvenile period of development. These fishes are relatively mature, and thus are not subject to the level of mortality that smaller larvae are.

At the other extreme, non-guarding pelagic spawners release eggs into the open water and provide no parental care at all. These eggs are typically small and very abundant. They suffer very high mortality rates. They are, however, often positioned in such a way as to facilitate their drift into productive nursery areas. This positioning is not only a spatial phenomenon, but a temporal one as well. Freshwater larvae are often larger and possess a somewhat greater capacity to relocate themselves to nursery sites. Fishes spawning in streams have the use of strong and fairly predictable current patterns to facilitate larval transport.

The spatial positioning or the process by which eggs are deposited in a location that facilitates larval growth and development is quite varied. Some species spawn in the site that will later be used as the nursery area. For example, many freshwater fishes spawn in the shallow, vegetated littoral zone of lakes and the young remain in the vicinity of the spawning area for many months during their larval and juvenile periods. Others, such as many pelagic marine fishes and stream-spawning freshwater fishes position themselves upcurrent from the nursery site and the young produced drift downstream to the nursery site. This is often referred to as a contranatant/denatant migration cycle (see Figure 7.1). Adult fishes migrate to the spawning ground which is located upcurrent relative to the nursery site.

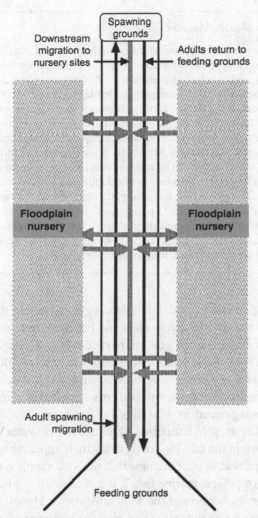

Figure 7.1 A schematic representation of the reproductive migration and subsequent larval drift of fishes in large tropical rivers. Spawning grounds are located upstream and nursery sites are downstream and lateral to the main channel. Nursery sites include flooded fields, forests, wetlands, small lakes, and connecting channels. Adapted from Bonetto (1981).

Spawning occurs and the eggs are deposited either in the open water or, in the case of stream spawners, on the bottom of the stream. This phase is the contranatant portion of the migration cycle, migration against the current to the spawning grounds. It is quite common in riverine fishes and pelagic-spawning marine fishes. The denatant migration begins as soon as the eggs or larvae become exposed to the current. In the case of the pelagic spawners, the denatant migration begins immediately, when the eggs and soon the larvae drift downcurrent. Many stream-spawning fishes produce demersal eggs and/or build nests, characteristics that detain the eggs at the spawning site until after hatching. Once the larvae are exposed to the current, they are subject to downstream drift and transfer to the nursery site. The life history of the Japanese sardine (*Sardinops melanostictus*), described in Chapter 11, follows this cycle. For species in which the adults spawn multiple times,

Box 7.3   The Member/Vagrant Hypothesis.

The issue of population regulation in marine fishes has long been of interest to fishery scientists. Sinclair (1988) proposed a hypothesis based on habitat (geographic settings) to explain population pattern, richness, and absolute abundance of fish stocks based on occupancy of retentive loops within which a species is able to complete its life cycle. He categorized individuals in the population as either members, if they are able to remain within a geographic setting that allows them to complete their life cycle, or vagrants, if they fail to reach an appropriate habitat or geographic setting. Vagrants, as a consequence of not reaching an appropriate habitat, generally do not complete their life cycle successfully. Membership requires that an individual be in the right place at the right time in its life cycle. The number and location of appropriate habitats where the life cycle of the species can be completed will establish the pattern and richness of the population within its range. The sizes of the habitats, particularly the spawning and nursery grounds, will determine absolute abundance of the population. A key element of the hypothesis is the importance of larvae in the life cycle and their ability to reach and remain within the nursery sites after spawning. The Member/Vagrant Hypothesis highlights the importance to populations of marine fishes of getting to, and remaining in, suitable habitat.

the contranatant migration is repeated, but for the eggs and larvae the denatant migration occurs only once in the history of the year class and becomes, in a sense, a property of that year class. What occurs during that migration, particularly as it relates to growth and mortality, will likely characterize the year class from that point on (see Box 7.3 and Chapter 4). As discussed in Chapters 2 and 6, it has become possible to chronicle such events through careful analysis of the otoliths using growth patterns or chemical signatures.

The temporal positioning occurs as a result of the synchrony between spawning time and the production cycle of the prey. In temperate zones, most freshwater fishes are either spring spawners or, if they spawn in the fall, the young delay their appearance until the spring. This insures that they will appear at or near the time that the food supply is beginning to build and approaching its peak. For pelagic marine fishes, the coordination of spawning with oceanic productivity cycles is the foundation of the Match/Mismatch Hypothesis. It is thought that the success of year classes is dependent on how well the timing of spawning and subsequent larval appearance matches the appearance of abundant planktonic prey (see Chapter 4). Whether the Match/Mismatch Hypothesis is a suitable explanation for variation in year-class strength is a matter of some dispute. It is obvious, however, that from a habitat perspective it would be of great survival value to a species if it were able to synchronize the appearance of its young with the appearance of large amounts of prey in a particular habitat.

As you would expect, fishes have developed a diverse array of methods to insure that sufficient larvae make it from the spawning site to the nursery grounds. I will outline four, which exemplify some of the major methods for accomplishing this important phase in their life history.

7.3.1   River-spawning fishes

Arguably, the young of river-spawning fishes have the simplest method of getting from the spawning grounds to suitable nursery areas. It is well known that many species of freshwater and anadromous fishes migrate upstream to spawn. The young produced in this way either stay in the general area of the spawning site if they are large enough and have the capacity, or they drift downstream with the current as eggs or early larvae. The latter is probably the more common method of distribution over all riverine fish taxa.

In tropical rivers with large floodplains, spawning fishes migrate upstream to good spawning sites, typically well upstream of the broad floodplain. Larvae drift downstream and move into the extensive shallow areas, wetlands, and small lakes associated with the flood- plain. Since these areas are very productive, they provide abundant food for the larval and juvenile stages. After spending time feeding and growing rapidly in the rich floodplain, the young either migrate or are forced out of the shallow water as the river level drops. Figure 7.1 summarizes the general pattern.

A regular pattern of appearance of different species in the drift is associated with the timing of adult spawning. The larvae from early spawners appear in the drift first, followed by larvae from later spawners, creating a regular chronological sequence of species that repeats itself year after year. In some species drift occurs primarily at night, in others there is no apparent diel pattern of drift, with larvae drifting throughout the day and night. Larvae tend to be found more commonly near shore and near the surface. But this changes somewhat as the larvae develop. Studies have shown that early stage larvae of clupeids (*Alosa*, similar to Figure 1.3m), carp (*Cyprinus carpio*, Figure 1.3f, g) and freshwater drum (*Aplodinotus grunniens*, similar to Figure 1.3r) are more common in the main channel, but later stage larvae are found more frequently near shore. Other investigators have noted a rapid depletion from the drift of larvae at about 8–12 mm in length with a concomitant increase in fishes of this size or larger in backwater areas of the river. This seems to imply that larvae up to 8–12 mm in length are subject to the vagaries of the current, but when they reach 8–12 mm and larger they are able to reposition themselves inshore.

The advantages of potamodromous migrations to riverine fishes may lie in locating spawning grounds upstream in a site better suited to survival of the eggs, with less silt and possibly higher oxygen concentrations, followed by downstream drift to more productive nursery sites. The migration is often timed to coincide with the beginning of high water so that the larval stages can be swept downstream into recently flooded wetlands. In rivers with large productive floodplains this provides an opportunity for the larvae to be carried into food-rich nursery areas. Rapid growth may then occur and upon reaching an appropriate size, the juveniles can migrate out into the main river.

In some cases, lake-dwelling fishes use tributary streams to spawn and the young thus produced drift downstream into the lake. Examples include white sucker (*Catostomus commersoni*), rainbow smelt (*Osmerus mordax*), some lake-dwelling salmonids, and walleye (*Stizostedion vitreum*) among others.

7.3.2   Open water marine fishes

One of the major issues in marine fishery research is the question of recruitment variability. Embedded in this issue is the question of how ocean-spawned eggs and larvae reach estuarine nursery areas. Many species of marine fishes spawn a considerable distance from shore and many hundreds of kilometers from their nursery grounds. Since the eggs and larvae appear to be incapable of traversing this distance on their own, the question arises as to how they get to the nursery area and what role oceanic currents and water movements might play. There are at least two phases to this process: open ocean transport and ingress and retention within the estuary.

The open ocean larval transport phase of the life history of marine fishes has been described for several different combinations of species and locations throughout the world. Each situation is somewhat different and is based largely on site-specific current patterns, weather, and hydrography, resulting in varying patterns of transport. One recent study, however, presents a scenario that provides a good example of the complexity of this issue. Hare & Cowen (1996) studied bluefish (*Pomatomus saltatrix*) that spawn in the South Atlantic Bight off the eastern coast of North America from Cape Hatteras to Cape Canaveral. Although some recruitment occurs in South Atlantic Bight estuaries, the bulk of it occurs to the north, in the Middle Atlantic Bight, which ranges from Cape Hatteras to Cape Cod. The question becomes: how do the larvae get from the southern spawning grounds to the more northerly nursery areas? Hare and Cowen divided the open ocean transport phase of bluefish into three parts: larval transport effected by the Gulf Stream, movement of larvae from the Gulf Stream to the edge of the continental shelf or across the slope sea, and crossing the continental shelf to the nursery areas.

Larvae are transported northward, out of the South Atlantic Bight within the Gulf Stream (Figure 7.2). As the Gulf Stream veers to the northeast, the more advanced individuals in the population (juveniles) swim across the slope sea to the continental shelf. Most larvae, however, are carried with the Gulf Stream and only break out of it when warm-core rings are formed, a common occurrence in this area. Warm-core ring streamers carry the larvae to the continental shelf, where they accumulate and eventually develop into juveniles. They remain on the shelf-slope front until the temperature front breaks up in the late spring or early summer. At that point, they are capable of swimming across the shelf to the estuaries.

For some species, small scale diel vertical migration provides a mechanism for accomplishing substantial horizontal movement through encounters with favorable currents. For example, larval Atlantic menhaden (*Brevoortia tyrannus*), less than 9 mm total length, have a diel activity pattern whereby they move toward the surface at night and into deeper water during the day. At about 10 mm, the larval swim bladder becomes functional and they swim to the surface at sunset to fill it. At dawn, they release gas from the swim bladder and swim to their normal daytime depth. It is generally assumed that larvae have sustained swimming speeds around 1–2 body lengths per second, which is too limited to enable them to make much progress swimming horizontally. But by moving vertically through the water column, it is thought that they may be able to take advantage of currents that would transport them to the nursery grounds.

Upon reaching the estuary many species of fishes utilize selective tidal stream transport (STST) to work their way into and up the estuary. This simply requires that the fishes move up into the water column on rising tides and stay near the bottom or out of the current on ebb tides. The net effect of such behavior is to facilitate movement into, and retention within, the estuary. For example, larvae of Atlantic menhaden have been shown to be most abundant in the water column at night on rising tides and rarely present at other times. During ebb tides they are thought to be close to the bottom, somewhat sheltered from the effects of tidal flow.

7.3.3   Coral reef fishes

Coral reef fishes have a bipartite life history divided between the benthic adult phase, closely attached to the reef, and the pelagic larval phase whose individuals drift in open water,

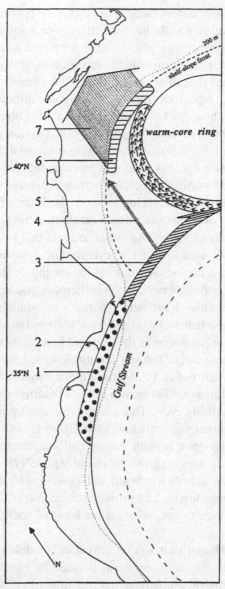

Figure 7.2 Diagram illustrating how larval and juvenile bluefish are transported from the South Atlantic Bight (SAB) spawning grounds to estuarine nursery grounds in the Middle Atlantic Bight (MAB). 1. Bluefish spawning grounds off the SAB. 2. After spawning, some relatively nearshore larvae may remain in the SAB and recruit to estuaries there. 3. The majority of the larvae remain in the Gulf Stream and are transported northeastward to the MAB. 4. More advanced individuals may actively swim from the Gulf Stream across the slope sea to the shelf-slope temperature front. 5. Warm-core ring streamers transport most larvae northwesterly across the slope sea. 6. Shelf-slope temperature front accumulates larvae and pelagic juveniles. 7. When the shelf-slope temperature front dissipates in late spring pelagic juveniles swim across the MAB shelf to estuarine nursery grounds (reproduced from Hare & Cowen [1996] with permission of the American Society of Limnology and Oceanography).

in some cases hundreds of kilometers away from any reef. Approximately, 95% of the families of coral reef fishes have such a life-history pattern. For many years, the commonly held view on recruitment from the pelagic phase to the coral reef was that pelagic larvae were derived from a variety of reefs, intermingled as larvae, and, after achieving an appropriate size, settled on the first reef that they encountered. Recruitment was generally thought to occur from fish spawning at upcurrent reefs. This left some difficult unanswered questions, not the least of which was, how does the most upcurrent reef obtain its recruits? Although research on this topic has been severely hampered by the physical and biological complexity of most reef ecosystems and the high cost of mounting appropriately scaled research studies, recent evidence has shed new light on coral reef recruitment. The major issue is the extent to which larvae are retained within the reef, in contrast to larvae recruited from elsewhere.

Using direct mark-and-recapture techniques on the damselfish (*Pomacentrus amboinensis*), Jones *et al.* (1999) demonstrated that a substantial fraction of the larvae on a reef near Lizard Island, Great Barrier Reef, Australia, are retained near the reef. Embryos were marked by immersion in a solution of tetracycline and returned to their original site. The tetracycline produced a distinctive mark on the otolith of the larvae. After hatching, the larvae became pelagic and dispersed. Upon settlement, young fish were collected and a subsample of 5000 fish examined for marks. Fifteen individuals had tetracycline marks indicating clearly that some fish were retained on their natal reef. By knowing approximately how many eggs were produced at the reef and how many were marked, the investigators estimated that the amount of fish self-recruiting varied from 15% to 60%.

In a similar fashion, Swearer *et al.* (1999) used natural marks based on a trace element signature found in the otolith as well as growth rates to determine where larvae were spending the pelagic phase of their life cycle. Those found in coastal water masses tended to possess a distinctive trace element signature and have higher growth rates, whereas those fish dispersing through less rich open oceanic waters had a different trace element signature and lower growth rates. By sampling the bluehead wrasse (*Thalassoma bifasciatum*) on St. Croix, US Virgin Islands, at both windward and leeward reefs they found that the wrasse recruited from two different sources. The primary source of recruits for windward reefs was pelagic larvae from the open ocean, whereas on leeward reefs, the primary source was locally retained larvae.

Finally, modeling has shown that simple advection models overestimate recruitment from open ocean populations to reefs, primarily because they underestimate losses to mortality and diffusion. To properly simulate actual recruitment rates, models may require the incorporation of some larval retention. The significance of this from a management point of view is great. Since the bulk (approximately 99%) of the mortality of a year class of coral reef fishes occurs during the pelagic dispersal phase, attempts to understand and manage recruitment variability in coral reef fishes must include knowledge of this phase. If recruits are coming from diverse sources, then this issue becomes very complex. But, if they are locally retained fishes, then the problem becomes more tractable.

7.3.4   Lake-spawning fishes

Up to this point, I have discussed examples in which eggs and larvae have been largely dependent on currents to transport them to the nursery grounds. For littoral spawning, lake-dwelling fishes currents seem a less likely means for moving from littoral to open water habitats, largely due to the fact that currents in lakes are relatively weak when compared to river or ocean currents and they are multidirectional.

There is good evidence that larvae of many species of lake-dwelling fishes undertake movements from their spawning grounds to the limnetic region of the lake. Documentation is available for brook silverside (*Labidesthes sicculus*), yellow perch and European perch (*Perca flavescens* and *Perca fluviatilis*, Figure 1.3j), bluegill sunfish (*Lepomis macrochirus*), and burbot (*Lota lota*) among others. The basic pattern is for adult fishes to spawn in shallow, nearshore areas. After hatching and yolk absorption, larvae move from the littoral zone of the lake to open water areas away from the vegetation. In productive lakes these two habitats are quite distinct. The limnetic zone is characterized by minimal structure, microscopic vegetation in the form of algae, abundant zooplankton, limited invertebrate predators, and relatively strong currents. In contrast, the littoral zone is characterized by a great deal of structure in the form of large macrophytes, reduced zooplankton, abundant invertebrate predators, and weak currents, if currents are present at all.

The mechanism by which larvae are able to move from the littoral zone to the limnetic zone has not been clarified, but may involve more active movement by the larvae than is the case with other larval habitat shifts. Based on the virtual absence of currents in dense macrophyte beds, currents would seem to be of less significance. In small lakes with well developed littoral zones, where this movement pattern has been shown to occur, currents are weak and negligible in the vegetated littoral zone. Upon reaching the limnetic zone, the dispersal of larvae may then be facilitated by currents. In large lakes where wind and currents are greater and the littoral vegetation is less dense, many of these species are more broadly distributed and the pattern is not as clear. But in small lakes where the distances are small, the possibility that the larvae are able to swim to the limnetic zone seems like a reasonable hypothesis.

The adaptive value of such a migration seems to be related either to finding more food or avoiding predators. Since predator and prey densities vary between habitats, larval fishes must balance their movement patterns to take into account available food and threats from predators.

7.4   Habitat recognition/habitat selection

Several studies have suggested that larvae or juveniles have the ability to recognize and select a habitat by moving into and remaining within it for an extended period of time, presumably because it provides for reduced mortality and/or increased growth rates. For example, juvenile *Lates angustifrons* and *Lates mariae* in Lake Tanganyika are able to differentiate between vegetated habitats. Both species have a pelagic early life phase, after which they move into aquatic vegetation. *Lates angustifrons* demonstrates a preference for *Vallisneria*, whereas *Lates mariae* is found more commonly in beds of *Potamogeton*. Juvenile summer flounder (*Paralichthys dentatus*) seem to have preferences for sand substrate. Preferences for sand have been demonstrated in the laboratory and several studies of the distribution of summer flounder in the field corroborate those results. Some coral reef fishes such as *Pomacentrus coelestis* have the ability to choose reef slopes in preference to lagoons, possibly using temperature cues. So the ability of early life stages of some species to recognize and select habitat appears to be well established.

For other species, however, there is doubt as to whether the larvae are making a choice or whether the observed distribution in the field is the result of increased survival in the optimal habitat. Atlantic cod (*Gadus morhua*, Figure 1.3p) and cunner (*Tautogolabrus adspersus*) showed no differences in settlement between habitats, even though post-settlement survival and ultimately juvenile densities were positively correlated with habitat complexity. This suggests the possibility of random settlement with ultimate distribution determined by differential mortality. If larvae are settling in a fairly random fashion and juvenile densities are determined by differential mortality, then habitat selection may not be operational. If, however, larvae are able to recognize and select a habitat based on a range of sensory cues, then more knowledge of this process could be helpful to fishery scientists.

A better understanding would have to begin with a clear outline of the development of sensory modalities that provide an ability to recognize appropriate habitat. The ontogeny of sensory systems in embryonic and larval fishes for several species has shown interesting correlations with behavior (see Figures 1.8 and 1.9). Others, however, have questioned the link between habitat recognition and sensory development. For example, Poling & Fuiman (1998) found very little correlation between sensory development in young sciaenid larvae (Figure 1.3r) and habitat. It was only later, when the larvae had reached an advanced larval stage that any correlation between sensory development and habitat could be demonstrated. They suggested that early larval stages are severely constrained by such "constructional restraints" at a small size. At small sizes, the eye and retinal area are very limited and this may control visual performance and the ability to recognize habitats.

The difficulty of understanding the sequencing of sensory capabilities and the impact it might have on habitat selection can be demonstrated by reviewing a study of a population of dwarf white suckers (*Catostomus commersoni*) in the Adirondack Mountains of New York. In this case, the process is complicated by the need to transfer sensory information across generations. Adult dwarf suckers spend essentially all of their time in the lake, moving into streams only to spawn. Eggs hatch in 4–7 days and 7–14 days later, at about the time of yolk absorption, the larvae move downstream to return to the lake. At this point they are 13–14 mm long. The interesting aspect of this is the fact that the adults return to the same stream to spawn each year. If they follow the salmonid model, then the larvae must be able to imprint on the odor of the natal stream at the time of outmigration as very small larvae. Do they have the structures necessary to do this? If they do not, then they may be using a model different from that of the better known salmonid model. There could be potentially two entirely different scenarios here:

(1)   larvae can smell sufficiently well prior to outmigration to imprint on the odor of the home stream upon leaving it for the lake, or

(2)   larvae cannot smell well enough and thus do not imprint. Instead, young adults imprint to the stream of first spawning and return consistently to it in subsequent spawning experiences.

Anatomical work by Werner & Lannoo (1994) has shown that the larvae do, in fact, have the basic structural elements required for olfaction, indicating that they may be able to imprint on the home stream at outmigration as the salmon do. Of course, the appearance of a morphological structure does not necessarily guarantee sensory capability, additional behavioral work would have to be done to answer that question. But clearly, the absence of a sensory structure or the incomplete development of the chain of structures necessary for a sensory function is a clear indication that the function is absent. With that in mind, a careful study of the ontogeny of sensory development in larval fishes would be a first step in trying to understand how larvae are able to recognize and select critical nursery habitat. Once the sequence of basic sensory capabilities is developed, then a search for sensory cues used in habitat recognition and selection can begin.

7.5 Ontogenetic changes in habitat utilization

The early life stages of fishes undergo very rapid changes in development and thereby quickly modify their ecological requirements and capacities. In Chapter 1, we outlined some of the things that make early life stages unique in terms of their relationship to their environment. These are:

(1) the development of anatomical structures and physiological capabilities that provide for relatively rapid increase in behavioral and ecological competence;
(2) an increase in size over many orders of magnitude; and
(3) the rapidity with which ontogenetic development and the increase in size occurs.

Significant and rapid changes in size, anatomy, physiology, and behavior lead quickly to changes in ecological requirements, which in turn, often demand or allow changes in habitat. Habitat suitable for an early stage may not be suitable for the larger and more capable later stages. Eggs may be deposited in a site that is suitable for embryonic development, but larvae may need to move to a different site, one that is better able to accommodate their need for food and their rapid development. As noted earlier, egg and early larval stages of fishes are dependent, for the most part, on forces other than their own swimming to get them from one habitat to another. As they grow, however, their ability to swim increases dramatically (Figures 1.10 and 11.31). I have already described patterns of movement, that lead to the occupation of appropriate nursery habitat by a variety of different species of fishes, both marine and freshwater. Such movements imply a kind of ontogenetic habitat shift, in which early life stages move from habitat to habitat as their development progresses.

For many marine fishes habitat shifts often occur at metamorphosis, when the larva undergoes a striking change from a pelagic, planktonic organism to a demersal one. Arguably, the most striking examples of this change occur among the flatfishes (Pleuronectiformes), where symmetrical pelagic larvae metamorphose into laterally compressed juveniles whose eyes have migrated to one side of the body. Such a dramatic change enables young fishes to take advantage of the new habitat in a variety of ways. Resting flat on the bottom with both eyes pointing upward allows the fish to spot prey against the background of diffuse light from above. The asymmetrical color pattern breaks up the outline of the juvenile and provides a kind of camouflage protection when the fish is resting on the bottom.

Ontogenetic habitat shifts are common among coral reef fishes. We have discussed the settlement of pelagic larvae as they recruit to the reef from the open water. The process may be somewhat more complicated, however, than simply settling on the first reef they encounter. For example, the Nassau grouper (*Epinephelus striatus*) occupies three habitats sequentially after it settles onto the reef. Early post-settlement fish are found exclusively on coral clumps that are covered primarily by the macroalga, *Laurencia*. Early juveniles move out of the algal mat but remain in close proximity to it. The larger juveniles tend to be found on patch reefs. This suggests a series of smaller scale habitat shifts following settlement on the reef. Presumably, the sequencing of each of these habitats is equally as critical in determining the survival of the year class.

Movement from one habitat to another during early life is well documented for freshwater fishes as well. The bluegill sunfish (*Lepomis macrochirus*) spawns in nests constructed in the littoral zone. Males guard the nest until the young hatch and absorb the yolk at about 6–7 mm total length. Sampling with fry traps in both the littoral and limnetic zones of a eutrophic lake has shown that the larvae can be found almost exclusively in the limnetic zone between 10 and 25 mm total length, but rarely in the littoral in these size ranges (Werner 1967). At 25 mm, the young have nearly completed the formation of scales and are then found much more commonly in the littoral zone. The littoral zone in many eutrophic lakes is occupied by a large number of invertebrate predators that prey on the relatively unprotected larvae. After a substantial increase in size and the development of scales while in the limnetic zone the bluegill becomes much less vulnerable to this source of mortality. They are, however, still vulnerable to predation from large fishes and thus may be able to reduce predation risk by taking advantage of the physical shelter provided by the littoral zone. In other lakes, the return to the littoral zone has been shown to occur at a smaller size (about 12.5 mm), suggesting the need to balance risk factors. Yellow perch (*Perca flavescens*) and the European perch (*Perca fluviatilis*) undergo ontogenetic habitat shifts as well. Adult perch produce long gelatinous strings of eggs that are typically deposited on vegetation in the littoral zone. The young, as with bluegills, remain in the littoral zone until the yolk is absorbed and then migrate to the limnetic zone where they remain for 4–8 weeks. At the end of that period, they either return to the littoral zone or become demersal.

Avoidance of predation may not be the only reason for changing habitats. Availability of abundant and suitable sources of food to sustain rapid growth may also be important. For example, Bear Lake sculpin (*Cottus extensus*) undertake a series of habitat shifts beginning with the deposition of eggs on a rocky inshore spawning site. After spawning, the larvae go through a pelagic dispersal phase during which they are distributed throughout the lake. At settlement they typically occupy one of two habitats: a food-poor, deep-water, profundal habitat or a shallower, food-rich, littoral habitat. Those fish in the profundal habitat have much slower growth rates and they soon make a unidirectional migration up into the richer littoral habitat. Those larvae that happened to settle in the food-rich littoral zone, of course, remain there. Thus, the movement from littoral spawning areas to the limnetic zone, followed by a return to the littoral zone, may be a widespread phenomenon among lake-dwelling fishes.

Chapter 1 identified some interesting correlates between habitat shifts and stage of sensory development. The correlation in time and developmental stage between a sensory capability and a change in habitat strongly suggests a link between the two. It is likely that the movement into a particular habitat may have to wait until the sensory capabilities have developed to the point that the habitat can be recognized and/or the predation risk is minimized.

The ecological advantages accruing to a fish undertaking ontogenetic habitat shifts were reviewed in detail by Earl Werner & James Gilliam (1984). They hypothesized that habitat

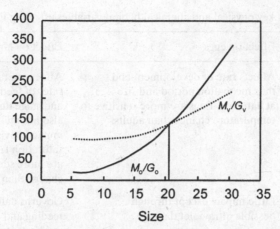

Figure 7.3 The ratio of mortality to growth rate for bluegill sunfish as a function of size for two different habitats, O – open water; V – vegetation. Point where curves intersect is the size at which a habitat shift should occur. Adapted from Werner & Hall (1988).

shifts occur to maximize growth rates and to minimize predation risk. This leads to the conclusion that larvae will attempt to occupy habitats that minimize the ratio between the habitat-specific mortality rate (M) and growth rate (G). The size at which a shift from one habitat to another occurs can be predicted by plotting habitat-specific M/G ratios against size. The intersection of the curves identifies the size at which the habitat shift should occur (Figure 7.3). Following a pattern of habitats that minimizes M/G leads to an optimal balance between the advantages of rapid growth and the risk of predation.

A study of the predation risk–growth rate model on bluegills in five different lakes in Michigan has shown that the limnetic zone is a much better foraging habitat for bluegills. But, if largemouth bass (*Micropterus salmoides*) are present, the predation risk is 40–80 times greater for small bluegills in the open water, implying that young bluegills should move from the limnetic zone to the littoral zone when bass are present. Recently, a cost-benefit analysis of ontogenetic habitat shifts of the Nassau grouper has shown that the pattern of habitat shifts was consistent with the "minimization of M/G" model, thereby broadening its application.

7.6 Putting it all together: how larvae are able to occupy suitable habitat to maximize growth and survival

Since the egg and larval stages of fishes are quite small relative to adult fishes and the anatomical structures and physiological capabilities are undergoing rapid development, the suite of habitat variables that influence eggs and larvae differs significantly from those affecting adults and often changes rapidly as larvae develop. The impact of physical and chemical parameters of the habitat is strongly influenced by the ratio of surface area to mass and the somewhat different physiology of eggs and larvae as compared to adults. The effects of key physical and chemical habitat characteristics on egg and larval stages of fishes are summarized in Table 7.1.

Table 7.1 Summary of key physical and chemical habitat variables affecting fish eggs and larvae.

Habitat variable	Effect on eggs	Effect on larvae
Temperature	Affects rate of development and thus incubation period and size at hatching. Eggs are more sensitive to temperature change than adults.	Affects metabolism and thus rates of feeding and growth and stage duration. Temperature also has indirect effects through impact on viscosity and oxygen saturation levels of water. Larvae are more sensitive to temperature change than adults.
Light	Little impact except through possible ultraviolet damage.	Governs daily duration of feeding and depth at which food can be captured. Available light strongly affected by turbidity and color.
Mechanical damage	Potentially serious source of mortality on eggs in shallow water, particularly for broadcast spawners.	Larvae in shallow water are subject to severe mechanical damage, if current or wave action present.
Shelter	Important as protection for eggs from predation and mechanical damage.	Important as protection for larvae from predation.
Current	Potential mortality source for unattached eggs. Provides a renewed oxygen supply to eggs. Can transport pelagic eggs toward nursery sites.	Can transport larvae to nursery sites. Strong currents generally avoided by larvae.
Oxygen	Respiration across chorion is relatively inefficient. More sensitive to low oxygen levels than adults.	The switch from cutaneous to gill respiration and the development of a functional blood oxygen transport system creates a very vulnerable situation during this time. Larvae are very sensitive to low oxygen levels.
Salinity	Affects buoyancy of eggs in marine environment.	Affects buoyancy of larvae in marine environment.
pH	Low pH has a major impact at hatching due to inactivation of hatching enzymes.	Low pH (below 5.5) can reduce larval survival.
Pollutants	Can impact eggs directly or through maternal accumulation. Embryos prior to gastrulation are particularly sensitive.	Yolk-sac larvae are most sensitive to pollutants.

Aside from suitable physical and chemical conditions in the spawning and nursery habitats, two additional requirements must be met: adequate prey densities and relatively low predator densities. The food supply must be great enough to allow for rapid growth so that the time spent in vulnerable stages is minimized and the suite of potential predators is reduced through increased larval size. In order for this to work, there needs to be a good fit between larvae and their prey, both spatially and temporally, as well as between prey size and larval gape dimensions. Since eggs and larvae have few defenses against predators, rapid growth is one of the best ways of minimizing the impact of certain sets of predators. Increased size, the development of sensory and locomotor skills, and the use of shelter will all help reduce the predation rate.

Once the physical, chemical, and biological requirements of spawning and nursery habitat are understood fishery scientists are faced with an even greater challenge. They must identify the sequence of habitats through which the population of interest needs to pass to produce a strong and successful year class. Optimally, each habitat should be able to satisfy the particular needs of the various ontogenetic stages in an appropriate sequence. This raises a number of questions. Are the larval and juvenile stages going to remain in the spawning area or do they drift into the nursery grounds from distant spawning areas? If so what is the pathway that the young will use to get from the spawning area to the nursery grounds? What is the optimal timing for arrival at each habitat in relation to production of prey and development of shelter? How is this movement accomplished and what are the factors that might disrupt the progress of the larvae? Are the disruptions predictable? Are they weather-related? If multiple sites are required, are they aligned in the proper sequence and is the timing such that they are in suitable condition when the larvae arrive? Once the pathway has been outlined, then the question of habitat quality can be addressed. Relative to other populations, is the habitat of excellent, moderate, or poor quality? Is there anything that can be done to improve the poorest quality habitat in the sequence so that it will be more suitable to the requirements of the larvae? As long as the poorest habitat is a critical part of the sequence it makes sense to focus on it first, since the probability of getting the greatest result with the least cost comes with working on the weakest link in the chain. Gathering all of this information together is admittedly a very large undertaking, but once it is completed, bottlenecks in the sequence of habitats occupied by the early life stages of a population of interest can be identified and appropriate management activity undertaken.

7.7 Summary

The unique contributions of early life stages with respect to habitat revolves around the fact that egg and larval stages are small and fragile and that their capabilities and requirements are changing rapidly. Because of this, the early life stages require a sequence of optimal habitats to service their changing needs. Fishery scientists must emphasize the importance of having the optimal habitat available at the right time and in the correct sequence in order to help insure a successful year class. Due to the large numbers of eggs and larvae produced by spawning populations, small changes in survival arising from careful study of habitat sequencing can yield substantial results.

Additional reading

Balon, E.K. (1975) Reproductive guilds of fishes: a proposal and definition. *Journal of the Fisheries Research Board of Canada* **32**, 821–864.

Blaxter, J.H.S. (1986) Development of sense organs and behavior of teleost larvae with special reference to feeding and predator avoidance. *Transactions of the American Fisheries Society* **115**, 98–114.

Blaxter, J.H.S. (1988) Pattern and variety in development. In: *Fish Physiology*. (Ed. by W.S. Hoar & D.J. Randall), pp. 1–58. Academic Press, New York.

Cowen, R.K., Lwiza, K.M.M., Sponaugle, S., Paris, C.B. & Olson, D.B. (2000) Connectivity of marine populations: open or closed? *Science* **287**, 857–859.

Crowder, L.B. & Werner, F.E. (1999) Fisheries oceanography of the estuarine-dependent fishes of the South Atlantic Bight. *Fisheries Oceanography* **8**(suppl. 2), 252.

Hare, J.A. & Cowen, R.K. (1996) Transport mechanisms of larval and pelagic juvenile bluefish (*Pomatomus saltatrix*) from South Atlantic Bight spawning grounds to Middle Atlantic Bight nursery habitats. *Limnology and Oceanography* **41**, 1264–1280.

Higgs, D.M. & Fuiman, L.A. (1998) Associations between behavioral ontogeny and habitat change in clupeoid larvae. *Journal of the Marine Biological Association of the United Kingdom* **78**, 1281–1294.

Hoar, W.S. & Randall, D.J. (1988) *Fish Physiology. Volume XI, The Physiology of Developing Fish. Part A, Eggs and Larvae*. Academic Press, Inc., San Diego, CA, 546 pp.

Jones, G.P., Milicich, M.J., Emslie, M.J. & Lunow, C. (1999) Self-recruitment in a coral reef fish population. *Nature* **402**, 802–804.

Leis, J.M. (1991) The pelagic stage of reef fishes: the larval biology of coral reef fishes. In: *The Ecology of Fishes on Coral Reefs*, pp. 183–230. Sale, PF 1991, Academic Press, New York.

Peters, R.H. (1983) *The Ecological Implications of Body Size*. Cambridge University Press, Cambridge, UK, 329 pp.

Poling, K.R. & Fuiman, L.A. (1998) Sensory development and its relation to habitat change in three species of sciaenids. *Brain, Behavior and Evolution* **52**, 270–284.

Post, J.R. & McQueen, D.J. (1988) Ontogenetic changes in the distribution of larval and juvenile yellow perch (*Perca flavescens*): a response to prey or predators? *Canadian Journal of Fisheries and Aquatic Sciences* **45**, 1820–1826.

Rothschild, B.J. (1986) *Dynamics of Marine Fish Populations*. Harvard University Press, Cambridge, MA, 277 pp.

Sinclair, M. (1988) *Marine Populations. An Essay on Population Regulation and Speciation*. University of Washington Press, Seattle, 252 pp.

Swearer, S.E., Caselle, J.E., Lea, D.W. & Warner, R.R. (1999) Larval retention and recruitment in an island population of a coral reef fish. *Nature* **402**, 799–802.

Werner, E.E. & Gilliam, J.F. (1984) The ontogenetic niche and species interactions in size-structured populations. *Annual Review of Ecology and Systematics* **15**, 393–425.

Werner, E.E. & Hall, D.J. (1988) Ontogenetic habitat shifts in bluegill: the foraging rate–predation risk trade-off. *Ecology* **69**, 1352–1366.

Werner, R.G. (1969) Ecology of limnetic bluegill fry in Crane Lake, Indiana. *American Midland Naturalist* **81**, 164–181.

Werner, R.G. & Lannoo, M.J. (1994) Development of the olfactory system of the white sucker (*Catostomus commersoni*), in relation to imprinting and homing: a comparison to the salmonid model. *Environmental Biology of Fishes* **40**, 125–140.

Chapter 8
Assemblages, Communities, and Species Interactions

Thomas J. Miller

8.1 Introduction

Much of the history of ecological thought has focused on understanding how communities or assemblages of plants and animals are structured and maintained. Researchers have asked how similar can two species be and yet still coexist? What is the role of predation in maintaining diversity in communities? Are diverse ecosystems necessarily more stable? How do ecosystems respond to disturbance? A substantial amount of work has attempted to determine whether there are indeed groups of species that are interdependent to the extent that they form communities, or whether coexisting species interact more loosely, thereby forming assemblages. During the last century, answers to these questions led to substantial increases in our theoretical and empirical knowledge of the processes that regulate the population dynamics of plants and animals and structure the assemblages in which they are found.

Research into larval fish or ichthyoplankton assemblages has a very different heritage. Perhaps differences should be expected as a result of the obvious differences between terrestrial and aquatic ecosystems. We experience terrestrial ecosystems directly. Distinctions between forest and hedgerow are clear. Thus, it was easier to recognize that different assemblages characterize different habitats. In contrast, we experience the aquatic realm remotely. Differences between habitats are less clear to us. It is only in the last 50 years or so that we have routinely entered aquatic ecosystems as divers and submariners. Prior to these inventions, all our knowledge of the aquatic realm was gleaned remotely through nets, grabs, and cores, which often obscure differences among habitats. As a result of this obstacle, the early work on ichthyoplankton assemblages in the late 19th and early 20th centuries concentrated on associating larval forms with the adult species type and delimiting when and where larvae occurred. Much of this work, fundamentally natural history, was published as wonderful texts full of drawings and descriptions. Sir Alistair Hardy's *The Open Sea*, published in 1965, was perhaps the last great example of the genre.

Another motivation for ichthyoplankton research can also be found in the early work – the need to understand recruitment, the central problem of fishery science. The landmark research by Johan Hjort, to which so many of the chapters in this book refer and which is thoroughly described in Chapter 4, identified the importance of early life stages of fishes in determining recruitment. Subsequently, fishery researchers have concentrated on the processes that contribute to survival of larval fishes (see Chapters 3 and 4). Research on ichthyoplankton assemblages has sought to understand why they occur in the places and at the times they do, and how these spatial and temporal distributions relate to patterns in recruitment. This has led to questions regarding the spatial and temporal overlap of ichthyoplankton with their prey, interactions with their potential predators, and life-history strategies that close life cycles. More recently, recruitment research has expanded to include variability at the individual level, focusing more intensely on processes that influence the characteristics of individual survivors (see Sections 4.7 and 11.11). This has led to renewed interest in the role of foraging ecology, predator–prey interactions, and the physics of dispersal.

8.2   What, why, and how

8.2.1   What is a larval fish assemblage?

A larval fish (or ichthyoplankton) assemblage is simply a suite of species whose eggs and larvae are collected in the same area and at the same time. There are no strict limits on the spatial or temporal scale of an assemblage. It is as correct to speak of the assemblage in a single season and region – for example, the spring ichthyoplankton assemblage in the North Sea – as it is to speak of the assemblage at specific times and locations – for example, in comparing ichthyoplankton assemblages among different estuaries in South Africa. A similar flexibility in the use of the term is found in all branches of ecology. There is, however, a fundamental difference in its use with respect to fish larvae compared to its classical ecological use. An ichthyoplankton assemblage is by definition transient. It is restricted to the egg and larval periods. An ichthyoplankton assemblage is a static snapshot of a collection of species whose eggs and larvae occur together, but whose juvenile and adult phases may not interact. Thus, ichthyoplankton assemblages cannot be thought of in terms of evolutionary communities in which tight trophic interrelationships among species have evolved over time. The fact that egg and larval stages of the species that form the assemblage occur together, however, can be taken to imply that these species have come to the same solution of closing their life cycles despite an aquatic environment that is often dispersive.

8.2.2   Why study assemblages?

If ichthyoplankton assemblages are simply a suite of species whose eggs and larvae co-occur in space and time, of what help are they in understanding recruitment variability and in improving fishery management? Studies of ichthyoplankton assemblages lead to insights into what regulates recruitment. If the strength of annual year classes and overall population dynamics are regulated during early life stages, as they appear to be, research into the sources of, and patterns in, early survival will improve understanding and predictability of recruitment. Moreover, the dynamics of species whose egg and larval stages occur sympatrically will be affected presumably by the same mechanisms. Hence, understanding why specific locations and times promote egg and larval survival may uncover the processes important to determining year-class strengths. Research that addresses why species "choose" to spawn at the times and in the locations they do, resulting in observed larval fish assemblages, may lead to insights into processes regulating recruitment. Understanding why a particular year or location did not produce strong year classes in individual populations may be equally valuable. If in any year particular assemblages lose coherence or are physically dispersed, overall survival and ultimately recruitment may be reduced. In turn, this suggests that understanding the forces that maintain assemblages may be insightful. If studies of ichthyoplankton assemblages do lead to improved understanding and prediction of recruitment, fishery management will benefit. The earlier year-class strength can be reliably predicted, the greater the opportunity managers and fishers have to respond to changes and trends in abundance.

Although improving recruitment predictability is perhaps the prime factor that has motivated research into ichthyoplankton assemblages, there are additional justifications. Co-occurrence of individual species in an ichthyoplankton assemblage suggests that they share common requirements during their early life. An alteration in the ecosystem that negatively affects one species will likely affect all the species in the assemblage negatively. One obvious example of such an alteration is global climate change. The broad scale changes in productivity or hydrography that global climate change might bring about will impact the entire assemblage and not just individual species. The impact on the assemblage may not be to reduce the abundance of all or any of the species of which it is composed, but rather, the impact may be to shift the geographic location or seasonal timing of the appearance of the assemblage. Additionally, there is an ongoing shift in the focus of fishery management from a single species approach to multispecies and ecosystem approaches. Obviously, these approaches require an understanding of the interrelationships among species and therefore of ichthyoplankton assemblages. Identification of which species interact during early life and the nature of these interactions may be important to understanding the response of the system to changes in patterns of commercial or recreational harvest.

8.2.3   *How to identify assemblages*

In order to recognize an assemblage one must compare the distribution of eggs and larvae between two, and often more, locations or times. The simplest attempts to identify assemblages involve distinguishing between assemblages based on the presence and absence of individual species. This is only a crude tool, however, since by their very nature ichthyoplankton assemblages are dynamic. It is important to include abundance in addition to simple presence or absence. The earliest attempts to use abundance information involved simply plotting abundance vs. time in simple bivariate plots. Assemblages are identified by overlap between the abundance curves of different species that occur in the same place at the same time and delineated from other assemblages by a lack of overlap in space and time. This approach is attractive not only because of its simplicity, but also because it is easy to include other factors in such analyses. For example, one can examine the influence of a suite of potential environmental or biotic factors that are believed to influence the assemblage by plotting them in the same bivariate space. These techniques, however, are limited in terms of the number of sites and/or times that can be examined and in the subtleties between assemblages they can identify. More sophisticated approaches are available.

There are two general classes of statistical analysis that have been used to examine patterns in the structure, duration, and extent of ichthyoplankton assemblages: cluster analysis

Box 8.1   Patterns of ichthyoplankton assemblages in the English Channel.

Alain Grioche and colleagues (1999) studied the ichthyoplankton assemblage during the spring of 1995 along the French coast of the English Channel. The ichthyoplankton assemblage was sampled at 45 locations in April. All samples were collected with a bongo net with 500-μm mesh towed obliquely from surface to the bottom. At each station a CTD profiler provided physical data. At least 21 different species were collected. A clupeid, *Sprattus sprattus* dominated collections during both months, accounting for approximately 50% of all larvae. Data from the ichthyoplankton assemblage were analysed in two ways. Relationships between the different stations and species were determined using cluster analysis. The analysis identified at least three different clusters of stations that represented nearshore and offshore assemblages, and a separation along the coast. Offshore and nearshore stations were most different, as indicated by the first level of branching in Figure 8.1a. Within each of these broad groups, there was a general south–northeast separation of stations. Cluster analysis also revealed that different geographic locations were often characterized by different species. Species in group A tended to dominate offshore stations, whereas species in group C dominated nearshore stations.

Subsequently, Grioche and colleagues used an ordination technique called multidimensional scaling to identify the biophysical features of the water masses in which the assemblages occurred. Multidimensional scaling separated stations with respect to fluorescence and depth along axis 1, and light transmission and fluorescence on axis 2 (Figure 8.1b). Nearshore stations had negative scores on both of the multidimensional scaling axes, indicating their association with estuarine water masses. In contrast, group 1 stations were more offshore in character, with high bottom salinity and low surface turbidity.

and ordination (see Box 8.1 for an example of each). Cluster analyses include a variety of statistical techniques that quantify the similarity among different samples. Statistical approaches that compare the level and variance of a single measure are termed univariate. Approaches that compare the levels and variances of multiple measures simultaneously are termed multivariate. Because the uses of cluster analysis to identify ichthyoplankton assemblages compare the pattern of abundances of many species at once, the approach is inherently multivariate in nature. Cluster analysis employs a variety of mathematical techniques to compare individual samples in pairwise fashion. The pairwise scores are then compared analytically to produce the most parsimonious arrangement of samples so that the most similar samples are closest together in a hierarchical tree (Figure 8.1a). Assemblages are identified because they share a common branching point on the tree. Clustering techniques have their roots in ecological studies of plants during the middle of the last century. The techniques have diversified greatly as they have been used in developing evolutionary relationships in recent years. These modern likelihood techniques have yet to be widely applied to ichthyoplankton studies.

The second group of statistical approaches to identifying ichthyoplankton assemblages is termed ordination. Ordination approaches share a common philosophy, that of reducing the number of variables in the dataset to a minimal set that still captures the overwhelming majority of the variability in the data. These techniques reduce the "dimensionality" of the dataset. The different algorithms reduce the dimensionality differently but all result in a reduced set of abstract variables or "components" that explain the data most efficiently. Often the reduction of the data permits the investigator to plot the species on the component axes to present the patterns in the data graphically (Figure 8.1b). Assemblages are

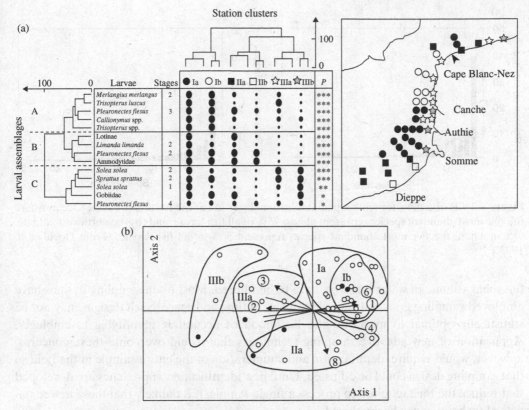

Figure 8.1 Examples of statistical approaches that are used to identify larval fish assemblages. (a) Cluster analysis of ichthyoplankton assemblages in the English Channel to identify groups of stations and taxa that form statistically identifiable groups. (b) An ordination of the same data presented in (a) to explore the physical factors responsible for the separation. See Box 8.1 for more detail. (Reproduced from Grioche *et al.* 1999 with permission of Academic Press.)

recognized formally by having similar scores on the component axes. Although the techniques do share a common philosophy, there are significant differences among them. The nature of the data available and the goals of the analysis can be helpful in identifying which of this class of techniques is most appropriate. Examples of ordination techniques that have been employed in ichthyoplankton studies include principal components analysis, canonical correlation, detrended correspondence analysis, and multidimensional scaling analysis.

There are common concerns over sampling designs used to identify and separate ichthyoplankton assemblages regardless of the analytical approach adopted. Assemblages are recognized by comparisons. Thus, most fundamentally, the sampling effort must be sufficient to ensure that an adequate number of samples is collected both from within and without the different assemblages. Clearly, without *a priori* knowledge of the distribution and extent of assemblages, this criterion can be difficult to meet. The situation is complicated further by the fact that not all species in the assemblage are equally abundant. Indeed many of the species that are vital for recognizing an assemblage are actually remarkably rare (Figure 8.2). This adds to the sampling demands, both statistically and practically, due to

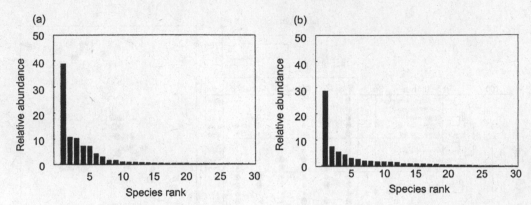

Figure 8.2   Rarity plots for ichthyoplankton assemblages in (a) the northeast Pacific Ocean where the five most abundant species represent almost 75% of all fish larvae, and (b) the northwest Atlantic Ocean where the five most abundant species represent 50% of all fish larvae. (From Doyle *et al.* 1993.)

the shear volume of water that may have to be filtered. Most often sampling designs have employed sampling grids distributed regularly in time and space. Such designs may not be statistically optimal to maximize the likelihood of accurately identifying assemblages. Application of new adaptive sampling techniques that would overcome these concerns, however, would require identification and enumeration of the entire sample in the field so that sampling design could be adjusted. Until new identification approaches are developed that reduce the time required to process a single sample, it is unlikely that these newer statistical techniques can be employed.

The statistical approaches outlined above can certainly identify ichthyoplankton assemblages, even when differences are subtle. These approaches by themselves, however, do not tell us why the patterns exist. Efforts to describe and understand ichthyoplankton assemblages must address the processes that affect the formation, maintenance, and disruption of the assemblage. I will extend the ideas of Boehlert & Mundy (1993) on this subject to include explicit considerations of scale (Table 8.1). This change reflects the idea that assemblage dynamics respond to evolutionary, biotic, and environmental forcing at different scales. For example, two categories of factors may influence the formation of assemblages but they operate at different scales. The primary factor is the "decision" by the adults of when and where to spawn. Larvae must hatch out into environments that will provide sufficient food for their growth and survival. Evolutionarily, this "decision" is likely driven by patterns in the productivity of the environment over long time scales and often broad spatial scales (see Section 8.2.2). The precise location of spawning, however, is likely driven by hydrographic processes that permit closure of the life cycle and that operate at mesoscales (see Section 8.2.1). Equally, ichthyoplankton assemblages are maintained by processes that operate at different spatial scales. Hydrography, which operates at mesoscales, ensures the continuity of the assemblage. But, maintenance of the assemblage also relies on the survival of the individual larva, which is governed by much smaller scale processes. For example, survival implies adequate growth and relies on the small spatial scales that govern encounters between larval fishes and their planktonic prey. Acting against the forces maintaining the assemblage are three principal scale-dependent factors

Table 8.1 Processes and scales of time (t) and space (s) important to the formation, maintenance, and disruption of ichthyoplankton assemblages.

Type of process		Macroscale ($s > 10^4$ m, $t > 1$ month)	Mesoscale (10^4 m $> s > 10^2$ m, month $> t >$ days)	Microscale ($s < 10^2$ m, $t <$ days)
Formation	Physical	Currents and circulation patterns (separation of water masses)	Convergent fronts (concentrating available larvae)	
	Biological	Spawning behavior (seasonal) Retention areas+	Spawning time (daily or lunar rhythms)	Lasker events (prey patch formation in stable environments)
Maintenance	Physical	Currents and circulation patterns (maintenance of water mass distinctions) Upwelling (maintenance of water mass distinctions)	Convergent fronts (increased local production) Gyres and eddies (limited advection and diffusion) Upwelling (uniform advection patterns)	Small scale turbulence (increases in feeding rates)
	Biological	1° and 2° production (provision of suitable prey field)	Larval growth and survival (condition-dependent distributions) Primary and secondary production (distribution of prey field) Swimming behavior (light, temperature, salinity preferences)	Foraging behavior (area restricted searching)
Disruption	Physical	Current meanders (breaking down water mass distinctions)	Divergent fronts, diffusion, and advection (disruption of patches)	Small scale turbulence (disruption of patches)
	Biological	Growth and development to metamorphosis (change to adult habitats) Seasonal declines in production (declines in prey availability)	Predation (mortality) Competition (reduction in prey availability)	

leading to the dissolution of the assemblage that are scale dependent. Biological and physical processes may disrupt assemblages. To avoid being eaten by their predators, larvae must follow a suite of processes that occur at slightly larger scales than larval foraging, but still operate at relatively small spatial and short temporal scales. Assemblages may be disrupted by hydrographic processes at mesoscales that disperse the water mass in which the assemblage exists. Larvae that grow and survive will ultimately metamorphose into juveniles and adults. These processes occur on time scales of months over the entire spatial domain of the assemblage. Ultimately, the assemblage is dispersed by its own success.

Now, I will review patterns in larval fish assemblages in time and space and the ecological processes that lead to assemblage formation, maintenance, and disruption. Clearly, the closed nature of many freshwater systems constrains the range of spatial and temporal variability in assemblages that may be observed. Thus, in the sections of the chapter dealing with spatial patterns, the focus will be on marine systems. The closed nature of many freshwater systems, however, does permit ecological processes to be studied more easily. Accordingly, the sections of the chapter dealing with ecological processes within ichthyoplankton assemblages will be biased toward freshwater systems. The differences in coverage reflect the very real constraints on the systems. It may be, however, that the differences in coverage also reflect subtle differences in the research heritages of scientists working in marine and freshwater environments.

8.3   Formation of larval fish assemblages

Both physical and biological processes regulate the formation of larval fish assemblages (Table 8.1). In general these processes operate at long temporal and large spatial scales.

8.3.1   *Physical processes affecting assemblage formation*

Physical features that affect assemblage formation occur over large spatial scales (Table 8.1). The multivariate statistical techniques used to identify ichthyoplankton assemblages commonly identify distinct larval fish assemblages that exhibit little spatial overlap at the scales of tens of kilometers. This separation can be most easily explained by differences in the physical characteristics of the environment.

Water masses of different densities will only mix when energy is added. In the absence of an input of energy, such as winds or tides, the different water masses will not be mixed. Thus, within any large body of water, different water masses can become isolated and distinct. In freshwater, density differences are driven by temperature, creating a warm, well lit zone above a thermocline (metalimnion) and a cold, dark zone below (hypolimnion). More rarely in large bodies of fresh water, thermal bars may separate inshore from offshore water masses. In the marine and estuarine realms, density differences result from both temperature and salinity. Often in marine systems, the separation of the water masses is reinforced by density driven currents. In general, different ichthyoplankton assemblages are often associated with different water masses. For example, Richards and colleagues (1993) identified both oceanic and shelf assemblages within the Gulf of Mexico. They linked the existence and distinctness of these two assemblages to the water masses associated with the

Florida Loop current. Working at slightly smaller scales, Thorrold & Williams (1996) identified three distinct assemblages of larval fishes (using clustering techniques) within the central Great Barrier Reef lagoon off Australia: a nearshore assemblage, a cross-lagoon group, and an outer lagoon assemblage. They postulated that the three assemblages could have been formed by a physical regime operating at a scale of at least 50 km, or perhaps by synchronized spawning behavior. The three larval fish assemblages in the English Channel (see Box 8.1) differed over approximately the same spatial scale as those in the Great Barrier Reef lagoon. Differences in water masses were the proposed mechanism responsible for the formation of these assemblages, although they too could not rule out the pattern of spawning as a causal agent.

Despite the occurrence of multiple assemblages at scales of tens of kilometers in the above examples, separate assemblages are not always present. In the western Gulf of Alaska, where the environment is characterized by a single water mass over a large area, only a single ichthyoplankton assemblage is found (Doyle *et al.* 1994). This lends further credence to the importance of differences among water masses in accounting for assemblage structure. While these examples have built a case for an association between distinct larval fish assemblages and different water masses, they leave unanswered the question of why this pattern occurs.

Physical processes may concentrate eggs and larvae, which promotes assemblage formation, even when adult spawning behavior does not place the larvae in the "right" place. In aquatic systems, interfaces between different water masses, known as fronts, have profound effects on both the distribution and survival of plankton. Convergent fronts, in which water from the two different water masses is brought together, are a concentrating mechanism, perhaps leading to assemblage formation. Govoni (1993) provided a detailed account of the role of convergent fronts to the ecology of ichthyoplankton assemblages. He studied assemblages associated with the Mississippi River plume front in the Gulf of Mexico and the Gulf Stream Front in the Atlantic Ocean off the United States. For both fronts, the assemblages contained species that spawned on the coastal shelf, beyond the front. Subsequently, larvae of the majority of these species must cross the fronts to reach nursery areas in estuaries and coastal bays. Both fronts clearly were convergent in nature with respect to concentrating larval fishes. But equally, mixing at the frontal boundaries was important in helping larvae of the estuarine-dependent species move shoreward. Thus, the role of fronts may not simply be one of concentrating larvae, they may also aid favorable transport. Although discussion of the importance of fronts has been restricted to marine and estuarine environments, they may be of equal importance in lakes. Recent research into the physical limnology of large lakes has identified shoreline fronts that are sites of increased productivity.

8.3.2   Biological processes affecting assemblage formation

Even if physical considerations are important in determining formation of assemblages, the ultimate cause of ichthyoplankton assemblage formation must be the spawning behavior of adults (Table 8.1). Fishes have evolved a wide range of life-history strategies in response to the diversity of environments they occupy (see Chapter 1). In many cases, early life stages are separated spatially from later stages. Thus, central to the evolution of any life-history

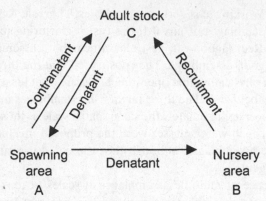

Figure 8.3   The Migration Triangle Hypothesis. In Harden Jones' (1968) original formulation, the three components of the population were spatially distinct. Completion of the life cycle required either active migration (adults) or hydrographically assisted movements (eggs and larvae).

strategy is the requirement that the life cycle be closed. Any offspring must be produced at a time and a place, from which they can grow, develop, and mature to join the spawning population. The range of times and locations of spawning that meets this central requirement may be limited. Species that are similarly restricted become members of the same ichthyoplankton assemblage. There has been a great deal of work into the relationship between the characteristics of spawning times and locations and assemblage formation. Two conceptual advances need to be recognized, however. Harden Jones (1968) formalized the widely recognized separation between spawning areas, nursery grounds, and adult grounds in many species in terms of a Migration Triangle Hypothesis (Figure 8.3). Spawning areas are those locations that deliver larvae to appropriate nursery grounds. Similarly, nursery grounds are those locations that permit survival and growth of juveniles so that they can enter the adult reproductive population at the appropriate place. Although fundamentally a spatial concept, Harden Jones' Migration Triangle Hypothesis also speaks to temporal variability. Successful spawning grounds must deliver larvae to the nursery grounds at the appropriate time. The Migration Triangle Hypothesis concept challenged and motivated the work of others.

Sinclair (1988) hypothesized that populations of species are maintained if and only if eggs and larvae are placed in locations in which they are retained and not dispersed. He proposed that the number of such retentive areas was related to a species' population richness, and the area of the retentive zone was related to that population's abundance (Figure 8.4). Sinclair's hypothesis does not presuppose spatial separation of the different life-history stages. It merely requires that the larval forms develop, grow, and survive to join the adult population. It may be that the entire life history is completed over a very limited spatial extent, as in several labroid fishes around Bermuda, or it may involve long distance dispersal of larval and juvenile forms, as in wreckfish (*Polyprion americanus*), which are carried around the North Atlantic basin during development. Several hydrographic features have been proposed as factors that may increase the chance of retention and survival of ichthyoplankton in favorable environments. Examples include gyres (from ocean basin scales to coastal banks), upwelling areas, convergence zones, and flow around isolated

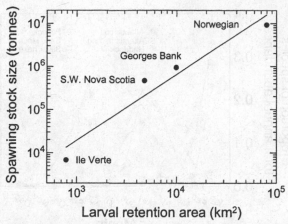

Figure 8.4 Relationship between the area of the larval retention area and estimated population size for several populations of Atlantic herring (*Clupea harengus*). (Reprinted with permission from Iles, T.D. & Sinclair, M. Atlantic herring: Stock discreteness and abundance. *Science* **215**, 627–633. Copyright 1982 American Association for the Advancement of Science.)

bottom topography. Given the essential requirement to close a life cycle, it is only to be expected that spawning locations have evolved to favor these hydrographic structures.

Fishes must spawn in the right place to ensure that their progeny hatch into a favorable environment. But equally, fishes must also spawn at the right time to maximize survival. Considerable effort has been expended in defining both the spawning and larval periods for a wide variety of fishes. In some cases, it is possible to find larvae of a particular species in nearly all months of the year, especially for species from lower latitudes. For example, a compendium of information on the seasonal distribution of larvae of approximately 200 genera in 61 families from the Gulf of Mexico north of latitude 26°N indicates that larvae of 31% of the species are present at least 9 months of the year. In addition, larvae of 12.5% of the species are present in all 12 months. In contrast, larvae of only two species are found in at least 9 months in the Middle Atlantic Bight (the Atlantic coast of North America from Virginia to New Jersey, inclusive). No species produce larvae for more than 9 months in the Gulf of Maine. The increased seasonality of the larval period in higher latitudes seems to be a general one, caused by the increasing influence of seasons at higher latitudes. Thus, ichthyoplankton assemblages tend to be more distinct seasonally at higher latitudes than comparable assemblages in lower latitudes. Interestingly, the high latitude ichthyoplankton assemblages from the Antarctic appear to deviate from this general pattern. For example, 47% of species reported from Antarctic waters have larval occurrences of greater than 9 months.

By themselves, patterns in seasonal occurrence are not sufficient to define ichthyoplankton assemblages. Abundance must be considered as well as diversity. While larvae of an individual species may be present over a number of months, only a few species may be common and they may be common in only a few months. The distribution of species ranked by abundance is termed a rarity curve, a concept that has been applied in ecology since the 1970s. Theoretically, the shape of the rarity curve expresses how a single or multiple key resources are partitioned in the assemblage. Accordingly, assemblages may be characterized by the

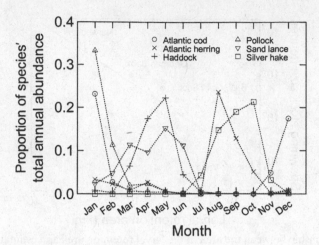

Month

Figure 8.5 The temporal distribution of the six most abundant species on the Scotian Shelf, based on samples collected from March 1991–May 1993 (T.J. Miller, unpublished data). Sampling was conducted using an $8 + 2\,m^2$ rectangular midwater trawl deployed at 45 stations during monthly surveys. Details of the sampling can be found in Miller *et al.* (1995). The abundance of larvae is expressed on a relative scale. Relative abundances are estimated as the proportion of the total annual abundance for each species that occurred in the specified month, which permits all species to be plotted on the same scale.

shape of their rarity curve. For example, on the Pacific coast of the United States, larvae from 33 species occur in the ichthyoplankton, yet the five most abundant species account for almost 75% of all larvae (Figure 8.2a). On the Atlantic coast, the ichthyoplankton assemblage is characterized by 50 species, and the five most abundant species account for 50% of the total assemblage (Figure 8.2b). Such rarity patterns are not unique to ichthyoplankton assemblages. If, the dominant species are sufficient to characterize the assemblage, these rarity patterns may ameliorate the sampling concerns that make adaptive statistical approaches difficult to apply (Section 8.2.3).

Despite these broad scale seasonal patterns, it is the specifics of individual species that really define an assemblage. For example, the ichthyoplankton assemblage on the Scotian Shelf, a series of banks that spread south from Newfoundland (Grand Banks) to Massachusetts (George's Bank), consists of three distinct groups that can be identified with respect to larval occurrence (Figure 8.5). The spring-spawning group, which includes haddock (*Melanogrammus aeglefinus*, Figure 1.3n), sand lance (*Ammodytes* sp.), and historically Atlantic cod (*Gadus morhua*, Figure 1.3p), is perhaps the best known. Larvae of the spring spawners feed on the abundance of plankton in the spring bloom and grow rapidly throughout the early summer. A second peak of abundance occurs in the summer. Summer-spawning species include herring (*Clupea harengus*, Figure 1.3m), yellow tail flounder (*Pleuronectes ferrugineus = Limanda ferruginea*), and silver hake (*Merluccius bilinearis*). This group may take advantage of production arising from the microbial loop of the food web. The final group is an autumn/winter spawning assemblage, which includes cod and pollock (*Pollachius virens*). This group uses winter production. As evident in Figure 8.5, however, shifts from one assemblage to the next are not instantaneous, nor complete. Yet despite this variability, the different groups are consistent from year to year.

The restricted seasonal patterns of abundance, such as those found on the Scotian Shelf are by no means atypical. Similar patterns of distinctly different ichthyoplankton communities during the course of the year have been reported often and are likely to be typical of ichthyoplankton assemblages generally. Indeed, it was the relationship between spring zooplankton production and the abundance of larval cod that was at the heart of David Cushing's Match/Mismatch Hypothesis, which suggests that the timing of larval production is tied directly to the timing of the secondary production that is necessary to support larval growth and survival (see Chapter 4). Cushing also suggested that in systems in which secondary production is pulsed, the timing of larval release would be pulsed to match production cycles, whereas in systems in which secondary production is continuous, larval production would be more continuous. Cushing suggested that cod populations in the North Sea were an example of pulsed systems and that sardine populations in the California Bight were characterized by more continuous production. Although support for the Match/Mismatch Hypothesis is equivocal, it remains the most influential concept relating the seasonality of larval production to assemblage formation.

The role of spawning behavior in assemblage formation is not restricted to seasonal time scales (Table 8.1). In many fish species, spawning behavior is synchronous and results in the formation of large patches of larvae. These patches can be followed, through the use of drogues, and sampled over time to yield information on advection, dispersal, and mortality. One such study of bluefin tuna (*Thunnus maccoyii*, similar to Figure 1.3s) larvae showed that a particular patch of larvae was formed by a single synchronous spawning event spanning only a few days and involving between 50 and 500 000 adult fish. Similarly, a patch of walleye pollock (*Theragra chalcogramma*, Figure 1.3o) larvae in Alaskan coastal waters was presumed to have resulted from a single synchronous spawning event. The synchronization of spawning that leads to assemblage formation may be promoted by two factors. Spawning synchrony in coral reef fishes, for example, may be related to the onset of favorable environmental conditions or by the need to swamp locally abundant predators. It has also been hypothesized that synchronous spawning increases fertilization success, although support for this hypothesis is meager.

In summary, ichthyoplankton assemblages form largely as a result of patterns in spawning behavior of adults, which have presumably evolved to match the timing of larval release with adequate abundances of the larva's principal food resources. Moreover, the location of release likely reflects locations of favorable hydrography from which the survival of larvae is sufficient to ensure adequate numbers survive to replace the spawning stock.

8.4   Maintenance of larval fish assemblages

As with their formation, assemblages are maintained by both physical and biological agents (Table 8.1). Whereas processes affecting formation typically occur at the macroscale (kilometers and months), processes affecting maintenance of the assemblage are more likely to operate at the mesoscale (meters and days) or the microscale (centimeters and seconds). The principal physical forces responsible for the maintenance of larval fish assemblages relate to mesoscale circulation patterns and density differences that retain larvae within particular water masses and microscale physical processes that promote growth and survival. The principal biological processes that help maintain larval fish assemblages relate to factors that promote growth and survival. It would be wrong, however, to draw sharp distinctions between the two categories, as they interact strongly.

8.4.1 *Physical processes maintaining assemblages*

Once developed, assemblages will only be maintained if they overcome the forces causing their disruption. These forces are principally diffusion, advection, and dispersal. There are two general ways in which this is achieved (Table 8.1). The first, somewhat counterintuitively, is that assemblages are often maintained in highly advective zones. Provided that the advection is sufficiently consistent and prolonged, the assemblage will move *en mass* with the water. Thus, the strong directional advection simply swamps physical processes that would lead to diffusive losses and so the assemblage maintains its integrity. The many upwelling-related ichthyoplankton assemblages are examples of such systems. For example, advection in the upwelling regime of the Benguela Current is critical to maintaining the highly productive ichthyoplankton assemblage along the west coast of Africa. Not all advective systems are driven by upwelling, however. Extensive research on the walleye pollock population that spawns in the Shelikof Strait off Alaska demonstrates that the strong currents in the strait are responsible for aggregating and moving patches of eggs and larvae through the strait. In addition, mesoscale eddies and gyres associated with variations in current velocity are also important for maintaining these patches once they are formed. In river systems, strong advection is universal. Thus, riverine assemblages are largely governed by advection. For example, the current regime in different reaches of the Missouri River (United States) and its tributaries is a good predictor of assemblage membership. Some species are associated with areas in which currents are stronger (more advective) and others are found only in more benign conditions. Thus, strongly advective systems can be thought of as a conveyor that carries species along with it and prevents other species from jumping on board along the way.

The second class of physical processes that help maintain ichthyoplankton assemblages are those features that concentrate passive particles (Table 8.1). These processes range from gyres and eddies, at the broadest scale, which limit outward diffusion, to convergent fronts, at smaller scales, which actually concentrate larvae and overcome both diffusion and advection. Gyre-like structures may be important to maintaining ichthyoplankton assemblages on coastal banks, isolated islands and sea mounts, and on coastal shelf slopes. In a modeling study, Werner and colleagues (1993) clearly showed the importance of a cyclonic flow field in retaining larval cod and haddock on the Georges Bank system. Werner and his colleagues seeded the model's flow field with "virtual" larvae and explored the implications of differing spawning locations on the probability that larvae would be retained on the bank. They found that spawning locations on the northeast peak of the bank were strongly associated with retention of larvae. Their model results indicated that larvae in the surface waters would be advected off the bank due to wind-driven flow. In contrast, larvae below the surface waters were transported by residual currents around the bank and were ultimately retained (Figure 8.6). In a field study on the Scotian Shelf, higher concentrations of cod larvae and zooplankton occurred within a gyre than outside it, and highest concentrations were associated with a convergent front near the edge of the circulation pattern. The front and gyre were acting to retain larvae in the region, overcoming diffusion losses (Figure 8.7).

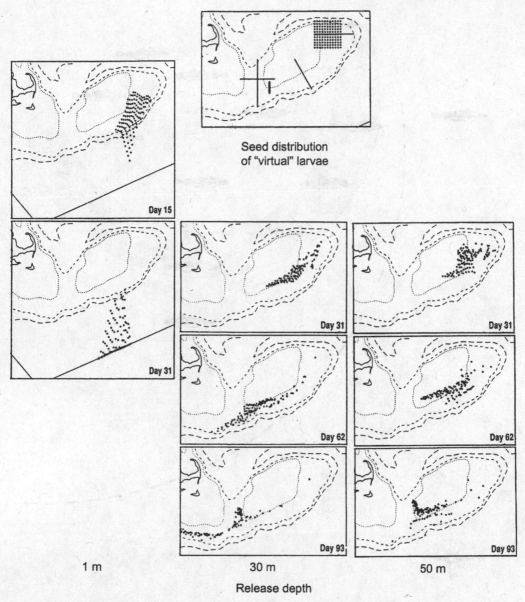

Seed distribution
of "virtual" larvae

1 m   30 m   50 m

Release depth

Figure 8.6 The fates of "virtual" larvae in a coupled biological–physical model of Georges Bank (from Werner *et al*. 1993). All "virtual" larvae were released in a grid of locations on the northeast corner of the grid. The model followed their positions over the subsequent 100 days. Results are shown for particles released at a depth of 1, 30, or 50 m. (Reproduced from Werner *et al*. 1993 with permission of Blackwell Scientific Publications.)

Frontal regions can also act to maintain assemblages. Convergent fronts may concentrate larvae, thereby overcoming diffusive and advective flows. Fronts have been shown to con-tribute to the maintenance of assemblages in Tokyo Bay, the Mediterranean Sea, and the North Sea. Fronts may act to maintain assemblages by promoting growth and survival.

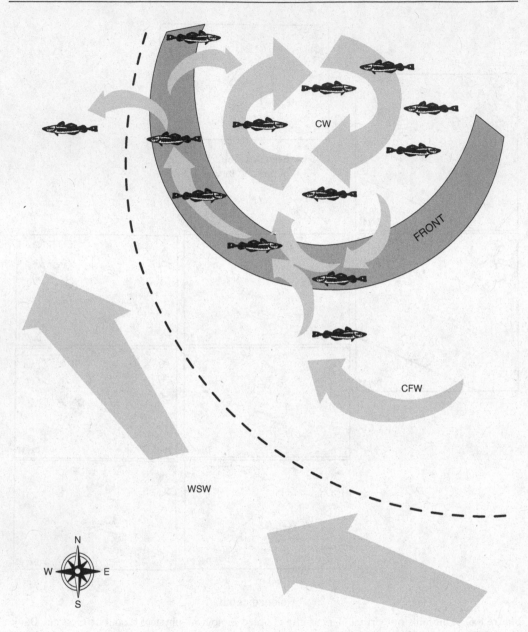

Figure 8.7 Conceptual diagram of the biological–physical interactions in a gyre-like flow around a shallow bank. Four different water masses are identified. Crest water (CW) rotates anticyclonically. Cold fresh water (CFW) occurs beyond the crest. The interface between these two water masses is a frontal feature at which larvae and zooplankton concentrate. Larvae mixed back into the CW are retained on the bank. Beyond these water masses is a third water mass consisting of warm salty water (WSW), which is moving northwestward. Larvae that become entrained in this water mass are likely lost from the system. (Reproduced from Lochmann *et al.* 1997 with permission of the National Research Council Canada.)

Riverine plumes, the frontal region between marine and estuarine water masses, are locally very important to the maintenance of ichthyoplankton assemblages because they promote retention, growth, and survival of fish larvae.

Two physical processes acting at microscales have been proposed as encouraging maintenance of assemblages through their presumed impact on growth and survival (Table 8.1). Following intensive study of the growth and survival of northern anchovy (*Engraulis mordax*, similar to Figure 1.3l), Lasker (1975, 1978) concluded that the establishment of a local concentration of their dinoflagellate prey, resulting from a "period of calm" (which later became known as a "Lasker" event), was crucial to the survival of newly hatched northern anchovy larvae. Conditions that disrupted these local concentrations of dinoflagellates produced lower survival and recruitment (see Section 4.5).

Small scale turbulence is the final physical force that may promote maintenance of assemblages. Small scale turbulence increases ingestion rates of larval fishes, suggesting that growth and survival may be higher in more turbulent regions. But to date no one has addressed the costs of growth in such environments, and so the overall impact of small scale turbulence on survival is uncertain. Moreover, the net effects of small scale turbulence are equivocal as it also disrupts local patches of prey, and this reduction in prey availability may offset the increases in potential feeding rates.

8.4.2   Biological processes maintaining assemblages

Clearly assemblages can be maintained only if the individual larvae that comprise the assemblage survive and grow. Thus, the principal biological processes supporting maintenance of assemblages are those that promote growth and survival. Support for several specific processes can be found in Chapters 2 and 3.

Larvae are often not randomly distributed in the environment and they are not passive particles, although they have only limited vagility. Recent research has indicated that at least some larvae may have greater swimming performance than previously believed, leading to speculation that larvae may have considerable ability to control their position in the water column, both horizontally and vertically. Therefore, ichthyoplankton assemblages may reflect behavioral responses of larvae to environmental signals. Fish larvae have been shown to respond to physical structures, currents, and chemical cues. For example, walleye pollock larvae respond to gradients of light and prey concentrations, preferring higher light environments and higher prey concentrations. Similar behaviors have been reported in other species, although not always relying on responses to visual stimuli. Red sea bream (*Pagrus major*), for example, responds to chemical scent trails in the water. Similarly, cod respond to the amino acid arginine, a common component of food organisms. These behavioral observations indicate that ichthyoplankton assemblages may be maintained, at least in part, by the expression of behavioral responses to common environmental signals.

Foraging behavior has often been proposed as a response that will tend to lead to aggregations. One of the first studies of swimming behavior associated with feeding demonstrated that larval northern anchovies alter their swimming behavior in response to prey density to maintain themselves in patches of prey. Walleye pollock larvae also respond to patches of prey, altering their behaviors to remain in the patch.

Feeding success may have subsequent consequences for the maintenance of larval fish assemblages. The position of larvae in the water is strongly influenced by their buoyancy. Eggs and larvae will remain in water in which they are neutrally buoyant. Feeding and growth change the biochemical composition of larvae, thereby affecting their buoyancy. Thus, there is the potential that feeding success will impact the vertical distribution of larvae in the water column. Larvae that have fed and are in good condition are denser than larvae that have not fed. In a model of the vertical distribution of cod larvae, Sclafani and colleagues (1993) demonstrated that buoyancy changes resulting from failure to feed caused cod larvae in poor condition to float to surface waters where they were subject to wind-driven currents and eventually lost from the system. In contrast, well fed cod larvae maintained a level of buoyancy that kept them in the middle region of the water column. This process implies that maintenance of larval fish assemblages may benefit from condition-dependent changes in buoyancy. Field evidence for such patterns is available from Georges Bank.

8.5   Disruption of larval fish assemblages

As with formation and maintenance of assemblages, both physical and biological processes, over a range of scales, can lead to the disruption of larval fish assemblages. Many of the processes that lead to disruption do so only because of the magnitude, timing, or location of their impact. For example, large scale patterns in currents and circulation help to form larval fish assemblages, but when these same currents meander off course, they can disrupt assemblages. Similarly, while processes that promote growth and survival are required to maintain assemblages, they ultimately lead to the disruption of the assemblage as survivors undergo metamorphosis and adopt the adult habitat and lifestyle, which are often different and separate from those of the larvae.

8.5.1   Physical processes disrupting assemblages

Physical processes that disrupt ichthyoplankton assemblages can be acute or chronic. Acute processes often involve entrainment of the assemblage in hydrographic features that move them away from favorable environments. In the northwest Atlantic, the formation of warm core rings, has been hypothesized to be so frequent as to affect recruitment patterns and ultimately ichthyoplankton assemblages. Studies by Kenneth F. Drinkwater and colleagues have provided evidence of a negative relationship between recruitment and the frequency of warm core rings for several commercially important species. Most recently, they presented clear evidence of the role of a warm core ring in entraining larval redfish (*Sebastes* spp.; Figure 1.1) on the Scotian Shelf. These investigators found that larvae within the ring were in poorer condition than those that remained on the coastal shelf and hypothesized that the entrained larvae would not survive to recruit to the population. These large scale incursions of currents into coastal areas that support high abundance of larvae are not restricted to the northwest Atlantic Ocean. Fish larvae, particularly cape anchovy (*Engraulis capensis*, similar to Figure 1.31), can be entrained into an Agulhas Ring in the coastal waters of Namibia, which leads to reduced recruitment.

Physical processes that lead to chronic dispersal of assemblages are more varied. Turbulent mixing generally disrupts patches in aquatic systems. In one of the first studies of its kind, Fortier & Leggett (1985) followed, and repeatedly sampled, a patch of capelin (*Mallotus villosus*) larvae in the St. Lawrence River. Their analysis indicated that diffusion of larvae out of the patch, by progressive mixing with deeper water, could account for approximately 6% of the observed reduction in larval abundance over 46 hours of monitoring. The majority of this "mortality" of larvae in the patch appeared to result from biological processes that coincided with yolk absorption. Clearly, by reducing abundances of larvae in patches, diffusion and advection can have a large effect on the accuracy of mortality estimates. In fact, up to half of the estimated mortality in patches of walleye pollock larvae may, in reality, be reductions in abundance brought about by diffusion. On the other hand, a study of a patch of bluefin tuna larvae (Figure 1.3s) over 6 days showed that larvae did not diffuse out of the patch as fast as would have been expected if they behaved as passive particles. Whether this was because of specific larval behaviors is uncertain. Overall, the mortality rate in the patch was still about 50% day^{-1}. Taken together, these studies suggest that physical mechanisms are important factors leading to the disruption of assemblages.

8.5.2 Biological processes disrupting assemblages

Three principal biological processes – metamorphosis, competition, and predation – may lead to disruption of larval fish assemblages. The first, survival of larvae to metamorphosis, leads to the disruption of larval fish assemblages by definition: after metamorphosis the fishes are no longer larvae and they typically settle out of the plankton. The rate at which this occurs will depend upon the productivity and seasonality of the environment. Environments that support faster growth will be associated with shorter times to metamorphosis and hence more transient ichthyoplankton assemblages. In contrast, environments of lower productivity will produce larvae that take longer to reach metamorphosis and hence larval fish assemblages will be of longer duration.

Competition among larvae may represent another mechanism leading to the dispersal of larval fish assemblages. Several hypotheses relating survival to metamorphosis rely on a relationship between food availability and growth. For example, the Match/Mismatch Hypothesis (Section 4.5.1) links the abundance of appropriately sized prey and early survival. This might suggest that competition for food among larvae within an assemblage may reduce food availability to levels that would reduce overall survivorship. But are the foraging rates of larvae in an assemblage sufficient to reduce the concentration of their prey? It may be that patterns of density-dependent dispersal of fish eggs and larvae are a life-history adaptation that ensures that larvae do not overgraze their food supply and compete. The limited foraging abilities of early stage larvae may prevent them from seriously affecting the density of their prey, but as they grow, these limitations disappear and larvae do gain the capability of influencing the concentration of food.

To date there is no clear consensus regarding the existence or importance of competition in larval fish assemblages. In freshwater systems, there appears to be strong evidence of interspecific competition in larval fish assemblages. For example, interspecific competition during early life explains the impact of the exotic white perch (*Morone americana*, similar to Figure 1.3q) on native yellow perch (*Perca flavescens*, Figure 1.3j) populations in Oneida

Lake, New York. Interestingly, the competitive effects occur late in early life, in agreement with the idea that early larvae are poorly equipped to alter prey density. The foraging activities of young-of-the-year threadfin shad (*Dorosoma petenense*) caused a marked decline of the zooplankton community which, in turn, led to lower growth and survival of young-of-the-year bluegill (*Lepomis macrochirus*). Again, this study supports the occurrence of competition, but only later in early life. Controlled mesocosm experiments have shown that gizzard shad (*Dorosoma cepedianum*) and bluegill larvae may compete for zooplankton prey. Moreover, competition may result in reduced growth and survival of larvae when the concentration of larvae is high. Unlike the previous examples, the competition can occur during early larval stages. Similarly, early stage larval bluegill in small impoundments can reduce the availability of zooplankton prey, thereby bringing about the potential for intracohort and interspecific competition. In other situations, however, larval bluegills were shown to have little impact on zooplankton populations.

Less attention has been paid to competition in marine systems, perhaps because of their open nature. Avoidance of competition has often been invoked to explain the lack of spatial and temporal overlap between different taxa in an assemblage. The lack of a tradition and the difficulty of conducting controlled experiments in the marine environment, however, has meant that the unequivocal demonstrations of competition in freshwater systems are lacking for the marine environment. Conclusions have been drawn largely from inferences based on calculated ingestion and production rates. Just such an approach suggests that the entire summertime larval fish assemblage in Conception Bay, Newfoundland, consumes less than 0.1% day^{-1} of the available prey, making competition extremely unlikely. In contrast, the predatory impact of the larval fish assemblage on Dogger Bank in the North Sea represents 3–4% day^{-1} of the standing stock of preferred prey. Given that the production and growth rate of the prey is 3–7% day^{-1}, the larval fish assemblage may indeed have the potential to overgraze their prey base, the first requirement for competition.

The results of studies on competition in ichthyoplankton assemblages do allow some conclusions to be drawn. The lack of evidence of strong competition implies that competitive exclusion is unlikely to occur in most ichthyoplankton assemblages, with the possible exception of certain assemblages in lakes. A reduction in abundance of the larvae of one species, for whatever reason, will not benefit its competitors. Similarly, it is unlikely that dramatic changes in patterns of adult abundances are the outcome of competition among larval stages. It is more probable that growth and survival of all members of a community will covary. In years when the production of food is above average, growth and survival of all members of the assemblage will likely be above average. In contrast, low production will produce poorer growth and survival throughout the assemblage.

The third biological process bringing about the disruption of larval fish assemblages is predation. The evidence for the importance of this process is unequivocal. Mason & Brandt (1996) provided one of the clearest examples of the importance of predation to larval fish assemblages in a study of embayments of Lake Ontario (see Box 8.2). They took advantage of a "natural experiment" in which yellow perch larvae occurred in two small embayments. A potential predator, alewife (*Alosa pseudoharengus*), was present in only one of the two embayments. Mortality rates of yellow perch larvae in the embayment containing the predator were always higher than in the predator-free pond. Larger fishes and invertebrates have been shown to be important predators of larval bloater (*Coregonus hoyi*, Figures 1.3b and 11.20), a

Box 8.2 Predation as a cause of disruption of larval fish assemblages.

Doran Mason and Steven Brandt studied two small embayments or ponds off Lake Ontario. The embayments are important spawning grounds for several species including yellow perch and alewife. The two ponds are connected by a narrow channel, and only one of them, North Pond, is connected to Lake Ontario. North Pond contains the same fish community as Lake Ontario, but the channel connecting the two ponds precludes passage of adult alewife into South Pond. Over these 3 years, natural climatic variability produced variation in the timing of migration of adult alewife into North Pond. Only in 1985 did the arrival of adult alewife in North Pond coincide with the peak hatching of yellow perch larvae. This variability allowed Mason and Brandt to compare years of different predation intensity. Larval yellow perch were sampled using a bongo net from April to June over 3 years and adult alewives were sampled using a trawl. From the collections of larvae, Mason and Brandt calculated the instantaneous mortality rate (Z) of the larval perch cohort in both ponds over the 3 years. The estimates of Z were:

Year	North Pond		South Pond
	North Basin	South Basin	
1984	0.026	0.196	0.221
1985	1.580	0.601	0.225
1986	–	0.094	0.283

Mason and Brandt assumed that the estimates from the South Pond reflected background, non-alewife mortality rates. In 1985, when the timing of yellow perch hatching coincided with the arrival of alewife, mortality rates in North Pond were 3–7 times higher than in South Pond. Mason and Brandt's study clearly indicates the potential for predators to dramatically alter ichthyoplankton assemblages.

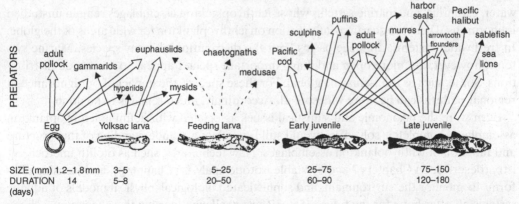

Figure 8.8 The generalized food web in which walleye pollock complete their life cycle. The thickness of the arrows corresponds to the importance of the trophic connections. Dashed lines indicate probable trophic pathways. (Reproduced from Brodeur & Bailey 1996 with permission of Swets & Zeitlinger Publishers.)

whitefish native to Lake Michigan. Even though it is considered to be an obligate planktivore, alewife was shown to be a voracious predator of larval fishes. For such animals, fish larvae may simply be particularly juicy "zooplankton."

Potential predators of ichthyoplankton assemblages are not limited to juvenile and adult fishes. Hydra, chaetognaths, ctenophores, and even other fish larvae have all been shown to be significant predators of fish larvae. In fact, larvae may face a gauntlet of predators as they grow and develop. A review of the suite of potential predators of walleye pollock larvae (Figure 8.8) indicates that between five and nine different families of predators may prey on each stage of pollock between the egg and late juvenile period. In summary, predation losses are probably the single largest cause of mortality in larval fish assemblages. In some cases, other factors, such as poor condition or disease may increase the susceptibility of larvae to predation, nevertheless predation remains the ultimate cause of death in most cases (see Chapter 3). Clearly, differential mortality among the species that comprise an assemblage will lead to variation in the composition of the assemblage.

8.6   Problems, issues, and future research

Research on ichthyoplankton assemblages has made substantial progress over the last several decades. In many cases, the original challenge of identifying the larvae of different species and thereby defining assemblage membership has been overcome. For many regions of the world (North America, Europe, northeast Pacific Ocean, Australia), atlases to aid larval identification of larvae are well advanced (see Chapter 12 for a partial listing). While there remain regions of the world for which this is not the case, many of the taxonomic challenges that once faced researchers interested in the early life stages of fishes are gone.

Identifying ichthyoplankton assemblages remains a challenge, owing largely to their dynamic nature and characteristic distribution of abundances that typify them (Figure 8.2, Section 8.3.2). Identification of ichthyoplankton assemblages requires intensive sampling efforts that present the practical challenge of filtering a sufficient volume of water, and the statistical challenge of characterizing infrequent events. The worldwide extent of freshwater, estuarine, and marine systems whose ichthyoplankton assemblages remain unstudied is enormous. There is a dearth of information on ichthyoplankton for wide areas of the globe. In freshwater systems, drainage basins limit the distribution of many species. Marine systems, however, are more open and many important species appear to have wide distributions. Some, such as tunas, may be global. For these species, the impact of our ignorance of regional patterns can be more widespread. Accordingly, much remains to be done.

Even were the taxonomic and statistical issues associated with defining ichthyoplankton assemblages completely solved, we would still lack adequate understanding of the structure and function of ichthyoplankton assemblages. New techniques, such as otolith microchemistry (described in Chapters 2 and 6), stable isotope analysis (Chapter 6), multisensing platforms to monitor the environment, and sophisticated biological–physical models of aquatic systems all offer hope for the future. The biggest challenge in using these techniques, however, will be to conduct sampling at the spatial and temporal scales appropriate to untangle mechanisms that regulate the formation, maintenance, and disruption of ichthyoplankton assemblages. Taggart & Frank (1990) provided an excellent account of some of the problems

of inference that arise when processes are not sampled at the correct scales. These considerations imply that we will have to combine large scale synoptic sampling that can define the
evolution of the assemblage over time, with smaller scale, intensive sampling focused on particular mechanisms that regulate the assemblage. There are examples of this type of
approach in several large studies of the dynamics of marine ichthyoplankton assemblages,
such as FOCI – Fisheries Oceanography Coordinated Investigations (see Section 4.8.2), but
the techniques have yet to be applied to any freshwater assemblage. While it is clear that
most freshwater systems work at smaller spatial scales than marine systems, both synoptic
and process-oriented studies are necessary to understand fully the structure of freshwater
assemblages.

Finally, considerable insight has come from comparative studies. Ichthyoplankton
assemblages reflect specific regional or local conditions. Often, these conditions are only a
subset of the possible range of abiotic and biotic factors that influence ichthyoplankton
assemblages. Thus, it is beneficial to examine many assemblages simultaneously, since they
may reflect responses to different constraints. Such comparative studies have provided considerable insight into the range of processes that structure ichthyoplankton assemblages
and will likely continue to do so in the future.

Additional reading

Bakun (1996) *Patterns in the Ocean: Ocean Processes and Marine Population Dynamics*. California Sea
    Grant College System, National Oceanic and Atmospheric Administration, La Jolla, CA.
Longhurst, A. (1998) *Ecological Geography of the Sea*. Academic Press, San Diego, CA.
Mann, K.H. & Lazier, J.R.N. (1996) *Dynamics of Marine Ecosystems*, 2nd edn. Blackwell Scientific,
    Cambridge, MA.
Moser, H.G., Smith, P.E. & Fuiman, L.A. (Eds) Larval fish assemblages and ocean boundaries.
    Bulletin of Marine Science **53**.
Moss, B. (1980) *Ecology of Fresh Waters*. Blackwell Scientific, Oxford, UK.
Sinclair, M. (1988) *Marine Populations. An Essay on Population Regulation and Speciation*. University
    of Washington Press, Seattle, 252 pp.
Taggart, C.T. & Frank, K.T. (1990) Perspectives on larval fish ecology and recruitment processes:
    probing the scales of relationships. In: *Large Marine Ecosystems: Patterns, Processes and Yields*.
    (Ed. by K. Sherman, L.M. Alexander & B.B. Gold), pp. 151–164. American Association for the
    Advancement of Science, Washington, DC.

Chapter 9
Fishery Management

Edward S. Rutherford

9.1 Introduction

The primary objectives of fishery management are to ensure long term sustainability of fish stocks, to prevent biological and economic overfishing, and to minimize disruption of ecosystems. Early life stages of fishes traditionally have played an important role in fishery management and promise to contribute significantly to supplementation and conservation of fish stocks in the future. In this chapter, I will briefly describe biological objectives, strategies, and regulations of fishery management, then review the unique contributions of early life stages of fishes to management theory, assessment, supplementation, conservation, and regulation of fish stocks.

A basic fishery management plan should include an assessment of the present state of development and exploitation of the fishery, objectives for managing the fishery, strategies for achieving those objectives, and regulations to be applied under various strategies. The relationship between stock assessment, management objectives, management strategies, and fishery regulations is depicted in Figure 9.1. Stock assessment is the estimation of vital rates and abundances to understand population dynamics and quantify current and potential yields. Management objectives may include restoration and conservation of fish stocks or maximization of economic or biological yield. Management strategies for achieving those objectives may include protection or supplementation of fish stocks and their critical habitats, or providing a minimum stock size or spawning stock to guard against recruitment overfishing.* Regulations to implement strategies may include input controls (limit fishing effort; restrict number, type, and size of fishing vessels or gear; restrict fishing areas or seasons), or output controls (limit weight, size, sex, or reproductive condition of the catch).

Study of the early life stages of fishes has contributed significantly to management theory and provided tools for assessment, conservation, supplementation, and regulation of fish stocks and their habitats. Variable survival rates and abundances in early life and their connection to recruitment have provided the theoretical basis for stock–recruitment and spawner-biomass-per-recruit analyses, which are central to management. Unique characteristics and

---

* Recruitment overfishing is a level of fishing that reduces adult stock enough to lower the probability of a successful recruitment. The case study of the Japanese sardine presented in Chapter 11 demonstrates methods for assessing recruitment overfishing.

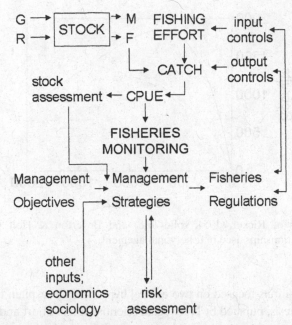

Figure 9.1 Relationships between stock assessment, objectives, strategies, and regulations in fishery management (from King 1995 with permission from Blackwell Science Ltd.).

abundances of early life stages may provide cost-effective means of indexing recruitment or adult biomass and estimating potential yield. Preservation or manipulation of critical spawning and nursery habitats, and supplementation from aquaculture production can stabilize or increase the variable abundances and survival rates normally experienced by fish populations during early life, and thereby conserve or enhance fish populations. Stocking of fish early life stages may help achieve management objectives of minimum or maximum sustainable yield. Early life stages are implicitly targeted by management regulations that affect spawning adults.

9.2 Fishery management theory

The ability to understand and predict fluctuations in recruitment and harvest of fish stocks is a desired component of any management plan. The relationship between adult spawners (stock or egg production) and their progeny entering the fishery (recruits), the so-called stock– recruitment relationship, is a cornerstone of fishery management theory that was developed from observed distributions and survival rates of fish early life stages. Studies of fish recruitment were first initiated in the early 1900s during investigations into the declining harvests of North Sea fisheries. By the early 1800s, signs of stock depletion, including declining fish sizes, catch rates, and harvests, were readily apparent to commercial fishermen in northern Europe. Scientists were slower to heed the warning signs and, as late as the 1880s, declared the seas to be an inexhaustible resource. By 1902, however, the problem of stock depletion could be ignored no longer, and the International Council for the Exploration of the Sea (ICES) was formed to investigate the causes of fishery fluctuations and declining harvest in the North Sea, thus beginning the modern era of fishery science.

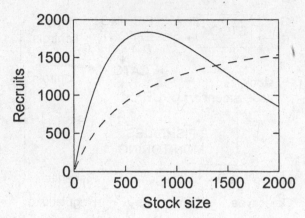

Figure 9.2 Examples of Ricker (1954, solid line) and Beverton & Holt (1957, broken line) stock–recruitment relationships used in fishery management.

The ICES investigations focused on two central hypotheses to explain the declines in harvest. The first hypothesis, pursued by Norwegian scientist Johan Hjort and colleagues, attributed fluctuating stocks and harvests to variable survival during the egg and larval stages. The alternate hypothesis, favored by most scientists of the day, attributed declining harvests to changes in oceanographic conditions that affected traditional migration routes of adults. Results of ichthyoplankton surveys conducted throughout the known distribution of adults and a time series of harvest data on known-age adults provided support for the first hypothesis. The results from the ichthyoplankton surveys indicated that spawning distributions were extremely localized compared to the distributions of adults. The harvest data collected from 1902 to 1914 for Atlantic herring (*Clupea harengus*, Figure 1.3m) demonstrated the presence of a very successful year class in 1904, suggesting that annual fluctuations in harvest were due to variations in year-class strength. Hjort suggested that variation in timing of larval production relative to that of their prey or variation in advection of larvae away from feeding areas were possible reasons for fluctuations in fish stocks. Importantly, he concluded that it was impossible to determine why year classes fluctuate and, instead, directed effort into monitoring year-class strength as a predictor of future yields. Hjort's ideas are presented more completely in Chapter 4.

Development of theoretical relationships to predict recruitment from spawning biomass has incorporated knowledge of vital rates and abundances during the early life of fishes. Two stock–recruitment relationships commonly used by fishery managers are the Ricker and Beverton and Holt curves (Figure 9.2). Both curves are non-linear, with a density-independent component of near proportional increases in recruitment at low stock sizes. The difference between the curves occurs at high spawner densities, when recruitment becomes density-dependent. At high spawner densities, recruitment declines in the Ricker curve, but increases at progressively slower rates to reach an asymptote in the Beverton and Holt curve.

Beverton and Holt noted that the shape of both relationships can be determined by events occurring during early life, between spawning and recruitment. They hypothesized that fish populations experienced two different stages of mortality during the period from spawning to recruitment: a period of high, density-independent mortality characteristic of the egg and larval stages, and a period of lower, density-dependent mortality characteristic of the juvenile

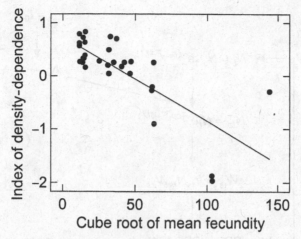

Figure 9.3 The relationship between the index of density dependence and the cube root of mean fecundity for 30 fish stocks. Adapted from Cushing (1971).

stages. The difference between the Ricker and Beverton and Holt stock–recruitment relationships is the time in life when density-dependence occurs. In the Ricker relationship, density-dependence is hypothesized to occur during the egg or larval periods, as a function of the initial number of spawning adults. For example, density-dependence can arise from high egg mortality due to superimposition of redds, from limited availability of feeding territories for larvae, or from cannibalism of eggs or young larvae by adults. Ricker believed that cannibalism by adults was the most common form of density-dependence. Salmonids and clupeids provide several examples of fishes whose stock–recruitment relationships are believed to exhibit this behavior. In the Beverton and Holt relationship, density-dependence is assumed to occur at any time during the pre-recruit period, after initial population numbers (eggs) are defined, as a function of the carrying capacity* of the environment. Beverton and Holt speculated that density-dependence may occur in the larval period through resource depletion, which would decrease individual growth rate and increase the amount of time larvae are exposed to high mortality. The Beverton and Holt stock–recruitment relationship has been fitted to harvest data for many fish species common to the North Sea for which recruitment is relatively stable over a wide range of stock sizes. In plaice (*Pleuronectes platessa*), for example, the mechanism believed to explain the asymptotic relationship was the limited carrying capacity of coastal nursery habitats for newly settled juveniles. After carrying capacity is exceeded, addition of subsequent juvenile production and recruitment is lost.

Variation in stock–recruitment relationships and the degree to which species exhibit density-dependence has been correlated with egg production. Cushing (1971) applied generalized regression models to stock–recruitment data for North Sea fish stocks and found an inverse relationship between stock density and fecundity. Stocks that had the highest values for the density-dependent coefficient in the stock–recruitment relationship had the lowest fecundities (Figure 9.3).

---

* Carrying capacity is defined here as the long term biomass of fish that a habitat can support.

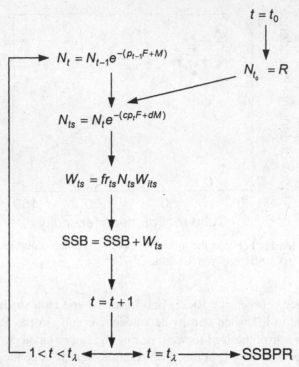

Figure 9.4 Schematic diagram of calculations of spawning-stock biomass per recruit, SSBPR (Gabriel *et al.* 1989). R = initial number in cohort; N_t and N_{t-1} = numbers in cohort at ages t and $t-1$ on 1 January, respectively; P_t and P_{t-1} = fractions recruited to gear at ages t and $t-1$, respectively; M = instantaneous natural mortality; F = instantaneous fishing mortality; t_0 = age at recruitment to grounds; t = oldest age in stock; N_{ts} = number in cohort of age t alive at time of spawning; c = fraction of fishing mortality within a year before spawning; d = fraction of natural mortality within a year before spawning; W_{its} = average weight of an individual at age t at time of spawning; fr_{ts} = fraction mature at age t at time of spawning; W_{ts} = weight of spawning segment of cohort at age t at time of spawning; SSB = accumulated total weight of spawning stock over the lifetime of a cohort.

One of the problems facing a manager using stock–recruitment relationships is the relatively poor fit of the relationships to observed data. The need for more accurate predictions of recruitment has stimulated research into mechanisms influencing survival and growth of early life stages (see Chapters 2 and 3). The stock–recruitment regression relationships have been improved by incorporating coefficients for factors believed to affect growth or survival of early life stages. As an example, Nelson and colleagues (1977) fitted a Ricker stock–recruitment model to data on Atlantic menhaden (*Brevoortia tyrannus*) year-class strength and found that deviations from the relationship were most strongly correlated with zonal Ekman transport, which acts to transport larvae via wind-generated currents from offshore spawning grounds to inshore nursery grounds. This transport accounted for 84% of the variation in year-class strength over a 15-year period. Madenjian and colleagues (1996) used a Ricker stock–recruitment model to explain 85% of the variance in recruitment of walleye (*Stizostedion vitreum*) in Lake Erie by incorporating warming rate during the egg and larval periods, and the biomass of prey available to spawners in the previous fall. Although these types of regression approaches indicate the potential for environmental factors to influence early life stages and recruitment in the years when data were collected, they often have failed to predict recruitment in subsequent years, in part, because of the potential for multiple environmental factors to influence egg and larval survival and recruitment success (see Chapter 4).

The poor relationship between stock and recruitment data has led fishery managers to believe there is little or no relationship between spawning biomass and recruitment. As a consequence, many fish stocks have collapsed from a combination of recruitment overfishing and poor environmental conditions. A recent meta-analysis of many harvested fish populations indicates that there is a weak positive relationship between stock and recruitment: good recruitments are more likely to occur at high spawning biomass, and low recruitments are more likely to occur at low spawning biomass (Myers & Barrowman 1996). As a consequence, fishery managers now commonly use spawning-stock biomass per-recruit (SSBPR), a variation of yield-per-recruit theory, as a metric to determine the effects of fishing on the reproductive potential of the stock, and by proxy, its recruitment potential. The analysis of SSBPR uses age and weight structures, maturity schedules, weight–fecundity relationships, and adult mortality and growth schedules to quantify the reproductive potential of the stock (Figure 9.4). A critical assumption of SSBPR is that proportional increases in spawning biomass will result in proportional increases in egg production and potential for recruitment, but as will be discussed below (Section 9.4), age and size structure, and condition may have non-linear effects on the number, quality, and viability of eggs.

9.3 Stock assessment

Fish early life stages have been used to estimate recruitment and adult abundance and to characterize unit stocks, the basic unit of management. The biomass or relative abundance of a fish stock can be estimated from the abundance of its spawn, thus providing a cost-effective alternative to sampling adult stages or using fishery-dependent data for estimating stock biomass. For species like the Atlantic menhaden, ichthyoplankton surveys may provide a better measure of relative abundance than fishery-dependent indices, such as catch-per-unit-effort (CPUE), which may be poorly correlated with fish abundance due to schooling behavior of fishes and the searching ability of fishers. Chapter 5 covers details of the egg-production approach to estimating adult biomass.

The presence or absence of fish eggs or larvae may be highly correlated with adult abundance, thereby providing a relatively inexpensive way to index trends in recruitment or abundance of widely distributed populations like the northern anchovy (*Engraulis mordax*) or bluefin tuna (*Thunnus thynnus*), or of depleted populations such as yellowtail flounder (*Pleuronectes ferrugineus = Limanda ferruginea*) or striped bass (*Morone saxatilis*). For example, egg and juvenile stages have been used to index the decline and recovery of the striped bass in Chesapeake Bay, USA. The striped bass population had declined severely by the late 1970s due to a series of low recruitments, overfishing, and environmental degradation. In 1984, the fishery was closed to protect the last successful year class until it reached maturity. The management plan for recovery of Chesapeake Bay striped bass populations linked harvest regulations to a juvenile recruitment index. Once the average recruitment index from three consecutive years exceeded the 50-year mean recruitment, the fishery was allowed to re-open under restricted harvest limits. Subsequent analyses indicated that the distribution and presence of eggs successfully tracked the decline and the recovery of the population.

Distributions, relative abundances, and unique physical attributes or characteristics of environmental histories experienced by early life stages may provide information about the unit stock. The unit stock is a discrete group of individuals which has the same gene pool, is self-perpetuating, and has little connection with adjacent groups that may be assessed as a discrete entity. It is the basic population unit, subject to fishing mortality and management. Molecular genetics techniques and microchemical analyses of tissue from early life stages are important tools for identification of unit stocks (some of these methods are described in Chapter 6). For example, analyses of genotypic variation in coral reef fishes, combined with knowledge of ocean circulation patterns and estimated dispersal rates of larvae, now permit identification of unit stocks for management purposes. The analysis of environmental histories recorded in fish otoliths through trace element and stable isotope composition of fish otoliths or scales can provide a natural tracer or fingerprint for identification of natal habitats and stock composition. Chemical tags (such as oxytetracycline or strontium) and thermal marks are now routinely applied so that identifiable marks appear on the otoliths and scales of hatchery-produced larvae and juveniles before release. These marks are utilized to determine the relative contribution of hatchery fishes to the harvest. These techniques are being used by fishery scientists around the world to identify, protect, or supplement fish stocks. Chapter 6 gives more details about stock identification based on early life stages.

9.4   Restoration, conservation, or supplementation of fisheries

Restoration and conservation of fish populations and optimization of fishery yields are prominent objectives of fishery management, since the majority of the world's stocks are fully exploited or overfished. The value of early life stages for achieving restoration or conservation, or maximizing yield, lies in the tremendously high fecundities of some species and the exponential rates of growth and mortality experienced during early life (see Chapter 3). Improvements in survival rates of these life stages may have dramatic effects on recruitment and adult biomass. Strategies for achieving restoration or conservation objectives using early life stages rely on maintenance of minimum population size or reproductive potential, protecting, improving, or increasing critical habitats, and supplementing population size through stocking. Strategies for maximizing yields using early life stages involve aquaculture and culture-based fisheries. A distinction should be made between intensive culture, in which fishes are raised until harvest under controlled conditions for all, or a major part, of their life history and extensive culture, in which control is exercised over a part of the life history and larvae or juveniles are transplanted or released for grow out in natural environments.

9.4.1   Maintenance of minimum stock size or reproductive potential

The relationship between spawning biomass, egg production, and potential recruitment forms the basis for management strategies of minimum stock size or reproductive potential. Management plans for most commercially harvested species incorporate target levels of

SSBPR to maintain or restore depleted stocks. The level of spawning biomass necessary to protect against recruitment failure is generally unknown for most fish stocks, but is believed to range from 30% to 60% of virgin, unfished biomass.

9.4.2 Supplementation of fish stocks

The practice of culturing early life stages to supplement or increase population size has increased in importance as wild fish populations have become depleted through overfishing and habitat destruction. Fishery products resulting from aquaculture now represent nearly 30% of global fishery production. Future increases in world fish production likely will come from aquaculture, as most capture fisheries have reached or exceeded maximum sustainable yields. Studies of diet, energetics, physiology, behavior, and ecology of early life stages have allowed fish culturists to significantly enhance stocks and harvests of many species.

The collection and culture of early life stages of fishes to supplement harvest is an ancient practice that originated in China nearly 1000 years ago. China and India traditionally have withdrawn large numbers of eggs and larvae (primarily of carps) from rivers to supply material for culture. These sources have decreased as exploitation and environmental disturbances have increased, and have been largely replaced by artificial spawning and rearing of larvae. In Indonesia, traditional culture of milkfish (*Chanos chanos*) has been practiced since the 14th century by farmers who obtain naturally produced eggs and larvae from rivers and coastal areas and grow them out to harvest size in specially prepared ponds. In Africa, by contrast, aquaculture production is a relatively recent phenomenon that began during the 1940s with stocking of tilapia and catfishes in ponds and impoundments.

By the 1950s and 1960s, development and spread of techniques for controlled, induced spawning, and development of artificial, nutritionally complete foods for larvae and juveniles, released farmers from having to depend on natural sources of eggs and larvae. The mass rearing of brine shrimp (*Artemia*) for prey now permits culturists to raise fishes beyond the first feeding stage and release them at larger sizes when they are less vulnerable to predation and starvation. As a result, survival rates from egg to metamorphosis for cultured fishes have improved to 30–65% (Welcomme 1998, Blaxter 2000), much higher than survival rates observed in nature (see Chapter 3). For milkfish farmers, the advent in the mid-1980s of hormonal preparation of broodstock now allows consistent production of larvae and, as a result, milkfish production has increased ever since.

Attempts to increase fish production through the release of early life stages (extensive culture) date back centuries for freshwater fishes and to the 1870s for anadromous fishes. Seed material for stocking traditionally came from natural spawning in rivers and lakes and more recently from aquaculture facilities. Early stocking of carp was also based on seed from natural sources. In Europe during the Middle Ages, carps were transported into ponds and lakes by traveling monks wishing to protect communities against famine. In Eurasian lakes and impoundments, coregonid and percid larvae (similar to Figure 1.3b, j) have been successfully stocked since the late 1800s to support commercial and recreational fisheries. In the United States, freshwater stocking programs were initiated during the late 1800s, to support sport fisheries. Eggs and larvae of carp (Figure 1.3f, g), trout (Figure 1.3d), and bass (*Micropterus* spp., Figure 1.3i) were released across the country with variable success until consideration was given to the size of individuals stocked and the carrying capacity of the receiving waters. In Japan, salmonid culture began during the late 1800s, but was largely unsuccessful until techniques of intermediate feeding and timely release of larvae were introduced in 1961. In Canada, hatchery production of salmonids began in the late 1800s and greatly increased during the 1960s when the development and application of spawning and egg incubation channels greatly improved egg and alevin survival.

Stocking of early life stages to supplement marine fish stocks has been less successful than freshwater or anadromous species. A noted exception is the centuries-old practice of valliculture, in which larvae of eels, mullets, and other seasonally migratory species are enclosed in coastal lagoons (valli) and reared to harvestable size in the Mediterranean. In Europe and the United States, hatchery production of marine finfishes began in the late 1870s when fish eggs and larvae of American shad (*Alosa sapidissima*), cod (*Gadus morhua*, Figure 1.3p), and flatfishes were fertilized and released extensively to increase fish stocks and harvests. Hatchery production increased dramatically in the period before World War I; by 1917, American hatcheries produced and released three billion larvae of winter flounder (*Pleuronectes americanus*), pollock (*Pollachius virens*), cod, and haddock (*Melanogrammus aeglefinus*, Figure 1.3n). But, the massive stocking efforts were largely futile. The numbers of larvae released were neglible compared to the production of a relatively few wild adults. Although North American and European countries released billions of hatchery-reared larvae in the years before World War II, there was little evidence that the plantings increased recruitment or adult stock. A number of factors impeded survival of hatchery larvae, including inadequate stocking numbers, susceptibility to environmental conditions, predation, advection away from local nursery areas, and insufficient ecosystem carrying capacity. Relative contributions of hatchery fish to total harvest were poorly known due to failure to adequately mark and recognize hatchery releases. Blaxter (2000) provided a comprehensive review of these efforts.

During the 1950s, 1960s and 1970s, increased knowledge of culture techniques and declining harvests from capture fisheries revitalized interest in extensive culture of marine fishes in Europe and North America. In Japan, the National Culture-Based Fishery Project succeeded in developing techniques for mass seed production and subsequent release of red sea bream (*Pagrus major*), sole, Japanese flounder (*Paralichthys olivaceus*), and yellowtail (*Seriola quinqueradiata*). In the United States, culture and release of red drum (*Sciaenops ocellatus*) juveniles in Texas significantly increased stock biomasses and catches by 20–30% in some years.

On a global scale, the volume of eggs, larvae, and juveniles currently stocked to supplement fisheries is impressive. Nearly 57 billion larvae and/or fingerlings are stocked annually. A key factor in the rapid growth of cultured finfish production is the increasing availability of hatchery-produced seed. The countries with the most active stocking programs include China, Japan, India, and other countries in southeast Asia, as well as the United States and Norway. The majority of finfishes cultured for release are freshwater species, principally carps, and tilapia.

Research on diet, physiology, energetics, genetics, behavior, and ecology of early life stages has allowed culturists to grow fishes to the sizes and quantities needed for commercial or recreational harvest. Factors contributing to stocking success include egg quality, development of planktonic diets (such as rotifers or brine shrimp), development of nutritionally complete, artificial diets for early juveniles, and sizes and behavioral quality of fishes for release. Fish nutritionists have lagged far behind other experts in fish culture. For many years, young fishes were fed on vegetables and insect larvae produced from organs and carcasses of livestock, with no regard for nutritional quality. The science of fish nutrition began in 1927 by United States scientists trying to develop an inexpensive food for brook trout (*Salvelinus fontinalis*). It was not until 1955 that the first dry, nutritionally complete, practical diet was commercially produced for salmon. Genetic manipulation of fish embryos now permits production of transgenic fishes that grow substantially faster than unmanipulated fishes, and can double or quadruple the weight of adult carp or salmon.

Ecological factors may be most critical to stocking success when individuals are first released and most sensitive to environmental conditions. Release experiments of marked red sea bream in a Japanese estuary demonstrated that larger (4 cm) individuals survived better than smaller (2 cm) individuals, probably due to the higher predation pressure on the smaller life stage. Initial releases of Japanese flounder from the 1970s to early 1980s were unsuccessful because of pigment abnormalities in released juveniles that made the flounder easy targets for predators. Success of hatchery releases improved in the 1990s when pigment deficiencies were corrected, but recapture rates are still low because these individuals spend more time feeding up in the water column compared to wild individuals, thus increasing their exposure to predators. Releases of otolith-marked striped bass larvae (Figure 1.3q) helped to determine the appropriate environmental conditions for successful larval stocking in a Chesapeake Bay tributary. Stocking of yolk-sac larvae was found to be feasible in years of poor to average natural recruitment, and was most successful under conditions of steadily rising water temperatures (Secor & Houde 1998).

Behavioral deficiencies in reared fishes may explain their lower survival rates in comparison to wild fishes, but innovative research on the use of behavioral keys for evaluating the quality of hatchery fishes suggest these problems may be corrected (see reviews by Blaxter [1976] and Masuda & Tsukamoto [1998]). In ayu (*Plecoglossus altivelis*), an amphidromous fish, jumping behavior at a waterfall is closely related to upstream migration, and thus recapture rate. In red sea bream, tilting behavior is used as a measure of post-release predator avoidance. Research on Japanese flounder indicates that newly settled juveniles must be able to bury into sediments and match their color to the background to survive predators. Reared juvenile flounder spend more time off the bottom while feeding than do their wild counterparts, thus exposing them to predators. Behavioral deficiencies in red sea bream are mitigated by extensive culture and in flounder by acclimation to semi-natural rearing conditions.

The increased ability of hatcheries to produce and rear early life stages has not always resulted in successful conservation or supplementation of fish populations. Despite the enormous releases of hatchery-reared salmonid larvae in the Pacific northwest, many wild salmonid populations are still threatened or endangered due to habitat degradation through land-use practices, obstruction by dams from upstream spawning habitat, changes in the ocean production cycle, predation mortality at release, out-breeding depression, or increased fishing pressure. For many Pacific salmon stocks, there is an inverse relationship between hatchery smolt production and ocean survival of wild fishes. Increases in harvestable populations resulting from hatchery supplementation have allowed managers to avoid the problems of declining nursery habitat resulting from damming and land use. Increased numbers of out-migrating hatchery smolts and stocking of non-native predators, such as walleye and largemouth bass, have increased predation pressure on out-migrating smolts. Ocean survival of released juveniles depends on abundance of food, which fluctuates with ocean production cycles that are regulated by decadal cycles in wind patterns and upwelling strengths. The period between the late 1970s and the mid-1990s corresponded to a period of favorable ocean production for juvenile salmon off Alaska, but an unfavorable period of production for juvenile salmon further south (British Columbia through California). Increases in adult salmon populations resulting from stocking have encouraged higher fishing rates on hatchery and wild individuals, resulting in declines of wild populations. The genetic and behavioral impacts of hatchery-reared adults on wild populations have been well documented. Hatchery-reared salmon tend to have lower survival rates in the wild and poorer homing ability than their wild counterparts. Thus, straying by hatchery adults and interbreeding with wild strains may result in dilution or loss of genetically inherited adaptations that enhance survival in their specific natal environments.

Documented failures of stocking programs to restore or enhance fish populations are not limited to Pacific salmon, however. Restoration of self-reproducing populations of lake trout (*Salvelinus namaycush*) in most of the Laurentian Great Lakes has been slowed by reproductive failures, despite the establishment of refuges around spawning and nursery habitats. Some of the factors hypothesized to block successful reproduction of lake trout include excessive fishing mortality that eliminates all but the youngest spawners, reduced egg viability from pollutants and toxins, high predation mortality on egg and larval stages, siltation of spawning habitats, and reproductive incompetence of hatchery strains. In the United States, only 5% of stocking programs for walleye have succeeded in enhancing fisheries or natural reproduction. The densities and life stages of hatchery-released walleye, and water temperatures and availability of natural foods at the time of stocking all influence the success or failure of stocking.

Future conservation and supplementation of fish stocks and increases in fishery production may only be realized with attention to environmental, behavioral, genetic, and economic factors influencing early life stages. The nature of the hatchery environment can affect behavior and viability of young fishes and lower their production in the hatchery and fitness for survival compared to wild fishes, particularly immediately after release. Differences in size (size hierarchies) often develop among larvae in raceways through differences in individual behavior, consumption, and growth (see growth depensation in Section 2.5.1), which can lead to cannibalism of smaller individuals by larger individuals or cause differential survival of released individuals, if mortality is size related. Larvae must learn cryptic and feeding behaviors to survive in the wild. The size of individual larvae, as well as season, location, time of day, depth, and density of release all may influence survival rate of hatchery-produced fishes. The carrying capacity (food and space) of the environment must be sufficient to support the number of larvae released. There must be adequate prey densities and adequate habitat to support released individuals. Hatchery-released larvae should supplement or coexist with wild fish, not displace them. Consideration should be given to minimizing the impact of hatchery releases on the wild genotype; broodstock should be taken from the same environment as the release site, and adequate numbers of adults (effective population size) should be collected to maintain genetic heterogeneity.

Benefit-cost analysis should be applied to improve economic efficiency of stocking programs. The effectiveness of stocking early life stages usually involves tradeoffs between costs of production and survival in the wild. Fish eggs and larvae are cheaper to produce and easier to release, but must be stocked in greater numbers than juveniles to attain an equivalent number of adults. The penalty of high mortality acts against stocking at too small a size, but production costs increase exponentially with increased size, especially with slow growing species (Masuda & Tsukamoto 1998, Blaxter 2000).

9.4.3 *Preservation and protection of critical habitats*

Alteration of fish habitats may serve to sustain early life stages and increase adult abundance, thereby achieving management objectives of stock conservation or supplementation. Certain alterations are undertaken to provide a competitive advantage for some species, while others are made to increase carrying capacity. Successful manipulation of habitats is based upon an understanding of habitat factors that limit production of desirable species (see Chapter 7).

Critical habitats for spawning and development of eggs and larvae have been restored or manipulated by managers wishing to improve reproductive success of desired species. For example, managers have manipulated aquatic macrophytes to provide shelter for eggs or larvae or substrate for prey organisms. Growth of macrophytes is limited by light and nutrients, and both variables are amenable to management through control of sediment and nutrient loadings. In Canada, hatchery workers have pioneered development and application of spawning and egg incubation channels that simulate high-quality natural spawning beds and provide favorable environmental conditions for survival and growth of salmonid eggs and alevins. In temperate lakes and reservoirs, cobble/gravel substrates or brush piles have been added to shoreline habitats to create spawning habitat for percids. Lake water levels may be manipulated to inundate wetland habitats and increase spawning areas favored by esocids and percids.

Fishery managers have also treated or manipulated spawning and nursery habitats to eliminate reproduction by undesirable species or to prevent their introduction from foreign environments. Water levels can be lowered or barriers erected to reduce spawning habitats. In tributaries to the Laurentian Great Lakes, fishery managers have used mechanical and electric weirs and maintained dams to prevent parasitic sea lampreys (*Petromyzon marinus*) from reaching favored spawning habitats. Managers routinely treat nursery habitats with a lampricide (TFM), which is toxic to larval lampreys, but less harmful to other species. Also in the Great Lakes, repeated invasions by early life stages of exotic species through ballast water in ocean freighters (see Chapter 10) will soon be treated through biocide applications of the ship's ballast water.

Physical characteristics of riverine habitats may be degraded by human activities, but can be manipulated to improve survival of early life stages. Spawning and nursery habitats for eggs and larvae may deteriorate from land-use patterns (logging, grazing, water diversion). Increased sediment loads resulting from land use or riparian development can fill shallow pool habitats and settle into gravel riffles, where they fill interstitial spaces and decrease egg or larva survival. Logging activities can increase stream runoff, lower infiltration into groundwater (making river flows more variable), increase stream bed scouring, and reduce invertebrate prey densities. Dams affect early life stages by blocking spawning migrations, impeding downstream migrations of larvae and juveniles or their prey, or altering flows that increase mortality of sensitive egg and larval stages (see Chapter 10 and the Danube River case study in Chapter 11).

Managers may mitigate land-use damage to egg and larva survival through habitat manipulation at various spatial scales. On a watershed scale, management can improve survival of early life stages by protecting riparian zones and encouraging land-use patterns that minimize nutrient and sediment loadings. On the scale of a stream reach, managers can improve instream feeding and refuge areas for early life stages by removing sand, stabilizing banks, and adding cobble/gravel substrates and woody debris. Managers may mitigate dam impacts on early life stages through installation of fish ladders, manipulation of flow volume and periodicity to more natural conditions, and manipulation of flow source via epilimnetic (top-draw) or hypolimnetic (bottom-draw) releases from the upstream impoundment.

In marine environments, managers have restored or added critical habitat for larvae and early juveniles to maintain ecosystem function and support fisheries. Seagrasses, mangroves, and wetland plants provide settlement habitat and refugia for early life stages and have been planted by managers to restore habitat or mitigate previously destroyed habitat. For marine reef-dwelling fishes, attempts to augment recruitment through placement of artificial reefs have been controversial and largely unsuccessful, perhaps because of the structural design and location of the reefs. Many of the initial reef structures were discarded ships, cars, and tires placed near natural reefs to augment potential harvest. The controversy until recently has focused on whether artificial reefs simply attract existing adult spawners and recruits from adjacent natural reefs, or actually produce fishes through settlement, growth, and reproduction of resident individuals. Recent studies indicate that properly designed artificial reefs can attract and produce fishes for harvest without intercepting recruitment from adjacent natural reefs.

9.5   Importance of early life stages for management regulations

9.5.1   *Input regulations*

Closed areas

Often, managers will close off an area or time of year to fishing, to protect spawning, settlement, and development of sensitive early life stages or to provide a source of eggs for adjacent overfished areas. The use of sanctuaries and protected areas has recently received considerable attention from fishery managers trying to conserve or restore depleted fish stocks. Protected areas or reserves can maintain productive fisheries by protecting a critical stock within their borders. These protected stocks may enhance catches outside the refuge through adults that grow larger in the reserve and then migrate to fishing areas, or through enhanced larval transport and recruitment to fishing areas through increased population fecundity within the reserve. Marine reserves may be most effective when used in conjunction with traditional harvest limits in adjacent harvest zones.

Studies of fish early life stages have been used to determine the potential location and size of marine protected areas (MPAs). The size of MPAs is defined by the habitat needed for spawning and settlement. The location of MPAs is determined by prevailing currents that influence dispersal patterns of larvae to areas with depleted populations. MPAs may serve as sources of larvae to adjacent depleted populations or sinks to retain larvae being dispersed from elsewhere. Fishes, such as serranids, that have sedentary adult stages, localized spawning aggregations, and dispersive eggs or larvae would be ideally suited for protection within a reserve. Development and application of molecular genetics techniques to identify source populations, and otolith microchemistry analysis to identify environmental histories of migrating larvae and young juveniles (see Chapter 6) may help resolve questions concerning the location and size of MPAs.

Recent research on ocean current patterns surrounding coral reefs and on genotypic variation in reef fish populations indicates that larvae are not passive particles, but must behave in a way to promote local retention. Gene flow among reef fish populations has been negatively correlated with larval stage duration (Figure 6.8), suggesting that rates of larval dispersal among populations could be much lower than previously assumed, even among species with long larval stage durations. Circumstantial evidence suggests that fish larvae may accumulate in offshore areas before settlement. Extensive retention of larvae may require major reassessment of fishery enhancement models for reserves that depend on larval export for their effects.

Spawning biomass and egg production may build up rapidly in MPAs as fishes are allowed to grow because fecundity is a cubic function of fish length. The relative success of MPAs is related, in part, to the stock–recruitment relationship. Species that mature late and increase fecundity after maturation – the periodic strategy described in Chapter 1 – will benefit more from MPAs than species that mature early with high fecundity, unless the adjacent stocks are severely depleted.

Closed seasons

Managers may close the harvest during a time of year when fishes are aggregated and more vulnerable to fishing. Species are often aggregated during spawning, thus protection from harvest not only conserves adult stocks but also protects egg production.

9.5.2 Output regulations (size limits)

Fishery management can influence the maternal age and size composition of a stock through size regulations. Managers often have used size limits to protect or harvest mature adults. Minimum size limits are set above the size at first maturity to ensure that adults spawn at least once. Alternatively, stock abundance may be controlled through maximum size limits to reduce egg production and encourage good growth and survival of the remaining individuals.

The effects of size limits on age and size structure can seriously alter the potential egg production and recruitment of a fish stock. Reproductive parameters that may vary according to parental reproductive history include fecundity, egg size, egg viability, sperm quality, and duration and time of spawning. A reduction in age and size at sexual maturity and the depletion of older members of fish stocks have the potential to influence not only total egg production, but the size and viability of eggs as well. Changes in size and age composition of adults can affect the time of spawning and abundance and quality of gametes. Smaller and younger adults resulting from size-selective fishing practices have lower reproductive potential than adults from a wider range of sizes and ages for an equivalent population biomass. Compared to larger, more experienced females, smaller virgin females spawn during a limited portion of the spawning season, produce fewer eggs, and their eggs and larvae are smaller and are less viable. A consequence of a shorter spawning season may be a decrease in the likelihood of temporal overlap between young, first feeding larvae, and the high levels of zooplankton prey abundances necessary to sustain larval growth and survival.

Maintenance of a balanced age/size structure may be important for long-lived fishes that spawn in turbulent nursery areas. Spawning behaviors that vary with size or age might be an effective means to compensate for environmental stochasticity. Secor (2000) analyzed the influence of adult demographics on spawning behavior and recruitment success in striped bass. Periods of high egg production are often poorly timed for optimal larval survival. Large, old females may spawn earlier than small and young individuals. Increased age diversity in the spawning stock may increase the temporal and spatial frequency of spawning (spawning dispersion), and thereby increase the probability that some offspring will encounter favorable conditions. When the striped bass population was severely depleted during the 1970s and 1980s and the fishery was closed, few spawner age groups remained, increasing the chances of a mismatch between timing of egg production and occurrence of optimal environmental conditions. Recruitment was extremely poor during the period; only one average year class was produced during an 18-year period compared to an average or good recruitment every 2–4 years at high spawner densities. Over the course of the stock's recovery from 1989 to 1998, Secor found positive associations between egg presence ratio, age diversity, and year-class strength, supporting the hypothesis that year-class strength was positively related to age structure of mature females due to its influence on spawning diversity.

The effects of altered age and size structure of spawners on the quantity, quality, and timing of progeny produced have important implications for recruitment and sustainability of fisheries. Many iteroparous fishes harvested by humans have long life spans, low adult mortalities, and high fecundities (periodic strategy described in Section 1.3.3). When environmental conditions are favorable, diversity in spawner size structure may generate dominant year classes that drive population growth. Conversely, reduction in diversity of adult sizes may render a population more vulnerable to recruitment failures and more dependent on year-class successes.

9.6   Conclusions

Early life stages contribute greatly to the theory and practice of fishery management and promise to play an increasing role in the future. The importance of adult size structure and condition to egg production and potential recruitment will improve management's ability to reduce the likelihood of recruitment overfishing. Technological advances will improve the capability to measure the factors affecting fish recruitment on the appropriate temporal and spatial scales, and thus improve predictions. Although it is obvious that recruitment is influenced by multiple factors, an increased understanding of recruitment dynamics can be used to generate probability distributions of recruitment for a given adult stock. Stock-assessment techniques that incorporate early life stages improve the understanding of life history, movement, and habitat use, and thus improve identification and conservation of critical habitats.

Perhaps, the greatest contribution of early life stages to fishery management will come from extensive and intensive culture of fishes for harvest and conservation. Most of the world's capture fisheries are fully exploited and, as human populations continue to rise, pressures to fully exploit the remaining species will increase. The demand for cultured fishes for specialty markets or for subsistence will increase along with human populations, thereby placing a greater emphasis on spawning, rearing, and nutrition of eggs and larvae of selected species. Transgenic manipulation of early life stages to achieve higher growth rates will become as routine for fishes as it is now for vegetables and livestock. But, as culture techniques improve for marine and freshwater species, and cultured fishes make up an increasing proportion of the world's fish harvest, the will to conserve habitats for early life stages of wild fishes may decline. Conservation of critical nursery habitats through hydropower dam removal or restrictions on land use may be less economic and less popular than supplementation using hatchery products.

Additional reading

Blaxter, J.H.S. (2000) Enhancement of marine fisheries. *Advances in Marine Biology* **38**, 2–54.

Bohnsack, J.A., Eklund, A.M. & Szmant, A.M. (1997) Artificial reef research: is there more than the attraction-production issue? *Fisheries* **22**, 14–16.

Conover, D.O., Travis, J. & Coleman, F.C. (2000) Essential fish habitat and marine reserves: an introduction to the second Mote Symposium in fisheries ecology. *Bulletin of Marine Science* **66**, 527–534.

Cowx, I.G. (Ed.) (1998) *Stocking and Introduction of Fish*. Fishing News Books, Oxford.

King, M. (1995) *Fisheries Biology, Assessment and Management*. Blackwell Science, Oxford.

Masuda, R. & Tsukamoto, K. (1998) Stock enhancement in Japan: review and perspective. *Bulletin of Marine Science* **63**, 337–358.

Myers, R.A. & Barrowman, N.J. (1996) Is fish recruitment related to spawner abundance? *Fishery Bulletin, U.S.* **94**, 707–724.

Richards, R.A. & Rago, P.J. (1999) A case history of effective fishery management: Chesapeake Bay striped bass. *North American Journal of Fisheries Management* **19**, 356–375.

Rothschild, B.J. (1986) *Dynamics of Marine Fish Populations*. Harvard University Press, Cambridge, MA, 277 pp.

Secor, D.H. (2000) Spawning in the nick of time? Effect of adult demographics on spawning behavior and recruitment in Chesapeake Bay striped bass. *ICES Journal of Marine Science* **57**, 403–411.

Sinclair, M. (1997) Prologue. Recruitment in fish populations: the paradigm shift generated by ICES committee A., In: *Early Life History and Recruitment in Fish Populations*. (Ed. by R.C. Chambers & E.A. Trippel), pp. 1–28. Chapman and Hall, New York.

Trippel, E.A., Kjesbu, O.S. & Solemial, P. (1997) Effects of adult age and size structure on reproductive output in marine fishes. In: *Early Life History and Recruitment in Fish Populations*. (Ed. by R.C. Chambers & E.A. Trippel), pp. 29–55. Chapman and Hall, New York.

Chapter 10
Human Impacts

G. Joan Holt

10.1 Introduction

This chapter will discuss anthropogenic changes to aquatic environments and how they affect fish populations through impacts on their eggs and larvae. The aquatic ecosystems in which fishes live are especially impacted by urbanization and development. The majority of fishes live in the narrow band of water along the margins of land masses. They must share this region with over half of the world's population (about 3.5 billion people) who live within 100 km of the shore. This juxtaposition of man and fishes sets the stage for the activities of man to influence how well fishes survive, grow, and proliferate.

Humans have traditionally concentrated near coastlines and large bodies of water because of the activities and economic opportunities associated with shipping, fishing, and tourism. But man's presence brings major changes to aquatic habitats through increased inputs of nutrients, sediments, and contaminants. Rivers are used as transportation corridors, waste disposal sites, and water sources. Hundreds of millions of tons of toxic chemicals, sewage, industrial waste, and agricultural runoff enter the world's seas each year. At the same time, overfishing has impacted more than 40% of the world's most important commercial species and biodiversity of fishes has been reduced. All together, 755 of the world's fish species are listed as endangered or threatened according to the IUCN Red List 2000 (see Box 10.1), but freshwater fishes are the most heavily impacted. For example, in North America, 35% of the freshwater fish species are threatened, nearly all because of alterations to lakes, isolated wetlands, rivers, and streams. Freshwater habitats are extremely vulnerable and species occurring in them are likely to face a much higher risk of extinction than those in marine environments. Although few marine species are listed (mostly sharks and rays), the status of 14 000 marine fish species is unknown. There are clear reductions in population size and distribution of many marine species. It is expected that the number of threatened fish species will increase significantly as more data are collected, particularly from tropical seas and coral reefs.

The early life stages of fishes are especially vulnerable to habitat loss, water quality changes, and pollutants. The unique contribution of early life stages to fish population dynamics amplifies these impacts since changes in hatching success, growth, and mortality (vital rates introduced in Chapters 2 and 3) determine whether a population grows or declines. Variability in annual recruitment, generated by the high and variable mortality during early life stages (see Chapter 4) is increased by man's impact. Such pressure on the

> **Box 10.1**   IUCN – The world conservation union.
>
> The International Union for the Conservation of Nature, founded in 1948, is a consortium of government agencies, scientists, and experts from 181 countries. The Red List produced by the IUCN is the world's most comprehensive inventory of the global conservation status of plants and animals. It uses a set of criteria, relevant to all species and all regions of the world, to evaluate the extinction risk of thousands of species and subspecies. There are eight Categories of Threat in the IUCN Red List system: Extinct, Extinct in the Wild, Critically Endangered, Endangered, Vulnerable, Lower Risk, Data Deficient, and Not Evaluated. A species is listed as threatened if it falls in the Critically Endangered, Endangered, or Vulnerable categories. Habitat loss is the main threat to 85% of the species on the threatened list.

brief, early life of fishes can strongly influence recruitment success and ultimately the long term dynamics of adult fish populations.

10.2   Categories of human impacts

Humans have made substantial alterations to habitats that are critical to many of the fishes of the world. These include changes to global and local climates, and to nutrient cycles that directly affect aquatic ecosystems. For example, depletion of the atmospheric ozone layer by chlorofluorocarbons has increased exposure to ultraviolet rays from the sun that are particularly dangerous for eggs and larvae that live near the surface of the water. Food production has disrupted the normal nitrogen cycle by generating excess fixed nitrogen. Increased greenhouse gases, mostly carbon dioxide, methane, chlorofluorocarbons, and nitrous oxide, from fossil-fuel combustion and deforestation contribute to global warming and increase the rate of climate change (Chapin *et al.* 2000). These human impacts act on early life stages by reducing survival rates either through direct mortality or through changes in behavior, development, or distribution that lead to increased mortality. Indirect effects that reduce growth and survival of larvae include food-web changes that alter predator–prey relationships. Reductions in survival of early life stages generated by abnormal development, reduced growth rate, increased predation, or any number of factors severely limit the number of pre-recruits. Such changes result from habitat degradation, introduced species, overfishing, or from chemical pollution. Each of these impacts will be discussed in detail in the following sections.

10.2.1   Physical modification of the habitat

Urbanization and associated recreational and economic activities have modified shorelines and waterways extensively. Because of the specific habitat requirements of fish eggs and larvae (Chapter 7), these changes have serious consequences for fish populations.

Shoreline development, channelization, and water diversions

Estuarine and marsh habitats are vitally important nursery grounds for larvae of many species of fishes. Seventy-five percent of the recreational and commercially important

---

Box 10.2   Types of wetlands.

Wetlands are areas where water is the primary factor controlling the environment and the associated plant and animal life. They occur where the water table is at or near the surface of the land, or where the land is covered by shallow water. Five major wetland systems are generally recognized: marine (coastal wetlands including coastal lagoons, rocky shores, and coral reefs); estuarine (including deltas, tidal marshes and mangrove swamps); lacustrine (wetlands associated with lakes); riverine (wetlands along rivers and streams); and palustrine (meaning "marshy" – marshes, swamps and bogs). Wetlands have fundamental ecological functions, as regulators of water regimes and as habitats supporting a rich biodiversity. Over two thirds of the world's fish harvest is linked to the health of wetland areas. Ensuring their wise use should conserve wetlands. Wise use is defined as sustainable use of a wetland so that it may yield the greatest continuous benefit to present generations while maintaining its potential to meet the needs and aspirations of future generations (from Davis & Blasco 1997).

---

fishes in the US depend upon estuaries and brackish water marshes for nursery grounds. Floodplains and freshwater marshes play similar roles for many freshwater fishes. Nevertheless, ports and harbors are built in important nursery habitats, primarily wetlands, marshes, and seagrass meadows (see Box 10.2, Chapter 7). The associated development removes shallow habitat, changes hydrology, and interferes with along-shore movement of fishes. Water diversions for flood control, irrigation, and drainage change the hydrology of environments and thereby alter larval fish transport, retention, and the quality of nursery habitats.

Nearly 50% of the wetlands and marshes in the US were lost between 1780 and the early 1950s. This occurred before we fully understood the importance of these habitats to fish production and how their loss would affect larval fish survival. Comprehension of this important value of wetlands led to the enactment of laws to protect the remaining habitat and since that time wetlands loss in the US has slowed considerably. At the same time, a global overview indicates that massive historical losses of wetlands have occurred worldwide, and the majority of the remaining wetlands are degraded or under threat of degradation. Generally, wetland loss is difficult and costly to reverse, although wetland restoration and wetland creation are increasingly popular applied science and conservation tools.

Land use changes in watersheds

The clearing of terrestrial vegetation in watersheds increases runoff, river flow, and sedimentation in a water body, consequently changing habitat characteristics important to eggs and larvae. Upstream development and channelization also increases water flow to the lower parts of a river. The increased flow removes vegetation, washes eggs from the substrate, and inundates backwater areas that are important for feeding by many larvae. The increased sediment loads and sedimentation rates can reduce survival of fish embryos. For example 98% of brown trout (*Salmo trutta*) eggs in streams with high siltation rates died during the incubation period, compared to control sites with only 15% mortality (Turnpenny & Williams 1980). A silt deposition rate of 1 mm per day resulted in 97% mortality of eggs of northern pike (*Esox lucius*) (Hassler 1970). Fish species that require clean, well-oxygenated substrates for depositing their eggs, such as paddlefish (*Polyodon spathula*), may be particularly vulnerable to increased sedimentation. Once widespread and abundant in the free-flowing rivers of the Mississippi River drainage in the US, paddlefish are now listed as a "species of concern" by the US Fish and Wildlife Service.

Clearing vegetation in watersheds also causes substantial changes in trophic dynamics. Sedimentation and the resuspended sediments increase turbidity, reducing light levels that are essential for photosynthesis by submerged aquatic plants and corals. Phytoplankton blooms often occur when nutrients are increased by the augmented runoff and sedimentation. If these processes continue, physiological stresses could eliminate the natural primary producers, changing the base of the food web and consequently its entire structure. This is especially important for fish larvae because, as we learned in Chapter 1, they are more prone to starvation than juveniles and adults. Their growth and survival depends upon encountering and capturing prey of the right size within a narrow time frame. As we will see later, land use changes in watersheds have had important consequences for fish populations in the Laurentian Great Lakes and in the Danube River (see Chapter 11).

Trawl and dredge fishing

Commercial fishing operations, such as shrimp trawling and clam dredging, physically alter benthic habitat. The continental shelves have been trawled and dredged so often over decades that there are probably no unaltered shallow continental shelf habitats anywhere in the world. In addition to the damage to benthic habitats, trawl and dredge fishing increases turbidity, which may negatively affect primary production as described above. Trawls and dredges directly affect demersal eggs and young fishes when these organisms are dredged or passed through the net.

10.2.2 Biological impacts on population and community structure

Biological impacts include commercial exploitation of fishes, the introduction of exotic species, and harmful algal blooms. These impacts often affect early life stages indirectly through changes in population and community structure that reduce the number of eggs produced or alter the population structure of predators or prey of fish larvae.

Fishing pressure

Many societies in the developing world depend upon aquatic ecosystems to provide their only source of protein and livelihood. About 16% of the total animal protein consumed by humans on a worldwide basis is derived from fishery resources, while in the developing world more than a third of the total protein comes from fish. However, most of the major fisheries of the world are in serious decline, and all have either reached or exceeded their maximum sustainable yield. The larger individuals of a species are often the target of commercial fishing, including some that have not yet spawned (for example, salmon) and others that have not yet reached their potential maximum fecundity. In general, early life stages are not directly affected, with the exception of roe fisheries (for example, sturgeon) and the collection of young fishes for aquaculture (for example, milkfish). The indirect impacts of fishing, acting through population and community level changes can be substantial. Reduced population size means that fewer eggs are produced, and changes in trophic structure often lead to reduced survival of the eggs and larvae that are produced.

Heavy fishing pressure on a single species, as is generally the case, dramatically alters community structure and food webs. For example, targeting a large predatory fish such as tuna releases pressure on its prey, allowing the forage species to increase in abundance. On the other hand, fishing on small forage fishes such as anchovies and sardines removes food important for predators (for example, cod and hake) and releases predation pressure on small invertebrates. This could result in increased grazing pressure on phytoplankton and ultimately a shift in primary production to other aquatic plants. The indirect effect on the ecosystem has rarely been investigated, but key changes in trophic and competitive interactions are likely to have serious consequences for fishes and their eggs and larvae. Research in the future will focus increasingly on understanding the mechanisms of ecosystem-level changes on the early life stages of fishes.

Exotic species

Species that flourish in a new geographic area as a result of man's activity are called exotic or introduced species. The introduction of species into new environments is a major contributor to depletion and extinction of indigenous fishes, second only to habitat loss. Large numbers of species are worldwide travelers and take advantage of the transportation system man has developed to move from place to place. Eggs and larvae are well adapted for dispersal, enabling organisms to invade new habitats and expand their range. Man-made devices, such as ships and canals, provide opportunities for dispersal into water bodies previously inaccessible to a species (see Box 10.3). For example, it is likely that some new invasions into the Great Lakes came in the ballast water of transoceanic ships passing through the St. Lawrence Seaway, a shipping channel completed in 1959 that connects the Atlantic Ocean with the Laurentian Great Lakes.

Exotic species are often deliberately introduced as sport fishes or unintentionally introduced when bait is discarded. Escaped aquaculture or released aquarium organisms also contribute to the exotic species pool. Frequently, exotic species have extraordinary effects

---

Box 10.3   Transoceanic shipping: a modern Trojan horse.

Ships take on water in special holding tanks to balance their loads. When they reach another port and take on more cargo, they release this ballast water into the harbor, bringing with it any organisms it contains. According to a recent study, at least 2.4 million gallons of ballast water arrive in U.S. harbors from foreign ports every hour. Life-history traits that facilitate ballast-water transport are planktonic larvae, wide salinity tolerance, and occurrence in estuarine waters where shipping traffic is frequent. The zebra mussel (*Dreissena polymorpha*) and more than 30 other exotic aquatic organisms, including such fishes as the ruff (*Gymnocephalus cernuus*) and the tubenose goby (*Proterorhinus marmoratus*), likely were introduced into the Great Lakes via ballast water dumping. The problem is not limited to the Great Lakes, however, but is of great concern worldwide in both freshwater and saltwater ports.

on an ecosystem because they can quickly dominate communities. For example, Nile perch (*Lates niloticus*) were introduced into Lake Victoria in Africa in the early 1960s to increase fish catches in the lake and within a decade there were catastrophic effects on the native cichlid populations. As a result of predation, competition, and food-web changes, two-thirds of the 300 species of cichlids once found in the lake are lost. Nile perch now make up 80% of the fish biomass in Lake Victoria.

Harmful algal blooms

Single celled algae that grow very fast, accumulate into dense, visible patches near the surface of the water, and have negative impacts are called harmful algal blooms. Some of these species of algae produce potent neurotoxins that can be transferred through the food web. These toxins may injure or kill higher forms of life such as zooplankton, fishes, and even humans that are exposed to them directly or through their diet. Harmful algal blooms affect early life stages of fishes directly through toxins that kill larvae or reduce feeding rates, or indirectly through trophic changes that affect larval feeding and predation. Intense blooms can lead to anoxia and mass mortality, while persistent blooms cause changes in trophic dynamics of the system. Trophic interactions and ecosystem structure are often radically altered in response to such blooms. For example, in South Florida algal blooms have contributed to the marked decline in extent and vigor of seagrass ecosystems that provide a vital nursery habitat for many species of fishes. The result has been a shift in dominance from the resident benthic and epibenthic species (rainwater killifish, *Lucania parva*; gold spotted killifish, *Floridichthys carpio*; and gulf toadfish, *Opsanus beta*) to pelagic planktivores (bay anchovy, *Anchoa mitchilli* and Spanish sardine, *Sardinella aurita*).

10.2.3 Pollution

Expanding human population growth, industrialization, and urbanization have increased the amount of pollution entering lakes, rivers, and estuaries. Included among these pollutants are excess essential nutrients (phosphorus, nitrogen, and ammonia) and toxic chemicals (pesticides, herbicides, and heavy metals). These pollutants may reduce egg production, increase egg or larval mortality, alter the physiology and behavior of larvae to reduce viability, and alter trophic structure.

Nutrients from urban and agricultural waste

Eutrophication of aquatic habitats is a well documented result of nutrient over-enrichment. Nitrogen and phosphorus are essential nutrients in aquatic productivity but over-enrichment stimulates excessive growth of algae. For example, phosphorus is often the limiting nutrient for primary production in lakes, but too much phosphorus causes massive growth of nuisance algae. Recent efforts to decrease phosphorus concentrations in discharges have dramatically reversed the eutrophic conditions in aquatic systems such as Lake Erie. The reverse condition, oligotrophication (that is, a significant reduction in productivity), is now an issue in the Great Lakes because the reduction in phosphorus was accompanied by a dramatic

---

Box 10.4   Oxygen-starved coastal "dead zones."

Researchers first documented hypoxic water pockets off the Mississippi River's mouth in the early 1970s. The zone gets its start early each year when melting snow and spring rains wash nutrients, including nitrogen and phosphorus, off the land into the rising river. The warmer, lighter river plume spills out into the Gulf of Mexico, sliding over the heavier, saltier ocean water to form a layer on top (see Section 4.5.4). Fueled by sunlight and the nutrients, massive blooms of algae thrive near the surface, attracting copepods and other organisms that graze on plankton. Dead algae and the grazers' fecal pellets sink to the bottom where oxygen-consuming bacteria devour them. Hypoxia occurs when oxygen levels in the bottom water drop below $2\,\mathrm{mg\,l^{-1}}$, a level too low to support most marine life. Anoxia occurs when the bacteria consume all of the oxygen, suffocating even themselves.

---

decrease in the abundance of phytoplankton and an associated decline in zooplankton biomass. Such changes affect feeding dynamics and can potentially reduce survival of pre-recruit stages of fishes.

Eutrophication induces blooms of macroalgae and filamentous epiphytes that reduce light availability to aquatic macrophytes that form critical nursery habitat for many fishes. This results in lower productivity by the macrophytes, habitat loss due to hypoxia, and eventual death of sensitive species. Loss of seagrass in Chesapeake Bay, for example, has been attributed to the deterioration of water quality caused by high nutrient input from agriculture and upland development. Atmospheric deposition of ammonia is a recently recognized source of excess nutrient input to aquatic systems. Ammonia released into the air from feedlots and other high concentrations of farm animals has contributed to changes in primary producers and food-web dynamics in aquatic systems far removed from the source. Runoff from agriculture has been suggested to play a major role in producing extremely large areas of anoxic water in the Gulf of Mexico near the mouth of the Mississippi River.

High turbidity caused by blooms of algae or other aquatic plants characterize eutrophic habitats. Decomposition of these plants reduces the water's dissolved oxygen content, adversely affecting fishes and other aquatic life forms. Low dissolved oxygen concentrations cause sustained periods of oxygen depletion (see Box 10.4). In the brackish Baltic Sea, marine species are able to utilize the higher salinity water found below the halocline in the deeper basins for reproduction. However, because of increased eutrophication in recent years, oxygen is often too low for the developing offspring. Larvae of pelagic spawners such as cod are most abundant below the halocline, where the oxygen concentrations are no longer suitable. Recruitment levels of Baltic cod have decreased continually since 1981, leading to serious declines in their populations.

Toxic chemicals from industry, agriculture, and urban runoff

Thousands of different toxic substances enter aquatic ecosystems via effluents from industrial activities and runoff from agricultural and urban areas. Toxic substances include heavy metals, acids, chlorine, pesticides, herbicides, and petroleum-derived hydrocarbons. These substances may be immediately lethal or cause physiological or behavioral changes that will eventually lead to death of the individual. Exposures that affect early life stages can occur in many ways: through maternal exposure, exposure of eggs and sperm during fertilization, exposure of the embryo, or through post-hatching exposure. Fish eggs and larvae are at least partly permeable to contaminants in water and several classes of contaminants, especially those that are lipophilic, are able to accumulate in the yolk as a result of maternal exposure. For example, selenium ingested by adult cyprinids accumulates in the yolk of eggs, causing morphological deformities during yolk absorption. Acidic precipitation with pH in the range of 3.5–4.5, common in the northeastern US and adjoining Canadian provinces, is lethal to yolk-sac larvae.

Sublethal exposure to contaminants may suppress the immune system or reduce anti-predator or feeding abilities. Contaminants have been shown to negatively affect escape responses of fish larvae and all aspects of the feeding sequence, including detection of the prey, prey capture, handling time, and ingestion of prey. In addition, food-web changes generated by toxic chemicals can alter the types and abundances of predators and prey of fish larvae.

10.3 Human impacts on early life stages

10.3.1 *Mortality due to changes in habitat structure*

Loss of nursery habitat and accessibility to remaining habitat has limited the number of recruits and seriously affected many fish populations. Alterations to aquatic habitats may interfere with transport and dispersal patterns and prevent larval fishes from reaching critical nursery habitat. Larvae of many fishes such as croakers and flounders (Figure 1.3r, t) are spawned in offshore or coastal waters and are transported to estuarine nursery grounds. Similarly, the spatially complex floodplains and backwaters of rivers are spawning and nursery habitats for freshwater fishes. Considering the importance of hydrological characteristics such as currents and tidal excursions for successful transport of larvae to nursery habitats (see Chapter 7), it is clear that changes in these characteristics can significantly impact survival. For example, increased stream flows and associated turbulence, dislodge brown trout eggs. High stream flows also reduce water temperature, which not only adversely affects rates of embryonic and larval development in American shad (*Alosa sapidissima*) but also reduce the density of zooplankton that serve as food for the larvae.

Structures that reduce or eliminate seasonal flooding, such as weirs and levees, can eliminate important refuge and foraging habitat for young fishes. Altered hydrology can also disperse aggregations of prey and subject fish larvae that depend upon concentrated zooplankton to starvation. Changes in streams flow are often accompanied by changes in temperature and light penetration. These factors regulate primary production and ultimately, the prey available for larvae. In general, loss or modification of wetland habitat means that suitable habitat for larvae may not be available. The backwater areas and tributary streams of large free-flowing rivers, with their dense concentrations of plankton, were important feeding and nursery areas for riverine species such as paddlefish. These habitats have been greatly reduced or eliminated through the modification of all the major rivers of the US and Europe. The case study of the Danube River in Chapter 11 documents some of these effects.

10.3.2   Mortality due to changes in water quality

Eutrophication may directly affect eggs and larvae. It has been suggested that algal exudates associated with eutrophication cause high embryo mortality in Baltic herring (*Clupea harengus*, Figure 1.3m). Certainly the high turbidity associated with eutrophication interferes with sensory systems used by larvae to feed, avoid predators, and to find and recognize suitable habitat. Suboptimal visual conditions caused by low light and high turbidity have been shown to reduce foraging success of Atlantic herring and bluegill (*Lepomis macrochirus*) larvae. Low oxygen levels associated with eutrophication can be lethal to eggs and larvae, or can be sublethal, causing reduced feeding as well as increased predation on the larvae. Denise Breitburg and colleagues (1997) discovered that changes in these predator–prey interactions result from differences among species (both fishes and invertebrates) in their physiological tolerance to low oxygen and how low oxygen affects their escape behavior. Low oxygen acts on swimming and feeding behaviors of the predators as well as the prey. Thus, low dissolved oxygen has the potential to cause major and perhaps predictable alterations in aquatic food webs.

Fish larvae are especially susceptible to toxins produced by harmful algae such as the "red tide" dinoflagellate *Karenia brevis* (formerly *Gymnodinium breve*). The Texas brown tide (*Aureoumbra lagunensis*) is toxic to newly hatched sciaenid larvae and significantly reduces feeding rates in first feeding larvae and in their zooplankton prey. Fishes and zooplankton avoid dense concentrations of certain harmful algal species, and laboratory studies indicate that some predators or grazers reject toxic species. Interestingly, the behavioral responses and toxin susceptibility of zooplankton and benthic grazers to toxic algae is often species specific. Ingestion of harmful algae has different consequences for different consumers, ranging from no effect to death. In addition harmful algal blooms cause severe light attenuation in estuaries that significantly reduces light available to submerged aquatic vegetation. Long-term or continuous bloom conditions (such as occurred with the brown tide in Long Island Sound) cause reductions in submerged plant biomass and eventually loss of crucial nursery habitat.

Larvae are the most sensitive life stage to many toxic chemicals in the environment, as evidenced by the wide array of morphological and behavioral responses reported in the literature. Sublethal exposure may cause reduced feeding ability, reduced predator avoidance, or suppressed immune response, all of which increase the probability of dying. Contaminant effects are often exacerbated by natural environmental factors. For example, the frequency of contaminant-induced malformations in larvae is altered by temperature.

At the community level, indirect effects of contaminants cause changes in the susceptibility of zooplankton to predation thereby altering food webs and larval feeding rates. When food webs are changed, the prey preferred by larval fishes at different ontogenetic stages may be diminished or eliminated. Particularly important, are increases in size of prey available for larvae as they grow. Trophic changes may also impact the prey available to adult fishes, resulting in inadequate nutrition and reduced egg production. In the Baltic Sea, reproductive failure in the commercially important species is linked to changes in their clupeid prey. High mortality and abnormalities seen in embryos of clupeids, major food items of cod and salmon, are attributed to toxic algae and pollutants.

10.3.3   *Exotic species and early life stages*

Early life stages can play a central role in expansions and invasions of species into new habitats. Some species that have invaded the Laurentian Great Lakes, such as the round goby (*Neogobius melanostomus*), may have been transported as adhesive eggs on the hulls of ships. This fast growing, territorial fish is tolerant of a broad range of environmental conditions and is therefore highly adaptable to new environments. It is suggested that the round goby feeds on the young of native species. In fact, they will raid nests and successfully prey on sunfish (Centrarchidae) eggs, sometimes even when the guarding male sunfish is present. It is not known how significant an impact round gobies could have on overall nesting success of sunfish. It appears, however, that successfully guarding the nest comes at a high energetic price for the male sunfish.

Profound changes in the ichthyoplankton of the Black Sea were caused in part by the ctenophore *Mnemiopsis*, a recent invader that is a voracious competitor for zooplankton. *Mnemiopsis* also feeds on the planktonic eggs and larvae of planktivorous fishes. Since the invasion of the ctenophore, the number of fish eggs have decreased by 2–4-fold and the number of fish larvae by 2–9-fold for different regions of the Black Sea. The large number of exotic species in western US streams has contributed to the threatened status of many native species. For example, introduced catfish and sunfish suppress native cyprinid populations by consuming their young.

Introduced species are known to act on early life stages of fishes by consuming food resources that native fish larvae would eat. The increased competition that may arise leads to reduced growth rates and increased mortality. Introduced species may also occupy or modify spawning habitats or nursery sites that are necessary for feeding, resting, or refuge from predators. They may also consume adults, eggs, and young of native fishes, spread parasites and diseases, or serve as prey to native species but lack certain essential nutrients, leading to death of native offspring.

10.3.4   *Multiple effects*

Human impacts are often not simple cause-and-effect relationships, but a combination of pressures on various life stages. An introduced species may prey on native species as well as modify spawning and nursery habitat. At the same time, the young of introduced species often consume prey that native fish larvae would eat, causing increased competition for food resources. As an example, declines in yellow perch (*Perca flavescens*) in the Laurentian Great Lakes are hypothesized to be due to declines in zooplankton attributed to heavy foraging by the introduced alewife and predation by round goby and alewife on egg and larval stages of the yellow perch (Figure 1.3j). Exotic species can further affect native populations by introducing diseases and altering primary production. Often human impacts are due to a combination of physical, biological, and chemical alterations. While these changes might be tolerated individually, they are often devastating when combined. Multiple stressors can lead to a cascade of changes throughout the ecosystem that make it difficult to decipher the important pathways that lead to high egg and larva mortality and the collapse of a fishery. The Early Mortality Syndrome found in salmonids in both the US and Europe (Box 10.5) exemplifies the complex interactions of multiple impacts,

---

Box 10.5   Cascading effects of multiple impacts on an ecosystem.

Salmonids in the Baltic Sea and in the Laurentian Great Lakes have recently had high incidences of yolk-sac larva mortality. This syndrome, variously called Early Mortality Syndrome (EMS), M74, and Cayuga Syndrome is associated with low thiamine concentrations in the eggs and appears to be caused by an unknown biochemical component in the adult diet that reduces the bioavailability of thiamine. In the Great Lakes, alewife (*Alosa pseudoharengus*) and rainbow smelt (*Osmerus mordax*), recent marine invaders, are the suggested culprits because they contain the thiamine-degrading enzyme thiaminase and they make up the major component of the contemporary diet of salmonids. These exotic species were able to flourish in the Great Lakes because they had few predators or competitors. Native species such as lake trout (*Salvelinus namaycush*), whitefishes (Coregoninae), and Atlantic salmon (*Salmo salar*) declined precipitously as a result of predation by the sea lamprey (*Petromyzon marinus*) and overfishing.

In contrast, the Baltic salmon and their most common prey, sprat (*Sprattus sprattus*) and herring (*Clupea harengus*), have coexisted for several millennia. Both of the clupeids also contain thiaminase. A major portion of Baltic salmon experience thiamine deficiency-dependent mortality. Yolk-sac larva mortality rates of 40 to 95% have been recorded along the east coast of Sweden. In affected Baltic salmon, the lowered concentration of thiamine is also accompanied by reductions in carotenoids and other antioxidants. These dietary deficiencies may be a result of alterations in community composition of algae and microorganisms in the Baltic Sea. Reduced concentrations of antioxidants implicate a general oxidative stress syndrome. Environmental contaminants may play a role since chloro-organics, such as DDT and PCB, have been linked to thiamine deficiency in mammals. There may be a contaminant–thiamine interaction such that contaminants increase the thiamine requirement of early life stages (from McDonald *et al.* 1998).

---

in this case introduced species, pollution, and overfishing, that severely reduce reproductive success through action on sensitive early life stages.

10.3.5   *Population changes due to reduced egg production and survival*

Human impacts that change life-history characteristics of adult populations can directly influence early life stages. For example, overfishing has altered life-history characteristics of fish populations including sex ratios and size and age at maturity. These changes are linked to egg size and fecundity, and have important implications for larval survival, as discussed in Chapter 1. For example, Baltic cod are rare as a result of a commercial fishery targeting large cod. The resulting reduction in spawning biomass drastically reduced egg production and subsequent recruitment of young fish. Some species respond to fishing pressure by maturing at a smaller size and younger age. But the biomass of eggs produced is positively related to body size, so spawning at a smaller size will reduce egg production. In contrast, larger fish often produce larger eggs with more energy for growth and development. This results in larger larvae that grow faster and have a better chance of survival.

The availability of spawning and nursery habitats for anadromous species has been greatly reduced by dams and levees on major rivers. Such structures also reduce access to backwater and floodplain nursery habitats for riverine species. Dams have interfered with natural reproduction of salmon, notably in the western US and Europe. Dams constructed on major rivers feeding the Baltic Sea have reduced spawning areas by two-thirds. As a result, recruitment of naturally produced salmon smolts has decreased by an order of magnitude.

Permanent reductions in spawning biomass have resulted from habitat loss and overfishing. The quality of spawning grounds for fishes that spawn on gravel, cobble, and rubble (for example, lake trout, lake whitefish, and walleye) is reduced by increased sedimentation that degrades benthic habitat. Such activities alter important spawning and nursery habitats for both demersal and reef fishes such as snappers and groupers. Alterations that reduce the preferred prey of adults can cause poor reproductive success since maternal nutrition affects fecundity, egg size, and hatching success.

The reproduction of fishes allows studies of both long term and acute effects on adult fishes and their sensitive early life stages. Exposure of adult fishes to pollutants can decrease reproductive output through reductions in the size of gonads, diminished sperm motility, alteration of spawning behavior, and reductions in the number of eggs per spawn and the number of spawns per female. An interesting study by Schaaf and colleagues (1987) looked at the effects of contaminants on fish populations using a matrix model with first year survival rate and age-specific fecundity (average number of eggs produced per individual at each age) as inputs. The goals were to assess chronic and acute effects of pollutants on fish populations and to compare relative vulnerability to pollution among stocks of fishes. For ten fish stocks, the predicted time for a population to return to equilibrium after an acute perturbation (a one-time mortality of 50% of early life stages) ranged from 4 to 18 years. Fishes were even more susceptible to chronic pollution (0.5% reduction in survival during each year), and heavily exploited populations were the most seriously influenced by the additional pollution stress. Population recovery varied among the species as a function of age-specific survival and fecundity. These results indicate that the model provides a fairly straightforward way to use previously published data on these two life-history characteristics to assess the effects of pollution on fish populations.

Using a different approach, Ken Rose and colleagues (1993) produced an individual-based model to evaluate the effects of toxic chemical exposure of eggs and larvae on recruitment success in striped bass (*Morone saxatilis*, Figure 1.3q). Chronic exposure of eggs and yolk-sac larvae had a greater impact on age-1 survival than did episodic exposure. But episodic exposure resulted in substantial reductions in age-1 survival if exposure occurred during a major spawning peak. They also noted that small effects acting on already depressed year classes could cause significant further reductions in recruitment.

10.3.6 Generational effects on eggs and larvae

Generational effects are anthropogenic impacts on the fishes that are expressed in their eggs and larvae. Adult exposure to pollutants can be detected by chemical changes in the eggs causing deformities or reduced viability. Lee Fuiman and colleagues have investigated the influence of parental exposure to various pollutants on the development of locomotor and sensory performance in larval fishes. They found that parental exposure to DDT resulted in reduced swimming speeds and startle responses in Atlantic croaker (*Micropogonias undulatus*, Figure 1.3r) larvae, which could impair the larva's ability to find food and escape predators (Faulk *et al.* 1999). Sublethal effects expressed as developmental or behavioral abnormalities in eggs or larvae may be followed through several generations. White and colleagues (1999) demonstrated that fathead minnow (*Pimephales promelas*) larvae two generations removed from benzo(a)pyrene (a potent mutagen) exposure, showed marked decreases in survival and as adults had significantly reduced reproductive capacity.

10.3.7   Summary

Mechanisms through which man's activities affect fish early life stages are reduced fecundity in adult spawning stocks and increased mortality of eggs and larvae.

Reduced fecundity is caused by:

(1)   reductions in spawning stock (via commercial fishing, exotic species);
(2)   low genetic variability due to limited gene pool or inbreeding (via exotic species or hatchery fish, overfishing);
(3)   reduced size/age of spawning females (via overfishing);
(4)   altered sex ratio (via size or sex targeted commercial fishing);
(5)   poor nutritional condition (via physical disturbance, pollution);
(6)   reduced gamete production, or impaired spawning activity (via pollution);
(7)   inability to locate or access spawning habitat (via physical disturbance);
(8)   loss of spawning habitat (via physical disturbance).

Increased mortality in eggs and larvae results from:

(1)   changes in parental care (via physical disturbance, exotic species, pollution);
(2)   reduced viability (via pollution, harmful algae blooms);
(3)   altered transport/migration conditions (via physical disturbance);
(4)   loss or interference with cues to locate nursery or settlement habitats (via physical disturbance, pollution);
(5)   altered or lost nursery habitat (via physical disturbance, pollution);
(6)   increased predation rates (via physical disturbance, exotic species, commercial fishing, pollution);
(7)   behavioral changes in predator avoidance (via pollution, harmful algal blooms);
(8)   decreased feeding rates at one or more ontogenetic stages (via physical disturbance, harmful algal blooms, pollution);
(9)   increased competition for food and/or nursery sites (via physical disturbance, exotic species).

10.4   The coral reef example

Coral reefs, with over a quarter of all marine fish species, contain the most diverse assemblage of fishes on Earth. Life-history strategies in most reef fishes include a larval stage that drifts in the ocean for varying lengths of time before settling on the reef (see Chapter 7). Pelagic larvae may occur a few kilometers offshore or thousands of kilometers off the reef in oceanic waters, and remain at sea for up to 17 weeks before settlement. There is often precise habitat selection at settlement, presumably facilitated by olfactory, auditory, or visual cues. Some species settle directly on the reef, while others use adjacent nursery habitat where they develop into juveniles before moving onto the reef. Lagoons, seagrass beds, and mangrove forests are important nursery habitats. Predation is high on reefs, so when competent larvae arrive at reefs they select sites with abundant cover. The structural complexity of shallow reefs provides ample shelter and high productivity, reducing predation and providing plentiful prey.

10.4.1 Habitat loss

Coral reefs and their associated mangrove, seagrass, and other habitats are the world's most biologically diverse marine ecosystems. Human impacts to coral reefs mirrors those found in other aquatic ecosystems. Approximately 10% of the world's reefs have been lost and another 60% are threatened according to the US Coral Reef Task Force. Pressures come from high human population density and the associated deforestation and shoreline development. Removal of mangroves has occurred worldwide, primarily for conversion of land to agriculture and aquaculture, firewood and charcoal production, and coastal development. Globally, over half of all mangrove habitat is already gone and the remainder is being reduced at a rate of about 5% per year. Drainage of canals in nearby coastal wetlands increases freshwater and sediment discharges onto reefs. Sedimentation is an important threat. Reefs near many tropical coastlines are being smothered with mud, which kills corals and fishes. All of these alterations are exacerbated by worldwide degradation of coral reef ecosystems through coral bleaching that is apparently caused by global warming (see Box 10.6).

The effects of coral reef habitat alterations on the early life stages of fishes are loss of spawning and nursery habitat, increased competition for settlement sites, and structural changes that reduce prey abundance and predator avoidance. All ultimately reduce larval fish survival rates.

10.4.2 Biological impacts

Overfishing reduces effective population size and ultimately influences egg production. Fewer eggs and larvae are available to produce subsequent generations, further reducing population size of the species. Impacts include fishing on spawning aggregations, over-harvesting of herbivorous fishes, and size-selective targeting of sequentially hermaphroditic species. Almost

---

Box 10.6   Coral bleaching and global weather.

Coral bleaching is a poorly understood general response of corals to stressful conditions. Corals expel the symbiotic algae (zooxanthellae), lose their color, and enter a starving stage, unable to grow or reproduce. Ultimately, weakened corals may die. Bleaching is a frequent symptom of pollution-induced stress, as well as a response to natural factors such as changes in water temperature. Massive coral reef bleaching events, first noticed by marine scientists in the 1980s, have increased in frequency and severity since that time. The widespread bleaching events are thought to be a consequence of steadily rising water temperatures driven by anthropogenic global warming in combination with El Niño.

50% of reef fish families contain hermaphroditic species (for example, wrasses, groupers, and porgies). Generally, sex change occurs when fish reach a certain age or size. Intensive size-selective fishing on hermaphroditic fishes can disproportionately remove one sex, biasing the sex ratio and potentially reducing spawning biomass and egg production.

Many commercial species including snappers and groupers migrate varying distances to spawning sites where they form large spawning aggregations. Some, such as Nassau grouper (*Epinephalus striatus*), have very long-term site fidelity, spawning in a specific location every year for decades. Obviously, overfishing on such sites can easily wipe out a spawning aggregation. Several Nassau grouper spawning sites have been lost in this way.

Destructive fishing methods that utilize dynamite or cyanide are employed on many reefs, especially in the Indo-Pacific region. Cyanide is used to stun and capture live coral reef fishes for the aquarium trade and to capture larger live reef fishes for sale to specialty restaurants in Asia. Divers crush cyanide tablets into plastic bottles of sea water and squirt the solution at fish on coral heads. Anesthetized fish float up and are captured, but attached sessile marine organisms are damaged or killed and corals are damaged (often causing loss of zooxanthellae and coral bleaching). The complexity of coral reef habitat that is so important to early life stages is further reduced by these impacts.

10.4.3 Pollution

Increased nutrients disrupt normal trophic structure and dynamics of coral reefs by encouraging algal blooms. While coral reefs have high productivity, the water is very low in nutrients. Thus eutrophication develops at extremely low levels of added nutrients. Sewage generated by coastal populations adds excess nitrogen and phosphorus that stimulate prolific growth of algae. The growth of phytoplankton leads to decreased water clarity and reduced light for coral growth. The increased phytoplankton also encourages the growth of filter feeding organisms, changing the trophic dynamics of the system. Additionally, excessive phosphorus concentrations weaken the coral skeleton thus making it more susceptible to damage from storm action. These losses of structural complexity and alterations of trophic structure especially affect fish early life stages.

10.4.4 Summary

Impacts on early life stages of fishes include:

(1) loss or degradation of nursery habitat for species that use mangroves, seagrass beds, and shallow lagoons;
(2) changes in trophic structure that increase predation or competition for resources, or reduce prey availability;
(3) alterations or masking of clues needed by settling young to find and settle into appropriate habitat;
(4) loss of settlement habitat through alterations in physical structure, or increased predation due to loss of structural complexity;
(5) interference with post-settlement movements, or with spawning migrations;
(6) loss of fecundity, due to overfishing, especially on hermaphroditic species;
(7) loss or degradation of spawning habitat.

10.5 Interpreting human impacts using early life stages

The more we know about how organisms adapt naturally, the better we can assess their ability to keep pace with the rate of man-induced change in their environment. For example, how are fish eggs and larvae affected by the changes in average global temperature caused by rising levels of carbon dioxide and other gases? What are the effects of nutrient inputs on nursery habitats in rivers, lakes, and coastal waters? How do global weather patterns modify local hydrology important to larval transport? Scientists are beginning to understand many of these processes that are important to fish early life stages but much more research is needed. Early life stages could serve as sentinels to alert us to changes at several levels of biological organization. Impacts on individuals result in death or physiological and behavioral changes that lead to death. At the population level, there are reductions in fecundity and early life survival (vital rates). Food-web changes and outbreaks of disease have community-level consequences. At the ecosystem level, changes in global weather patterns can interfere with successful retention and transport processes that have evolved to insure larvae recruit to appropriate habitats. Reductions in numbers of eggs and larvae or their absence could indicate an environmental problem that has ecosystem-wide repercussions.

10.5.1 Understanding the effects of urbanization

Urbanization brings about many changes in aquatic ecosystems but it is often difficult to interpret the significance of those changes. Can one group of organisms be used to interpret the effects of urbanization? Probably not, but there are several reasons why fishes, specifically early life stages of fishes, might function as indicators of aquatic health. Eggs and larvae are especially vulnerable to habitat alterations, water-quality changes, and pollutants. Although controls on population size and community structure operate at many stages in the life history, the early life stages dominate. Moreover, fishes have social and economic value to man that provides an incentive for investigating impacts on fish populations.

In a study that provides an example of how fish early life stages can serve as indicators of anthropogenic stress, Limburg & Schmidt (1990) examined the capacity for egg and larval density to reflect the degree of urbanization along the Hudson River (USA). The river was divided into 16 watersheds and an urbanization gradient was defined as the percentage of the watershed that was in urban land use. Their aim was to determine if fish spawning success would be a quantifiable response to the level of urbanization. Eggs and larvae of anadromous and resident species were collected in each watershed along with water-quality parameters, physiography, and land use. Multiple regression analysis was used to examine the relationship of egg and larva density to the many variables measured. The strongest relationship was between numerical density of eggs and larvae and the index of urbanization. None of the other variables (for example, pH, dissolved oxygen, watershed area) alone was as strongly related, reinforcing the idea that the impacts of urbanization are complex and suggesting that the response of fish eggs and larvae can integrate these impacts.

10.5.2 Essential fish habitat

Habitat complexity plays a significant role in many of the processes that regulate population size, including egg production, larval dispersal, and survival, as well as post-settlement processes, such as growth, competition, and predation. Thus, degradation and loss of habitat are key factors in the decline of fishery resources. The US has begun to focus on the loss of fish habitat because fishes support a significant national economic resource.

The reauthorization of the Magnuson-Stevens Fishery Conservation and Management Act in 1996 acknowledged that habitat loss in the US has reduced the capacity to support sufficient fish populations. In fact, the act specifically finds

"One of the greatest long-term threats to the viability of commercial and recreational fisheries is the continuing loss of marine, estuarine, and other aquatic habitats. Habitat considerations should receive increased attention for the conservation and management of fisheries resources in the US."

Fish eggs and larvae play a major role in defining essential fish habitat since fish population size depends upon egg production and survival of the young. Both of these processes require specific spawning, nursery, and foraging habitats that are often impacted by man's activities. It is, however, no longer sufficient to assess only the impacts; the Magnuson-Stevens Act anticipates that problems of the past will be fixed. So there is a mandate to restore and recover the aquatic populations that have been affected and to rectify the problems created over many decades of use. This should provide a unique opportunity to look more critically at larval fish transport and habitat requirements, and especially such issues as the carrying capacity and the quality of the habitat. Ultimately, these studies will be used to evaluate and possibly restore specific nursery habitat found to be crucial to fishery resources.

10.5.3   Indicators of pollution

Early life stages of fishes are used in other ways to indicate environmental health. Eggs and larvae are very sensitive to pollutants and are especially useful as early warning indicators of environmental alteration. Small amounts of contaminants, low pH and dissolved oxygen levels, or low concentrations of toxic algae can be detected in the environment as mortality in newly hatched fish larvae, abnormal development, or by reduced feeding and growth of young larvae. More specific sublethal responses of embryos and larvae to toxic chemicals include:

(1)   egg abnormalities, including increased egg fragility, altered buoyancy, embryonic deformities, abnormal embryonic development, modified incubation period, decreased hatching success, increased mortality;

(2)   physiological changes during larval development, including altered yolk absorption, change in osmoregulatory capability, poor equilibrium, altered buoyancy, changes in respiratory and heart rates, retarded growth, and increased larval mortality;

(3)   behavioral changes resulting from impaired locomotor and sensory development that affect swimming performance, predator avoidance, or feeding rate;

(4)   suppressed immune system that can lead to lowered disease resistance; and

(5)   development of specific chemically induced deformities in larvae.

Integration of effects of sublethal pollution (that is, reduced feeding, impaired development, and deformities) occurs very quickly in fish larvae, making them a model indicator of environmental health and valuable in toxicity studies.

The early life stages of fishes are the most sensitive to environmental contaminants such as crude oil, pesticides, and heavy metals. Thus, these stages are often used in damage assessment studies following large scale spills and other significant sources of pollution. Chemicals such as selenium induce very specific deformities in fish larvae during yolk absorption. Adult dietary intake of excess selenium replaces sulfur in proteins that are formed and deposited in the developing egg. When larval fish rapidly absorb the contaminated yolk for building new tissues, characteristic deformities in the hard and soft tissues occur. A method for using these bio-indicators in fishes as a basis for evaluating impacts of selenium contamination was proposed by Dennis Lemly in 1997. He developed an index based on the relationship between selenium loads in larvae, the prevalence of deformities, and larva mortality. Collecting fish in different areas and identifying and enumerating deformities provided data to calculate a population-level mortality. The index is more useful than merely measuring selenium in the water because it provides a conclusive cause–effect link between the contaminant and the fish. Another widely known example of the use of fish eggs and larvae to evaluate a major contaminant is the Exxon Valdez oil spill that occurred in Prince William Sound, Alaska (see Box 10.7).

10.5.4   Use of eggs and larvae in toxicity testing

Toxicity tests are used to quantify both acute and chronic effects of pollutant exposure. Acute toxicity tests involve the short term exposure, commonly 96 h or less, of test animals to several concentrations of a chemical or mixtures of chemicals. Acute toxicity tests have been criticized because the criterion for effect is death, and the more pervasive and subtle effects of sublethal exposure that diminish long term survival may not be detected. Chronic or long term exposure to lower levels of the same pollutant may affect the development, growth, physiology, and behavior of individuals, ultimately reducing their potential for survival and putting a population at risk.

Common chronic or sublethal toxicity tests include the life-cycle, partial life-cycle, and early stage toxicity tests. Life-cycle toxicity tests measure the effects of a pollutant on growth, survival, maturation, and reproduction by following a species through an entire reproductive cycle or "from embryo to embryo." These tests, however, are both costly and time consuming, requiring six or more months to complete. Early life stage (ELS) toxicity tests decrease test duration by starting with exposures of embryos and larvae and terminating at the larval or, at most, early juvenile period. These tests establish toxic levels by quantifying pollutant stress in terms of hatching success, larval growth and survival, and occurrence of morphological deformities. The use of short term embryo/larva tests in studies of toxicants has been shown to be more sensitive than longer life-cycle tests for screening a large number of toxic compounds. Thus early life stages are often the best measure of the long term effects of contaminants in aquatic ecosystems.

10.5.5   Use of early life stages to mitigate human impacts

The culture of fishes in captivity can increase our understanding of man's impact on aquatic systems and has the potential for alleviating some of these problems. Stocking of young fishes that have been produced in captivity is one of the oldest and most common fishery tools. It is based on the premise that fish population size can be increased if more young are

---

Box 10.7 Exxon Valdez oil spill.

On March 24, 1989, the Exxon Valdez oil tanker ran aground on Bligh Reef, spilling 42 million liters of crude oil into Prince William Sound, making it the largest oil spill in US waters. Over the next several days, winds and currents distributed the oil throughout Prince William Sound and into the northern Gulf of Alaska, contaminating approximately 500 km of shoreline. Scientists working for federal and state resource management agencies as well as those at several universities immediately began planning a comprehensive assessment of oil spill damage to the Prince William Sound ecosystem. The spill occurred just prior to the annual spawning of Pacific herring (*Clupea pallasi*) and the emergence of pink salmon (*Oncorhynchus gorbuscha*) from gravel spawning beds. An estimated 40% of the shoreline used by spawning herring and much of the nearshore habitat important for salmon larvae and juveniles was contaminated by oil from the spill. Because both species comprise important commercial and subsistence fisheries in Prince William Sound, considerable attention was focused on their early life stages.

Immediately following the spill, scientists worked at correlating Exxon Valdez oil concentrations at specific sites within Prince William Sound with observed physical abnormalities and mortalities in Pacific herring and pink salmon. Increased incidence of premature hatching, morphological defects, reduced growth, and genetic damage in larval herring were observed in oiled vs. non-oiled areas. Reductions in embryo survival and larva and juvenile growth rates were reported for pink salmon in oil-exposed areas. The effects on embryo survival continued for several years, presumably because of persistent oil residues in the gravel. Further, eggs from female pink salmon returning to oiled areas demonstrated lower survival than eggs from fishes returning to unoiled areas, even when reared in clean water. Although the effects of Exxon Valdez oil on the early life stages of both species are well documented, the long term effects at the population level are harder to quantify. Mortality of early life stages at oiled sites may have resulted in loss of more than 40% of the total production of herring from Prince William Sound in 1989. As a result, recruitment of that year class into the spawning population was greatly reduced, and by 1993 the returning 1989 year class was one of the smallest on record. Scientists also estimated that almost two million pink salmon failed to return to Prince William Sound in 1990. Thus, both populations appear to have suffered important consequences of the exposure of their early life stages to oil following the spill. Despite obvious effects immediately following the spill, after 10 years both populations had increased in size and were thought to be recovering from the negative impact of the spill (from Rice *et al.* 1996 and Peterson 2001).

---

placed in the system. Initially, eggs or newly hatched larvae were released into the sea, but that proved to be unproductive. Subsequently, older and larger larvae, or even juveniles have been raised in captivity before release to avoid the heavy mortality of the early larval stages. In Japan, an immense number of fishes has been released since the 1960s in response to fishery declines. These stocking efforts appear to have contributed to sustained fishery landings. If habitats for critical life-history stages are altered, stocking young fish will not help a population recover. Stocking programs are most successful when overfishing has reduced stocks to a level where there are recruitment limitations. Therefore, stocking may be most beneficial as an aid in the recovery of an over-exploited fishery, along with implementation of other management tools that include a reduction in fishing pressure. Chapter 9 discusses this role of early life stages in stock enhancement more thoroughly.

A novel approach to resolving the problem of incidental damage associated with the collection of fishes for the aquarium trade is the culture of marine ornamental fishes in captivity. The cultivation of coral reef fishes for the aquarium trade conserves natural reef resources by offering alternatives to wild-caught organisms. Although many of the fresh-water tropical species sold to the public are cultured, the vast majority (over 90% as of 2001) of ornamental marine organisms are collected from the wild. Survival through the early life stages has proven to be the major obstacle to large scale production of a wide variety of marine ornamentals in captivity. Difficulties with providing appropriate prey for first feeding larvae and the lack of information on ecophysiological requirements during the pelagic larval stage need to be overcome for successful cultivation of many of the 700 species in the tropical fish trade. Recent interest by government entities bodes well for increased emphasis on captive culture of marine ornamental fishes.

Culturing fishes will also provide valuable details about life history, including size and age at spawning, fecundity, morphological descriptions of early life stages, age at first feeding, stage durations, and growth rates. These life-history traits are unknown for a large majority of fishes of the world, yet this information is critical for understanding fish population dynamics. In addition, data on ontogenetic changes in larval feeding and physiological responses to environmental parameters would be invaluable for interpreting the response of fishes to human impacts. Thus, increased understanding of fish early life stages at all levels is crucial to understanding and mitigating anthropogenic impacts.

10.6   Conclusion

Man's impacts on fish early life stages are clear, as is the need to reduce these impacts to maintain or regain vigorous fish populations. The role of fishes as an economic resource means that they can be used to highlight human impacts on the environment. Since habitat destruction, overfishing, and pollution reduce egg production and increase egg and larva mortality, a troubled fishery would need a reversal of these impacts to rebound. Such an argument was used to clean up badly deteriorated aquatic systems such as Chesapeake Bay and Lake Erie and will continue to play a crucial role in defining essential fish habitat. The value of fish early life stages as indicators of anthropogenic stress will increase as we learn more about the mechanisms that underlie the response of eggs and larvae to man-induced changes in their environment.

Additional reading

Benaka, L.R. (Ed.) (1999) *Fish Habitat: Essential Fish Habitat and Rehabilitation*. American Fisheries Society Symposium, No. 22, Bethesda, Maryland.

Boesch, D.E., Anderson, D.A., Horner, R.A., Shumway, S.E., Tester, P.A. & Whitledge, T.E. (1997) Harmful algal blooms in coastal waters: options for prevention, control and mitigation. NOAA Coastal Ocean Program, Decision Analysis Series, No. 10, *Special Joint Report with the National Fish and Wildlife Foundation*, February 1997.

Breitburg, D.L., Loher, T., Pacey, C.A. & Gerstein, A. (1997) Varying effects of low dissolved oxygen on trophic interactions in an estuarine food web. *Ecological Monographs* **67**, 489–507.

Bryant, P.J. (1998) Biodiversity and conservations: a hypertext textbook on biodiversity and conservation. (Online) Available: http://darwin.bio.uci.edu/~sustain/bio65/Titlpage.htm [February 20, 2002].

CRC Reef Research Centre (2001) Townsville, Australia. (Online) Available: http://www. reef.crc.org.au [February 20, 2002].

Fuiman, L.A. (Ed.) (1993) *Water Quality and the Early Life Stages of Fishes*. American Fisheries Society Symposium, No. 14, Bethesda, Maryland.

Hunter, J.R., Kaupp, S.E. & Taylor, J.H. (1982) Assessment of effects of UV radiation on marine fish larvae. In: *The Role of Solar Ultraviolet Radiation in Marine Ecosystems* (Ed. by J. Calkins), pp. 459–497. Plenum, New York.

Jennings, C.A. & Zigler, S.J. (2000) Ecology and biology of paddlefish in North America: historical perspectives, management approaches, and research priorities. *Reviews in Fish Biology and Fisheries* **10**(2), 167–181.

Jennings, S. & Kaiser, M.J. (1998) The effects of fishing on marine ecosystems. In: *Advances in Marine Biology*, Volume 34. (Ed. by J.H.S Blaxter, A.J. Southward & P. Tyler), pp. 203–314.

Kaufman, L. (1992) Catastrophic change in species-rich freshwater ecosystems: the lessons from Lake Victoria. *BioScience* **42**, 846–858.

McKim, J.M. (1995) Early life stage toxicity tests. In: *Effects, Environmental Fate, and Risk Assessment*, 2nd edn. (Ed. by G.M. Rand), pp. 974–1011. Fundamentals of Aquatic Toxicology. Taylor and Francis, Washington, DC.

National Oceanic and Atmospheric Administration. Coral reef. (Online) Available: http://www. coralreef.noaa.gov [February 20, 2002].

National Sea Grant Nonindigenous Species Sites, National Sea Grant College Program. (Online) Available: http://www.sgnis.org [February 20, 2002].

Rabalais, N.N. (1998) Oxygen Depletion in Coastal Waters, NOAA's State of the Coast Report, National Oceanic and Atmospheric Administration, Silver Spring, Maryland. (Online) Available: http://state-of-coast.noaa.gov/bulletins/html/hyp_09/hyp.html [February 20, 2002].

Rabalais, N.N., Turner, R.E., Justic, D., Dortch, Q., Wiseman, W.J. & Sen Gupta, B.K. (1996) Nutrient changes in the Mississippi River and system responses on the adjacent continental shelf. *Estuaries* **19**, 386–407.

Ramsar Convention on Wetlands (2001) Gland, Switzerland. (Online) Available: http://www.ramsar.org [February 20, 2002].

Sala, O.E., Chapin, F.S. III, Armesto, J.J., Berlow, E., Bloomfield, J., Dirzo, R., Huber-Sanwald, E., Huenneke, L.F., Jackson, R.B., Kinzig, A., Leemans, R., Lodge, D.M., Mooney, H.A., Oesterheld, M., Poff, N.L., Sykes, M.T., Walker, B.H., Walker, M. & Wall, D.H. (2000) Global biodiversity scenarios for the Year 2100. *Science* **287**, 1770–1774.

Vitousek, P.M., Mooney, H.A., Lubchenco, J. & Melillo, J.M. (1997) Human domination of Earth's ecosystem. *Science* **277**, 494–499.

Warner, R.R., Swearer, S.E. & Caselle, J.E. (2000) Larval accumulation and retention: implications for the design of marine reserves and essential fish habitat. *Bulletin of Marine Science* **66**, 821–830.

Zaitsev, Y.P. (1993) *Impacts of Eutrophication on the Black Sea Fauna. Fisheries and Environment Studies in the Black Sea System*. United Nations, Food and Agricultural Organization, Rome, pp. 64–86.

Chapter 11
Case Studies

Resurgence and Decline of the Japanese Sardine Population

Yoshiro Watanabe

11.1 Life history

The Japanese sardine *Sardinops melanostictus* is a small pelagic fish inhabiting the temperate waters around Japan. Adult fish larger than 17 cm in length spawn in the waters along the Kuroshio Current in the Pacific and along the Tsushima Current in the East China Sea and the Sea of Japan. Egg-production data show that the spawning season extends from November to May but is concentrated between February and March (Figure 11.1). This species shows a contranatant/denatant migration cycle, as described in Chapter 7. Hatched larvae are transported northeasterly by the currents and juveniles migrate to northern subarctic waters in summer for feeding. In the Pacific, young-of-the-year (YOY) sardines migrate south to the waters off Boso Peninsula in the fall to overwinter. The following spring they migrate north to the subarctic Oyashio waters for feeding again. After accumulating energy in summer for spawning, adult sardines migrate to the southwestern part of their migration range and spawn in winter at the age of 2 years (Figure 11.2). Although not genetically differentiated (Okazaki *et al.* 1996), the Japanese sardine population is divided into the Pacific stock and Tsushima Current stock for management purposes. The following account is for the Pacific stock, which is better known.

11.2 Population decline

The Japanese sardine experienced two population peaks in the 20th century, one in the 1930s and the other in the 1980s (Figure 11.3). Total catches of the two stocks of the sardine in Japan reached 1.6 million tonnes in 1936 and 4.5 million tonnes in 1988. During the years around the second peak, the sardine catch constituted 30–40% of the total production of fisheries and aquaculture of fishes, shellfishes, and seaweeds in Japan. Biomass estimated by virtual population analysis (VPA) exceeded 38 million tonnes in 1988 for the Pacific stock and 11 million tonnes in 1989 for the Tsushima Current stock (Japan Fisheries Agency 1997). The catch of the Pacific stock in 1989 was as large as 2.9 million tonnes, but still less than 10% of the total biomass. Major fishing grounds of the Pacific stock were established in the northern part of the migration range off Hokkaido Island during the

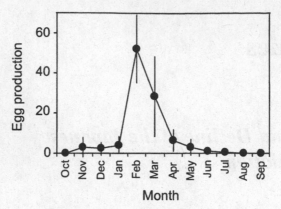

Figure 11.1 Eighteen-year (1978–1995) average (±s.d.) of monthly egg production as a percentage of annual production. Spawning season starts in November and ends in May. (Data from Mori *et al.* 1988, Kikuchi & Konishi 1990, Ishida & Kikuchi 1992, Zenitani *et al.* 1995, Kubota *et al.* 1999.)

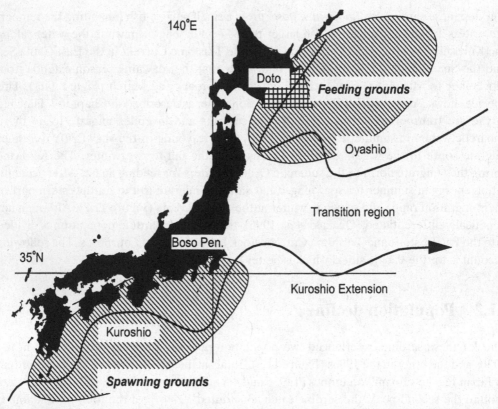

Figure 11.2 Schematic illustration of spawning grounds along the Kuroshio Current off southern Japan and feeding grounds in the Oyashio area off northern Japan. The Doto region (meshed area), located off the east coast of Hokkaido Island, constituted a major fishing ground in the 1980s. Immatures and adults migrate north to the Oyashio area in summer for feeding. Immatures migrate down to the waters off Fukushima and Boso Peninsula for wintering. Adults migrate further south to the spawning grounds in winter for spawning. Spawned eggs and hatched larvae are transported northeasterly by the Kuroshio Currents.

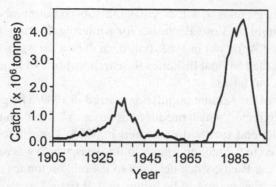

Figure 11.3 Long term fluctuations of the annual sardine catch in Japan. Two peaks, one in the 1930s and one in the 1980s, were dominant.

years of the second population peak in the 1980s. The Japan fishing regulation delimited this area of about $40\,000\,km^2$ as the Doto region (Figure 11.2). The catch in the Doto region exceeded 1 million tonnes in 1983–1988. In these years, fishermen were able to catch about 100 tonnes or 1 million sardines with one operation of a purse seine (Figure 11.4). The 24 purse seiners permitted to operate in the Doto region produced 10000 tonnes of sardines every day during the fishing season from July to October.

It was the purse seine fishermen who first noticed an unusual change in the sardine stock in the fall of 1988. They usually caught YOY sardines of 12–14 cm in the fishing grounds off northern Honshu Island in October and November, but in 1988 they did not see any young

Figure 11.4 A large purse seiner operating in the Doto region in summer. Twenty-four purse seiners were permitted to operate in the region for 4 months from July 1 to October 31 and produced more than 1 million tonnes or 10 billion sardines in the mid 1980s.

sardines in the fishing grounds at all. Normally, a substantial number of YOY sardines immigrated to the waters off Boso Peninsula for wintering, forming rich fishing grounds between November and April, but in 1988, only a small number was found. In spring 1989, a research vessel from the National Fisheries Research Institute of Japan tried to locate the young sardine schools but failed.

The drastic decline of the sardine population started in 1989. The purse seine fishery of the sardine in the Doto region, which produced as much as 1.2 million tonnes of sardines in 1988, collapsed rapidly and totally disappeared in 1994. The estimated biomass of the Pacific stock diminished to 0.7 million tonnes in this year, which was about 2% of the peak in 1988. The catch of the Pacific stock declined to 0.4 million tonnes in 1994, thus around 50% of the sardine stock was removed by fishing in that year.

11.2.1   Growth overfishing?

Marine fish populations replenish themselves every year through maturation and spawning of adults and survival and growth of young, which result in recruitment of a new generation to the population. A sustainable fishery harvest is possible by thinning out an appropriate proportion of a population. When a rapid decline of exploited fish populations or stocks occurs, it is generally understood that the cause of the decline is overfishing, a condition of a population or a stock which diminishes reproductive capacity to a level that cannot compensate for the losses due to fishing. Two different processes of overfishing have been recognized: growth overfishing results in a decrease in mean fish size with increasing fishing intensity, and recruitment overfishing results in a reduction of recruitment with increasing fishing intensity (Cushing 1977).

Age composition of the sardines caught in the Doto region changed greatly during the mid 1980s to the early 1990s. In 1984–1988 when the population was at its peak, each of 1, 2, 3, and 4+ year-old fish evenly occupied some 20–30% of the catches, with interannual fluctuations (Figure 11.5). Occurrence of the YOY (age 0) was occasional and variable in

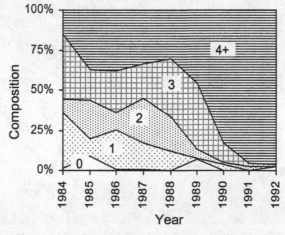

Figure 11.5   Age composition of sardines caught in the Doto region during 1984–1992. More than 95% of the catch came from fish older than 4 years in 1991 and 1992. This skewed composition resulted from recruitment failures in the 1988–1991 year classes. (Modified from Watanabe *et al.* 1995.)

this northern-most of fishing grounds. In 1989, 1-year-old fish (the 1988 year class) did not occur in the catch. One year later, the 1988 year class was still absent (2-year olds) and the 1989 year class was scare. This resulted in a large proportion of 3+ fish. One-year-old fish in 1991 (1990 year class) were virtually absent, and sardines younger than 3 years old constituted only 4% of the catch in this year. The age composition of the population became remarkably skewed to old age groups. In 1992, 1-year-old fish (1991 year class) did not occur and more than 96% of the catch was composed of fish older than 4 years (Watanabe *et al.* 1995).

If growth overfishing had caused the population decline, older age groups of large body size would have been the first to disappear from the population and a reduction in the average age of the population would have resulted. Most of the fish caught after 1989, however, were of the 1987 and earlier year classes. The average age of the catch increased greatly in the 4 years from 1989 to 1992 and older fish of large body size represented a large proportion of the catch. Thus, growth overfishing did not explain the decline of the Japanese sardine.

11.2.2   Recruitment overfishing?

The rapid increase in the composition of older age groups implies that recruitment of new generations to the sardine population was minimal after 1988. The number of age 1 immature sardines immigrating to the Doto region in summer was found to be a good index of recruitment and year-class strength of the sardine (Wada 1988, Figure 11.6). Estimates of the number of immatures during 1980–1987 had fluctuated by about one order of magnitude, from 2.1 billion in 1984 to 16.3 billion in 1983 (Figure 11.7). The large year classes that appeared in 1980, 1983, and 1986 put the sardine population up to the maximum biomass of 39 million tonnes in 1988. Recruitment in 1988, however, was as small as 0.1 billion and virtually no recruitment was found in 1990 and 1991. Severe recruitment failure occurred in the four consecutive years from 1988 to 1991. This was the cause of the rapid increase in the

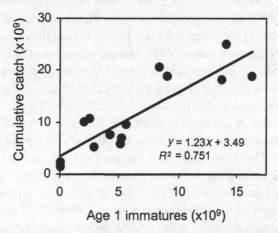

$$y = 1.23x + 3.49$$
$$R^2 = 0.751$$

Figure 11.6   Positive correlation between estimated number of age 1 immatures immigrating to the Doto region and cumulative catch from age 1 to 4+ in 14 year classes from 1975 to 1988 in the Doto region, indicating that the number of age 1 immature sardines immigrating to the Doto region is a representative index of year-class strength. (Data for age 1 immatures from Wada [1998], those for cumulative catch from Annual Reports from Kushiro Fisheries Experimental Station 1975–1992.)

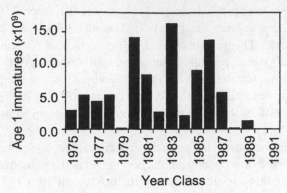

Year Class

Figure 11.7 Number of age 1 immatures in the 1975–1991 year classes. A dominant year class appeared every 3–4 years in the 1980s but recruitment was extremely low in the four consecutive years from 1988 to 1991. (Data from Wada 1998.)

proportion of older fish (Figure 11.5) and the precipitous decline of the sardine population (Figure 11.3). Intense fishing activity that exploited millions of tonnes of sardines a year might have reduced the spawning-stock biomass and resulted in low egg production in the years of recruitment failure after 1988. If this was the case, recruitment overfishing would be responsible for the population decline of the Japanese sardine.

Egg and larval censuses of small pelagic fish species such as the sardine, anchovy (*Engraulis japonicus*, Figure 1.3l), mackerels (*Scomber japonicus* and *Scomber australasicus*), and round herring (*Etrumeus teres*) have been conducted systematically since 1978 in the Pacific waters off Japan. The data from the census are used for estimating spawning-stock biomass and determining the spawning grounds and season of each species. These species produce pelagic eggs which are dispersed in the water column when spawned. Eggs of these species are immobile and quantitatively sampled by plankton net tows. A cylindrical–conical plankton net of 45 cm in mouth diameter and 0.33 mm in mesh aperture is used in the census. Data from 4000–6000 net tow samples per year were compiled in a spatially and temporally stratified database (see Watanabe *et al.* [1995] for details of the sampling and data compilation). Monthly egg production of the sardine was estimated from the monthly egg abundance data, and annual egg production was the sum of egg productions from October of the previous year to September (Figure 11.1). A map of the spatial distribution of egg production was constructed for each month and year (Figure 11.8).

Annual egg production of the sardine along the Kuroshio Current increased from 450 trillion in 1979 to 6660 trillion in 1990 with an exceptionally high production (8990 trillion) in 1986 (Figure 11.9). The production then declined sharply to 170 trillion in 1996. The mean annual egg production in the 4 years of recruitment failure from 1988 to 1991 was calculated as 4610 ± 1370 (mean ± s.d.) trillion eggs, which was greater than the production (1450 ± 480 trillion eggs) of the previous years from 1980 to 1987 exclusive of 1986. This result indicated that spawning adults of the sardine produced enormous numbers of eggs in the years of recruitment failure. The extensive egg and larval survey of the sardine clearly demonstrated that the recruitment failures from 1988 to 1991 were not owing to the decline of the reproductive output from the spawning adults. Recruitment overfishing did not explain the recruitment failures of the sardine from 1988 to 1991.

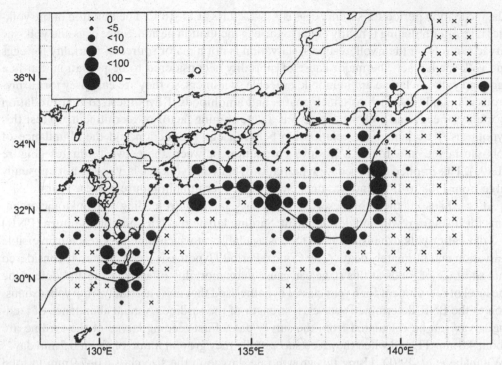

Figure 11.8 Sardine egg distribution along the Kuroshio Current (solid line) in the Pacific waters of Japan in 1990. Each closed circle indicates annual egg abundance in trillions in a 30′ × 30′ square of latitude and longitude.

11.2.3 Mass mortality in the early life stages?

The time series of the egg production of the sardine demonstrated that enormous numbers of eggs were spawned in the years from 1988 to 1991. Nevertheless, recruitment failed in these years. Mass mortality was suspected to have occurred after eggs were spawned.

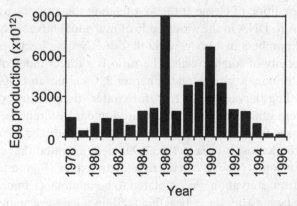

Figure 11.9 Time series of annual egg production in the Pacific waters of Japan. The production increased from 450 trillion in 1979 to 6600 trillion in 1990, then declined to 170 trillion in 1996. Exceptionally high production (8900 trillion) was recorded in 1986. (Data from Mori *et al.* 1988, Kikuchi & Konishi 1990, Ishida & Kikuchi 1992, Zenitani *et al.* 1995, Kubota *et al.* 1999.)

The incubation period for sardine eggs is 3.2–2.2 days at 15–18°C. The duration of the yolk-sac stage after hatching is nearly equal to the egg incubation period. Since eggs and yolk-sac larvae can rely on their yolk reserves, starvation is not a major source of mortality for eggs and yolk-sac larvae. The major cause of mortality is considered to be predation. Sardine eggs and yolk-sac larvae are immobile or feeble swimmers, so they are easy prey for carnivorous zooplankters, such as chaetognaths and amphipods. I hypothesized that predation mortality of eggs and yolk-sac larvae was a determining factor of recruitment. To test this hypothesis, I examined the relationship between egg production and the abundance of yolk-sac larvae and found a positive correlation between these two estimates (Figure 11.10a). This implied that predation mortality of eggs and yolk-sac larvae was not unusually large from 1988 to 1991 and was unlikely to cause recruitment failure in these years.

After exhausting their yolk reserves in 2–3 days, larvae start feeding on planktonic organisms for further growth and development. Sardine larvae reach a point of no return (PNR) 2–3 days after yolk exhaustion if no food is available, at which time they reach irreversible starvation and ultimately death. The Critical Period Hypothesis (see Chapter 4) considered that shortage of foods at this stage can cause mass mortality of early larvae and determine the level of recruitment for a year class. I hypothesized that mass mortality from yolk exhaustion to the PNR determined year-class strength of the sardine. Annual abundance of feeding larvae can be calculated from the survey data. First feeding larvae of the sardine are about 4.8 mm TL and when they feed successfully they grow at a rate of about 0.5 mm day^{-1} (Watanabe *et al.* 1991). Using this growth rate, larvae in the size class 6.0–7.9 mm in total length are calculated to be 2.4–6.4 days beyond first feeding and, therefore, should be past the critical first feeding stage of mass mortality by starvation. To test the hypothesis, I examined the correlation between abundances of yolk-sac larvae and feeding larvae. Annual abundances of feeding larvae were positively correlated with abundances of yolk-sac larvae (Figure 11.10b). This result implied that mortality at the first feeding stage by starvation and predation were not so variable as to disrupt the positive correlation in abundances between yolk-sac larvae and feeding larvae. In the years of recruitment failure, feeding larvae were abundant in proportion to egg production and yolk-sac larva abundance. The Critical Period Hypothesis was not applicable to the recruitment failure of the sardine population.

The nutritional condition of larvae at the first feeding stage can be assessed by measuring the ratio of RNA to DNA in the whole body of individual larvae. The amount of DNA is a function of cell numbers in a body, while that of RNA is directly proportional to the protein synthesis activity of somatic cells. The ratio is a direct index of protein synthesis activity and therefore somatic growth rate (Chapter 2; Clemmesen 1996). The ratios at the PNR for first feeding larvae were 1.17 for water temperature <17.5°C and 1.32 for >17.5°C, and were obtained from larvae reared under different feeding conditions in the laboratory and used for interpreting the ratio of wild larvae. The results from about 400 larvae collected in the Kuroshio area in 1993–1995 demonstrated that virtually no sardine larva was in poor nutritional condition at the first feeding stage (Figure 11.11) and the number of larvae dying from starvation was estimated to be minimal (Kimura *et al.* 2000b).

Although feeding larvae after the critical first feeding stage were abundant in 1988–1990, recruitment in these years failed. No correlation was detected between the abundance of feeding larvae and age 1 immatures through the 12 years of recruitment success and failure from 1979 to 1990 (Figure 11.10c). Some years, for example 1980 and 1983, produced large

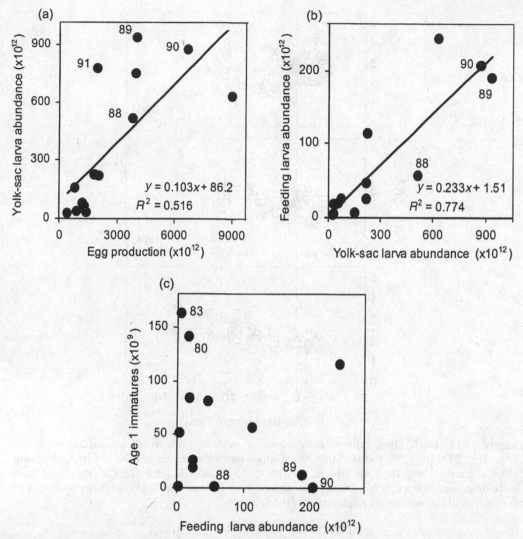

Figure 11.10 (a) Positive correlation between egg production and yolk-sac larva abundance for 15 years from 1978 to 1992. Mass mortality was not detected in the egg and yolk-sac larva stages in the 4 years of recruitment failure (1988–1991). (b) Positive correlation between yolk-sac larva and feeding larva abundances for 13 years from 1978 to 1990, indicating that interannual variability of mortality in the first feeding stage was not large enough to disrupt the correlation. (c) No correlation between feeding larva abundance and the number of age 1 immatures for 13 years from 1978 to 1990. Dominant year classes were established from low abundance of feeding larvae in 1980 and 1983, while recruitment failed from extremely high abundances of feeding larvae in 1989–1990. Numerals attached to the plots indicate year classes. (Data for egg production from Mori *et al.* 1988, Kikuchi & Konishi 1990, Ishida & Kikuchi 1992, Zenitani *et al.* 1995, Kubota *et al.* 1999.)

numbers of immatures from a relatively small amount of egg production (Figure 11.9) and less abundant feeding larvae. In contrast, an extremely small number of immatures resulted from an enormous amount of egg production (Figure 11.9) and feeding larvae in the years of recruitment failure. This suggested that variability in mortality after the first feeding stage

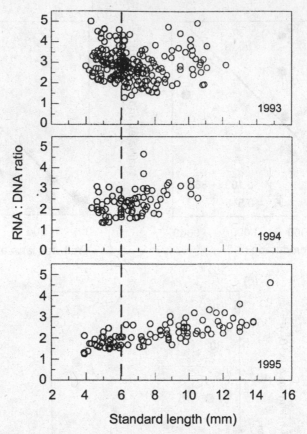

Figure 11.11 RNA:DNA ratio of early larvae of sardine collected off central Japan in 1993 (*n* = 213), 1994 (*n* = 78), and 1995 (*n* = 98). Dashed line indicates upper limit of first feeding stage (SL = 6.0 mm). Using the PNR criteria, 1.17 for <17.5°C and 1.32 for >17.5°C, virtually no larvae were diagnosed in starving condition. (Reproduced from Kimura *et al.* 2000b with permission of the International Council for the Exploration of the Sea.)

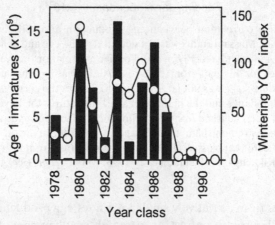

Figure 11.12 Interannual fluctuation of the YOY index (circle) and estimated number of age 1 immatures (column) during 14 years from 1978 to 1991. (YOY index from National Research Institute of Fisheries Science [1992], data for age 1 immatures from Wada [1998].)

determined the size of recruitment, unlike the prediction of the Critical Period Hypothesis. High survival after the first feeding stage in 1980 and 1983 produced large recruitment and dominant year classes. On the contrary, mass mortality after the first feeding stage destroyed the 1988–1991 year classes and led to recruitment failure.

The number of YOY sardines immigrating to the waters off Boso Peninsula for wintering has been used as the wintering YOY index, based on the catch of age-0 fish in the waters during November to March. The index agreed with the number of age 1 immatures migrating into the Doto region the following summer (Figure 11.12) and is used as the earliest reliable estimate of recruitment for a year in the sardine fishing forecast meeting in Japan (National Research Institute of Fisheries Science 1994). Size of recruitment is determined by the mortality suffered by fish from the first feeding to wintering YOY stages 8–12 months after hatching.

The results so far obtained lead to the conclusion that the magnitude of sardine recruitment is determined in the larval and juvenile stages up to about 10 months of age (about 13 cm in length). Neither predation on eggs and yolk-sac larvae nor starvation of the first feeding stage is specifically crucial to the recruitment-determining process in this case. Ecology of larval and juvenile sardines in the first spring and summer of their life needs to be investigated further.

11.3 Spawning and early life ecology

The data from egg and larval surveys (for example, Figure 11.8) provide us with information on the distribution and size of spawning grounds of pelagic fish species. Spawning grounds of the sardine changed in location and size through the process of population fluctuation (Watanabe *et al.* 1996). In the late 1970s when the population was below maximum, the spawning grounds were located in the coastal waters on the continental shelf off the Pacific coast of Japan. In the years of the population peak, the grounds expanded offshore to the oceanic waters along the Kuroshio Current axis where sea floor depths were some thousands of meters. The center of the spawning grounds was located 20–80 km inshore from the Kuroshio Current axis until 1984 but moved to the vicinity of the axis after 1985 (Figure 11.13). In the middle 1990s, egg production was greatly reduced and the grounds constricted to the coastal waters along the Pacific coast, similar to the distribution in the late 1970s (Kubota *et al.* 1999). During the period of population increase, the spawning grounds of the California and Chilean sardines expanded latitudinally along the west coast of North and South America (Schwartzlose *et al.* 1999). The offshore expansion in the Japanese sardine differs from the alongshore expansion of the California and Chilean sardines.

The dominant Kuroshio Current, the velocity of which is as much as 2 knots ($3.6 \, km \, h^{-1}$), transports large volumes of water northeasterly from the southwestern portion of the sardine spawning grounds. Pelagic eggs and larvae distributed around the current are swiftly transported. Stationary average flow fields at three levels of the water column (10, 50, and 100 m deep) in the waters off the Pacific coast of southern Japan were constructed from acoustic Doppler current profiler (ADCP) data and used to drive a particle-tracking model representing the dispersal of sardine eggs and larvae (Heath *et al.* 1998). When the model was used to simulate the transport of particles released in the waters ranging from the coast to the Kuroshio

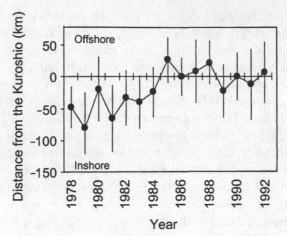

Figure 11.13   Latitudinal shift of the spawning ground center relative to the Kuroshio Current axis during 15 years from 1978 to 1992. The center was located in the inshore waters of the axis until 1984, but moved to the offshore waters in 1985 and stayed around that axis until 1992.

Current axis in southern Japan (first case), approximately 50% of egg production was exported to the waters off Boso Peninsula in 2 weeks, where the Kuroshio Current breaks away from Japan's coast and turns eastwards into the central North Pacific and is known as the Kuroshio Extension (Figure 11.2). A major proportion of the remaining 50% was transported to Pacific coastal waters. In a second case, when eggs were released only in the coastal waters, 5% of the egg production was exported to the Kuroshio Extension in 3 weeks and the remainder went to the coastal waters. Distribution of the spawning grounds during 1985–1992 was much more weighted to the Kuroshio area than in the first case, therefore more than 50% of the egg production was likely to have been transported to the Kuroshio Extension areas after being spawned in the years of offshore spawning.

Growth rates of early feeding larvae (determined from analysis of otolith daily increments) were positively correlated with ambient food concentrations (Oozeki & Zenitani 1996). Larval growth rates are variable and depend on environmental factors over a wide geographic range within the spawning grounds around the Kuroshio Current and transportation fields in the Kuroshio Extension area. In the spawning grounds, growth rates of early feeding larvae ranged from 0.3 to 0.9 mm day^{-1} (Watanabe *et al.* 1991). Early larvae transported to the offshore waters of the Kuroshio tended to grow more slowly and suffered high mortality (Zenitani *et al.* 1996) under low food availability (Nakata *et al.* 1995). Larvae transported to the Pacific coastal area formed concentrated shoals and grew at 0.6–0.8 mm day^{-1} (Watanabe & Kuroki 1997). Concentrations of copepod nauplii, food organisms of larval sardines, were distributed at 6–12 nauplii l^{-1} over the Kuroshio Current area (Watanabe *et al.* 1998).

The transition region lies to the north of the Kurosiho Extension between the Kuroshio Front (14°C at 200 m depth) and the subarctic Oyashio Front (5°C at 100 m depth, Figure 11.2). Although water temperature is relatively low, the Kuroshio–Oyashio transition region provides late larval and early juvenile sardines with higher food availability (12–17 nauplii l^{-1}) and constitutes a favorable nursery area (Watanabe *et al.* 1999). Early juvenile sardines grow at rates of about $1.0\,\text{mm}\,\text{day}^{-1}$ in the Kuroshio–Oyashio transition region (Kimura *et al.* 2000a), faster than the $0.7\,\text{mm}\,\text{day}^{-1}$ in the Kuroshio Current area off central Japan (Watanabe & Saito 1998). A recent study on Japanese anchovy (*Engraulis japonicus*) found that larvae that grow at a rate greater than $0.8\,\text{mm}\,\text{day}^{-1}$ successfully metamorphose into juveniles and eventually constitute adult populations (Takahashi 2001). Environmental variability in the transition region may be a determining factor of larval growth and eventual recruitment. The key processes seem to lie in late larval and early juvenile stages of the sardine inhabiting the transition region.

11.4 Implications for management

Regulating fishing by controlling the size of a total allowable catch (TAC) has been in effect since 1997 in Japan. Before 1997 the regulations were based on the number of fishing boats permitted to use a particular fishing gear and fishing area. For example, 24 large size purse seiners were permitted to operate in the Doto fishing region. Under this regulation, the fishermen caught as many fish as they could as long as the costs and gains were beneficial to them. Therefore, total catch or catch per unit effort of fishing could be used as an index of the density of fish in the fishing grounds. Under the TAC regulations, however, the central or local government allocates a total catch in weight per year to each fishing boat before the fishing season. Therefore, total catch cannot be used as an index of fish density. Because fishing activity is commercially operated, the cost-and-benefit rule works through fishing activities. If larger fish are more profitable than smaller fish on a weight basis, fishermen try to search for and catch fish of larger size and of older age. The size and age compositions of the catches can be greatly biased under this condition and are difficult to use in VPA estimates of fish biomass.

As discussed in Chapter 5, the size of a spawning stock can be estimated from the number of eggs it produces. Egg-production estimates of pelagic fish species are derived from egg and larval surveys and have the advantage of being free of biases arising from commercial fishing activity. One of the most sophisticated applications of this idea is the Daily Egg Production Method (DEPM, see Section 5.2), which was developed for the northern anchovy (*Engraulis mordax*) in the California Current region (Lasker 1985). The species-specific fecundity may be variable within and between spawning seasons, depending on environmental factors and/or spawning biomass. In this method, parameters of spawning activity are measured in the field together with the egg distribution during a spawning season. Methods based on egg production require more or less detailed analysis of adult spawning ecology and distribution and mortality of spawned eggs. A rough estimate of spawning biomass may be obtained by measuring the area of spawning grounds of a stock. In the Japanese sardine, the area is a positive function of the egg production (and therefore, spawning-stock biomass) (Figure 11.14). The area in this example was calculated by integrating areas of squares 30 min of latitude and longitude on a side in which sardine eggs were collected (Figure 11.8).

Recruitment of a new generation to a fish stock is a function of egg production by adult fish and mortality in the larval and juvenile stages. Studies on early life stages of fishes have the potential to establish a method of recruitment forecasting for a fish stock. A recent study on Japanese anchovy in the Kuroshio and Kuroshio Extension areas demostrated that late

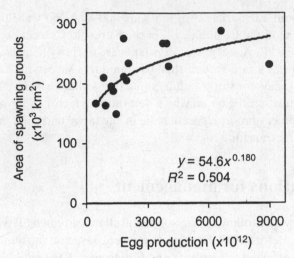

Figure 11.14   Regression of the area of spawning grounds against egg production of sardine along the Pacific coast of Japan. Rough estimates of egg production, and therefore spawning-stock biomass, can be obtained from the regression curve. (Data for egg production from Mori *et al.* 1988, Kikuchi & Konishi 1990, Ishida & Kikuchi 1992, Zenitani *et al.* 1995, Kubota *et al.* 1999.)

larvae (20–30 mm in length) with growth rates less than 0.7 mm day^{-1} have a low probability of successful metamorphosis to the juvenile stage and subsequent recruitment to the adult spawning stock (Takahashi 2001). We may be able to predict recruitment level of a coming cohort when the cohort is at this early stage of its life history. Recruitment variability is the most important component of fish population dynamics and it is one of the major subjects in fishery science. When established, recruitment forecasting will constitute essential information for management of fish stocks in the near future.

11.5   Summary

(1)   The Japanese sardine experienced two population peaks in the 1930s and 1980s and has been declining since the end of the 1980s. Age composition of the population became remarkably skewed to older age groups, implying that growth overfishing did not explain the recent population decline.

(2)   Time series of egg production estimated from egg and larval census data showed that even though enormous numbers of eggs were spawned in the years from 1988 to 1991, recruitment failed in those years. Recruitment overfishing was not the reason for the population decline of the sardine.

(3)   Recruitment failure from 1988 to 1991 was found to be owing to natural mortality after the enormous number of eggs were spawned. Detailed examination of egg and larval census data revealed that predation mortality on eggs and yolk-sac larvae was not unusually large in the years of recruitment failure. Starvation mortality in the first feeding stage was not critically high in these years. RNA : DNA data suggested that virtually no larvae were in poor nutritional condition at the first feeding stage.

(4) The number of age 1 immatures was not correlated with the abundance of feeding larvae. This suggested that mass mortality after the first feeding stage, probably up to 8–10 months of age, was responsible for recruitment failure from 1988 to 1991.

(5) Spawning grounds expanded offshore over the strong flow field of the Kuroshio Current in proportion to the amount of egg production, and presumably, spawning-stock biomass. A particle-tracking model of dispersal of eggs and larvae demonstrated that a large proportion of egg production was transported from the spawning grounds to the downstream Kuroshio Extension area in 2–3 weeks.

(6) Food availability for larvae and early juveniles was higher in the waters north of the Kuroshio Extension than in the spawning grounds. Otolith-based daily growth rates in early juvenile sardines were higher in the Kuroshio Extension area than in the Kuroshio Current area. The Kuroshio Extension area is thought to constitute a nursery of late larval and early juvenile sardines. Transport of eggs and larvae from the spawning grounds to the nursery area is an important component of early life history of the sardine.

(7) Spawning-stock biomass of sardines and other pelagic egg spawners can be estimated based on daily or annual egg production. A rough estimate of spawning-stock biomass may be obtained from the area of the spawning grounds. Biomass estimates based on egg and larval surveys are independent of commercial catch data and, therefore, one of the important pieces of information for management of fish stocks.

Cascading Effects of Human Impacts on Fish Populations in the Laurentian Great Lakes

James A. Rice

11.6 Introduction

The biological communities of the Laurentian Great Lakes have undergone dramatic changes over the last 200 years, many induced directly or indirectly by human behavior. Overharvest of fisheries and the establishment, both intentional and accidental, of exotic species have dramatically altered the food web and community structure of the lakes. Forest clear-cutting, shifts in land use, and changes in nutrient loading have altered the productivity of the Great Lakes system. Chemical pollution and waste disposal practices of the past continue to affect ecosystem functions today. These changes have presented extraordinary management challenges, and continue to do so.

Early work on fishes of the Great Lakes, dating back to the 1920s and 1930s, provided a basic foundation of knowledge concerning species distributions, early life histories, and basic biology. Unfortunately, over much of this time fish early life-history studies were not a major factor in fishery research and management. Historically, adult fishes were the primary focus of attention; larger fishes naturally drew people's interest and were the target of subsistence and commercial fisheries, which provided data from harvest records. They were more obvious than eggs and larvae and, in many respects, easier to sample. Even stock–recruitment analyses emphasized the role of spawning adults in determining recruitment.

More recently, fishery biologists have come to understand that events occurring during early life often are the primary forces that dictate the dynamics we observe in adult populations and communities. Over the last 20 years, an understanding of fish early life history has played a significant role in almost every major Great Lakes fishery management issue. How did this change in perspective come about, and how has it changed the way we approach fishery science and management? The Great Lakes story, far too rich a history to relate in full detail here, offers a wealth of opportunities to illustrate many of the concepts presented in this book. In the following pages I will share several examples that show how early life-history studies have been critical to understanding and managing the dynamics of Great Lakes fisheries, and have fundamentally changed our thinking about fishery management in general.

11.7   First, some background

The emergence of early life-history studies as a force in Great Lakes fishery management is best understood in the context of the history of Great Lakes fisheries. Until about 200 years ago, the fish communities of the Great Lakes were relatively stable. Lake trout (*Salvelinus namaycush*) and burbot (*Lota lota*) were the dominant predators, and a substantial portion of the fish population consisted of endemic species. Commercial fishing began in the early 1800s and was a major industry by 1850. Harvest continued to increase throughout the 19th century as better nets and improvements in fishing technology, such as the steam engine, allowed fishermen to range farther afield and harvest fishes more efficiently.

During this same period a growing lumber industry was deforesting much of the Great Lakes region. The reduced forest canopy increased water temperatures in tributaries, and increased erosion smothered spawning grounds in silt. Massive amounts of sawdust waste from sawmill operations were disposed of in rivers and floated 30–50 km out into the lakes, sinking and covering feeding and spawning grounds. The development of major cities and ports along the shore of the lakes led to increasing pollution and loss of wetlands. Construction of dams on tributaries impeded spawning migrations. By the late 1800s these impacts had already led to major declines in Great Lakes fisheries.

With further development of shipping came a new invasion, still going on today – colonization of the lakes by exotic species. In 1829 the Welland Ship Canal was built around Niagara Falls to allow ships to reach the four upper Great Lakes. Shortly after its enlargement in 1919, sea lampreys (*Petromyzon marinus*) began their march through the Great Lakes. These ancient parasites attach to large fishes and suck their blood, often until their victim dies. With no natural predators to control them, lampreys decimated populations of lake trout, lake whitefish (*Coregonus clupeaformis*), and other large species by the 1950s.

The rise of sea lamprey populations paved the way for the advance of another anadromous invader, the alewife (*Alosa pseudoharengus*). First reported in Lake Michigan in 1949, alewife populations exploded in the vacuum created by the decimation of the lakes' piscivorous predators. From 1960 to 1967 alewives increased from 8% of the fish biomass in Lake Michigan to a stunning 80% of the fish biomass, culminating in a massive die-off in the spring of 1967. The alewife invasion, in concert with other stressors, resulted in the extirpation or extinction of several common and endemic planktivorous fish species.

By the middle of the 20th century, the Great Lakes and their fisheries were in total disarray. These events galvanized previously unsuccessful efforts to initiate a comprehensive management strategy for the Great Lakes fisheries, culminating in the formation of the Great Lakes Fishery Commission by the United States and Canada. Their first mission was to find a way to control the sea lamprey.

11.8   In search of a vampire slayer

Clearly, it would be impossible to control lampreys in the large, open lake environment. The key to controlling lamprey lay in understanding their early life history, and, in particular, the concept of ontogenetic habitat shifts (Chapter 7). Because sea lampreys are anadromous, they move into small streams to spawn, depositing their eggs in clean gravel. After the young hatch, they drift downstream to muddy areas, burrow into the sediment, and spend 3–17 years as filter-feeding larvae called ammocoetes. Eventually the ammocoetes transform into parasitic adults and move into the lake where they feed on body fluids of host fishes for 1 or 2 years before returning to spawn.

It was in the spawning or larval stage, when sea lampreys and their young are concentrated in streams, where the battle would have to be joined. Early efforts to block lamprey spawning migrations using mechanical and electrical barriers failed. The focus then shifted to identifying a way to kill lamprey larvae, which were not only confined to streams, but were in the life-history stage when many fishes are most sensitive to toxic substances (Chapter 10). After testing more than 6000 chemicals over 5 years, US Fish and Wildlife Service biologists finally found the chemical TFM (3-trifluoromethyl-4-nitrophenol). It could be dripped into spawning streams to selectively kill lamprey larvae without significant harm to populations of other aquatic organisms. (They later identified a second chemical, Bayer 73, that worked more effectively in still or slowly moving water.) Stream treatments with TFM radically reduced lamprey populations, paving the way for efforts to re-establish lake trout, the primary native piscivore.

11.9   Bring back the lake trout

With lamprey populations greatly reduced, efforts to re-establish naturally reproducing lake trout were begun in the 1950s and 1960s, by rearing millions of lake trout (Figure 1.3d) in hatcheries and stocking them in the lakes. Lake trout populations rebounded as these stocked fish matured, but mysteriously the stocked lake trout failed to reproduce successfully. Restoration efforts are still continuing today, though with limited success except in Lake Superior.

Many hypotheses being explored for this limited success focus on early life stages. For example, naive hatchery fish may deposit eggs in unsuitable habitats because they do not home to historical spawning sites. Predation on eggs and larvae has been observed and may limit survival. There is also concern that chemical contaminants may inhibit egg viability and survival (Selgeby *et al.* 1995).

In recent years, another potentially important factor, called Early Mortality Syndrome (EMS), has been identified. This syndrome results in high mortality in salmonid larvae between hatching and first feeding due to a deficiency of thiamine, a B vitamin, in the eggs. Evidence suggests that this problem stems from the prominence of alewives, which contain high levels of thiaminase (a thiamine-degrading enzyme), in the diets of Great Lakes salmonids (see Box 10.5; Fisher *et al.* 1996, Fitzsimons *et al.* 1999). In some years, some salmonid populations in the Great Lakes Basin have experienced up to 90% mortality due to this deficiency. EMS also appears to be more common in contaminated ecosystems, and when spawning females exhibit deficiencies of vitamin E, an antioxidant. Fortunately, EMS can be prevented in the hatchery by treating eggs with thiamine during incubation.

When it became clear that lake trout populations were not going to rebound rapidly, managers turned to two other exotic species to help fill the void at the top of the food web. They stocked millions of chinook salmon (*Oncorhynchus tshawytscha*; Figure 1.3c) and coho salmon (*Oncorhynchus kisutch*), natives of the Pacific Ocean. These predators flourished, but not before the alewife population, free from predatory control, exploded. The alewife's rise to dominance would prove to have one of the longest lasting impacts on the Great Lakes fish community.

11.10   The secret lives of alewives

As alewives became abundant first in Lake Ontario, then in Lakes Huron and Michigan, native planktivore communities in each lake followed a similar pattern of dramatic decline. For example, by the 1960s all but one of the seven native species in Lake Michigan's coregonid deepwater cisco complex were either extinct or severely reduced. Only bloater (*Coregonus hoyi*), the smallest species, remained relatively abundant, but it too declined in the late 1960s. The emerald shiner (*Notropis atherinoides*) had been extremely abundant in Lake Michigan until the late 1950s, but disappeared in the early 1960s. While fishing and lamprey predation likely contributed to declines in the larger species, alewives were strongly implicated in the decline of all of these species.

Alewives are primarily planktivores, so their main impact was thought to occur via competition. Changes in the plankton community supported this argument; in Lake Michigan, mean zooplankton size decreased sharply from 1954 to 1966 in response to intense planktivory by the increasing alewife population, then increased again following the alewife die-off in 1967 (Wells 1969). These impacts on plankton size structure put some native species at a disadvantage. For example, bloaters feed on individual prey and are not able to filter-feed on smaller zooplankton, as alewives can (Crowder & Binkowski 1983). Juvenile bloaters also grow more slowly on small zooplankton than on large zooplankton, even when the total biomass of prey is the same (Miller *et al.* 1990).

Certainly alewives had competitive interactions with native planktivores, but in 1980 Larry Crowder made the case that predation on eggs and larvae, rather than competition,

Table 11.1 Most native Lake Michigan species with semi-pelagic eggs or larvae that overlapped spatially and temporally with alewife became rare or extinct after alewives became abundant, whereas native species with demersal eggs and larvae persisted after the alewife invasion (Crowder 1980).

Status after alewives became abundant	Eggs or larvae semi-pelagic and available to alewives?	
	Yes	No
Rare or Extinct	9	0
Persistent	1	11

Bloater was identified as the one species with semi-pelagic larvae (Figure 1.3b) that survived the invasion, but it also showed strong declines during periods of alewife abundance.

was the main mechanism by which alewives led to the decline or extinction of native species. He pointed out that all the native species that had become rare or extinct in Lake Michigan had eggs or larvae that were pelagic or semi-buoyant and co-occurred with alewives, while most of the remaining coexisting species had demersal eggs and larvae, and sometimes parental care (Table 11.1). Other researchers had suggested that alewives might prey upon fish eggs and larvae, and several observations of alewife cannibalism and predation on larval fishes had been reported in the early 1970s. Until Crowder's 1980 paper, however, the importance of alewife predation on fish early life stages had not been widely recognized.

In hindsight, Crowder's idea makes obvious sense. Fish eggs and larvae (ichthyoplankton) are larger than most other plankton species, and alewives, like most other planktivores, tend to consume the largest planktonic prey available (Brooks & Dodson 1965). Today, the importance of predation by alewives, and predation on larvae in general, is widely accepted. In fact, recent lab and field studies have shown that predation by alewives on lake trout fry may be inhibiting the recovery of lake trout populations (Krueger *et al.* 1995).

Empirical evidence of predation in the field is rare because larvae are digested so quickly. Doran Mason and Stephen Brandt, however, were able to document that alewife predation was a major source of mortality for newly hatched yellow perch (*Perca flavescens*, Figure 1.3j) in Lake Ontario embayments when the onshore spawning migration of alewives coincided with yellow perch hatching (see Box 8.2 and Section 8.5.2). Thus, it seems plausible that predation by alewives on larval yellow perch may be an important factor underlying the generally inverse relationship observed between yellow perch year-class strength and alewife abundance over the last half century.

Predation plays an important role, but it is not the whole story. The alewife–yellow perch interaction is a classic example of the importance of understanding how ontogenetic niche shifts (Chapter 7) influence the outcome of interactions between species. While alewives shift from being a competitor to a predator of young yellow perch as they grow, perch also change roles. Small yellow perch feed primarily on zooplankton, but become more piscivorous as they grow, including both alewife eggs and young-of-the-year alewives in their diet. Indeed, yellow perch predation on alewife eggs has been implicated as a factor in the decline of alewives in Lake Michigan during the 1980s (Hansson *et al.* 1997). Thus, as these ontogenetic shifts occur, the nature of competitive and predatory interactions between species often completely reverses. Similar dynamics underlie the Lake Erie yellow perch–walleye (*Stizostedion vitreum*) oscillation, where these species are competitors as juveniles, as well as both predator and prey for each other at other life stages. Perhaps one of the most enduring lessons of the alewife invasion of the Great Lakes will be the awareness that fully understanding the interactions between species requires that we take into account the dynamics occurring at all ontogenetic stages.

11.11   Bloater recruitment and the characteristics of survivors approach

Throughout the 1970s there was a growing awareness of the importance of events occurring early in life for determining survival and year-class strength of fishes. Researchers, however, were often frustrated in their efforts to pinpoint the mechanisms governing annual recruitment variation. What were the most important factors? Starvation? Predation? Environmental effects? Despite our growing knowledge of fish early life dynamics, the list was getting longer, rather than shorter. Simultaneously collecting information on all the potentially important factors affecting recruitment of a particular species was virtually impossible.

Two developments converged to allow a radical change in our approach to understanding recruitment variability. The first was Gregor Pannella's (1971) discovery of daily growth increments in the otoliths of larval fishes, which allowed us to collect detailed information on the history of individual fish (Chapter 2). The second was the development of techniques for rearing larvae in the lab, which was an essential prerequisite for conducting process-oriented experiments on the factors affecting larval fish growth and survival. These tools allowed us to shift our focus from the vast majority of larvae, which die, to the small fraction of larvae that survive. Rather than trying to account for all the factors determining mortality, we could instead ask, "What is unique about survivors?" and use the answer to help us identify what is important in determining survival and recruitment. This concept, pioneered by Richard Methot (1983), has since been dubbed the "characteristics of survivors" approach (Chapter 4).

In the early 1980s, Larry Crowder, Fred Binkowski and I applied the characteristics of survivors approach to try to decipher the mechanisms underlying the poor recruitment of Lake Michigan bloaters in the 1970s. Binkowski had developed techniques to reliably rear bloater larvae (and subsequently alewife and yellow perch larvae) in the lab, which allowed us to conduct experiments to determine what information we could get from their otoliths (Rice *et al.* 1985). We found that bloaters begin depositing growth increments at first feeding (about 3 days after hatching), and we could determine the date of first feeding within 6 days for bloaters up to 5 months old (Figure 11.15). Using this age estimate we could also determine their growth rate. Lastly, we found that periods of low ration would leave noticeable stress marks on the otolith (Figure 11.16).

We applied this ability to interpret bloater otoliths to larvae that were collected from the field in 1982 and 1983. Bloaters spawn on the bottom in deep water, where the eggs incubate for several months (Wells 1966). Larvae spend 1–10 days in the hypolimnion before moving to the surface (Wells 1966, Crowder & Crawford 1984). We used data on bloater

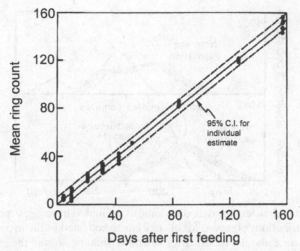

Figure 11.15 Relationship between the number of rings (increments) on the otolith and the number of days after first feeding for lab-reared bloater larvae. The regression line and 95% confidence interval are shown. The regression equation is: Ring Count = 0.331 + 0.961 × days ($R^2 = 0.99$).

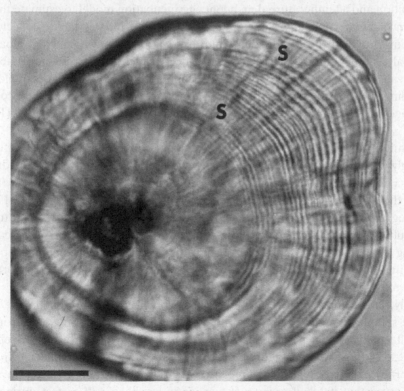

Figure 11.16 Otolith of a 13.1-mm lab-reared bloater larva 35 days after first feeding, showing evidence of stress resulting from two 5-day periods of low ration (indicated by S). Scale bar represents 20 μm. (From Rice *et al.* 1985.)

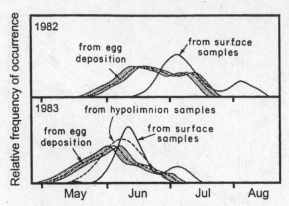

Figure 11.17   Expected first-feeding date distributions estimated from egg deposition, and observed first-feeding date distributions observed for bloater larvae collected in the hypolimnion shortly after hatching or at the surface about 1 month after hatching. Shading around the expected distributions reflects uncertainty in the position of these curves on the X-axis due to possible variation in when ripe females deposited their eggs. Hypolimnetic samples were not collected in 1982. (After Rice *et al.* 1987a.)

egg deposition and a temperature-dependent egg-incubation model to estimate the temporal pattern of bloater hatching dates. Otolith analysis of bloater larvae collected in the hypolimnion provided the first-feeding date distribution of newly hatched bloaters, and otolith analysis of bloaters collected at the surface provided the first-feeding date distribution of larvae about one to two months old. We compared the characteristics of older larvae (first-feeding date distributions, growth rates, stress-mark patterns) with larvae collected at earlier stages to determine how survivors differed from the average individual (Rice *et al.* 1987a).

Our results suggested that events during egg incubation and the first month after hatching strongly influenced patterns of bloater survival. Larvae with early hatching dates were much less common in hypolimnion samples of recently hatched larvae than we expected based on the distribution predicted from observed egg deposition (Figure 11.17). Early-spawned eggs likely experienced higher total mortality because cooler temperatures early in the incubation period extended their incubation period by about two and a half weeks, prolonging their exposure to sources of egg mortality. Because environmental conditions in the hypolimnion of Lake Michigan are very stable during the winter, egg mortality was likely due to biotic factors such as predation, rather than abiotic stressors.

Similarly, larvae that hatched early were less common among older larvae collected in surface samples than expected from the distribution of first-feeding dates for younger larvae collected in the hypolimnion, while larvae hatched later were over-represented among these older survivors (Figure 11.17). Growth rates of larvae collected in the field were generally as high or higher than growth rates of larvae reared in the lab on *ad libitum* rations. Larvae that hatched early, however, grew less than half as fast during the first 3 weeks of life as larvae that hatched later in the season. Furthermore, larvae hatched early were significantly more likely to show evidence of brief stress periods on their otoliths; 43% of larvae with stress marks first fed early in the season when survival was less than expected,

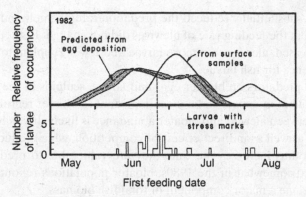

Figure 11.18 First-feeding date distributions of bloater larvae in 1982 estimated from egg deposition and observed for larvae collected at the surface about 1 month after hatching (top panel), compared to the first-feeding dates for larvae in surface samples that had stress marks on their otoliths. While only 13% of all surviving larvae fed first early in the season when survival was less than expected (to the left of the vertical dashed line), 43% of larvae with stress marks fed first during this period. (From Rice *et al.* 1987a.)

while only 13% of all larvae first fed during this period (Figure 11.18). Stress marks all occurred at a common point in ontogeny, 5–10 days after first feeding, rather than on common dates, suggesting the stress was associated with migration from the hypolimnion to the surface rather than with temporal phenomena, such as storms.

Taken together, these patterns suggest that mechanisms operating during the first month or so after hatching have an important effect on bloater survival and recruitment. The relatively high growth rates of all larvae, including those with stress marks, suggested that starvation was not an important factor affecting bloater survival. Subsequent lab experiments lend support to this conclusion; starved bloater larvae took 25 days to reach 50% mortality (Rice *et al.* 1987b)! There was also no evidence to suggest that short term environmental variability, such as storms, played a significant role in determining bloater recruitment. Rather, the patterns we observed implicated sources of mortality that were size- or growth-rate dependent, particularly during the first few weeks of life. We hypothesized that predation was the most likely mechanism; predation is typically size-dependent, and variations in growth rate would determine how long larvae remain susceptible to gape-limited predators.

The characteristics of survivors approach allowed us to eliminate some hypotheses and narrow the scope of our research to the most likely mechanisms. In subsequent research we evaluated the impact of potential predators on bloater eggs and larvae (Luecke *et al.* 1990). Potential hypolimnetic predators included deepwater sculpins (*Myoxocephalus thompsoni*) and slimy sculpins (*Cottus cognatus*), adult alewives, and adult bloaters. Of these, only sculpins proved to be significant predators on bloater eggs and yolk-sac larvae under hypolimnetic conditions (darkness and low temperature). Rainbow smelt (*Osmerus mordax*), yearling alewives, and yearling bloaters overlap with bloater larvae once they move to the surface. All three of these predators consumed bloater larvae under epilimnetic conditions, in a strongly size-dependent manner, with capture success falling from about 85% on 13.2-mm larvae to about 5% on 31.5-mm larvae (Figure 11.19). The presence of zooplankton as alternative prey substantially reduced the predation rate of smelt and yearling bloaters on bloater larvae, but the feeding rate of alewives did not change. Rather, as the density of bloater larvae increased, alewives ate more larvae and fewer zooplankton, suggesting that they have a preference for fish larvae.

Of the potential predators on bloater eggs and larvae, sculpins in the hypolimnion and alewives in the epilimnion appear to be most important to bloater recruitment. The negative relationship between alewife and bloater abundance is likely driven by direct effects of alewife predation, as well as indirect effects of competition, which reduces growth rate of bloater larvae and thus increases the time they are vulnerable to predation. As alewife abundance declined somewhat in the 1980s, bloater populations rebounded dramatically and once again became a major component of total fish biomass.

The characteristics of survivors approach proved to be a powerful tool in our effort to decipher the mechanisms governing bloater survival. It not only helped us understand the dynamics of bloater recruitment, but suggested broader generalities that might be applicable to other species as well.

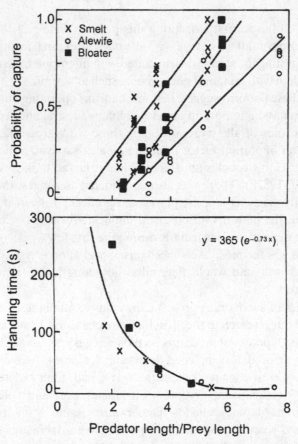

Figure 11.19 Size-dependent capture success (top) and handling time (bottom) of yearling rainbow smelt, alewife, and bloater preying upon bloater larvae. (Reproduced from Luecke *et al.* 1990 with permission of NRC Research Press.)

11.12   All larvae are small (but some are smaller than others!)

At the time we began our work on bloater recruitment, there was no unifying conceptual framework that reconciled conflicting information among species about which factors, such as starvation, predation, or abiotic conditions, played the key role in governing survival. In fact, Reuben Lasker, one of the most prominent scientists working on the early life stages of fishes, stated in his plenary address to the 1986 Larval Fish Conference that recruitment "depends on a species' behavioral and physiological response to the biotic and abiotic environment and that generalities with respect to recruitment mechanisms cannot be made."

Our experience with bloater larvae suggested that body size might provide such a unifying framework. We realized that while bloater larvae seemed quite small to us (having worked mostly with adult fishes), they are quite big in comparison to other species (Figure 11.20). We had been surprised to find that 50% of newly hatched bloater larvae survived 25 days without food – very different from results with anchovies and sardines, which are much smaller at hatching! In fact, about 90% of all fish species hatch at a length less than 9.8 mm, the size of bloaters at hatching (Chapter 1). Our subsequent analysis of the available published data (Miller *et al.* 1988) indicated that fishes from diverse taxa and habitats share common size-dependent relationships, suggesting that body size does indeed provide a good starting point for understanding survival and recruitment mechanisms of larval fishes. As we saw in Chapter 1, such scaling relationships can be further refined by accounting for differences among species in their size at comparable stages of development.

The accumulating studies on early life stages of Great Lakes fishes created an ideal opportunity to test the relative importance of species vs. body size. Thomas Miller chose three species from different families with very different morphologies and hatching sizes (alewife, 3.8 mm; yellow perch, 5.5 mm; bloater, 9.8 mm). He conducted a series of experiments with all three species at several common sizes ranging from 10 to 40 mm, evaluating the effects of size and species on foraging ability, including prey detection, pursuit, attack, and capture. For most of these traits, body size accounted for much more of the variation than did species differences. Size explained 71–91% of the variation in parameters of the Holling Type II

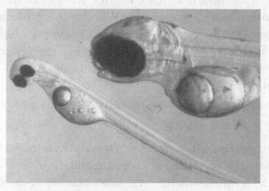

Figure 11.20   A newly hatched bloater larva (9.8 mm in length) in comparison to a newly hatched rainbow smelt larva (4.4 mm in length). The differences in size and development are pronounced, even though smelt are even larger at hatching than about half of all fishes.

functional response relationship (attack constant and handling time), whereas the effect of species was not significant (Miller *et al.* 1992). Similarly, body size explained 72–78% of the variation in measures of visual ability critical for foraging (histological acuity, reactive distance, visual angle), while species did not have a significant effect (Miller *et al.* 1993).

In contrast, both size and species significantly affected swimming speed and the distance from which larvae initiated attacks on prey. Capture success was not closely related to absolute size, but was more dependent on the predator–prey size ratio. In longer term experiments with these same three species, Benjamin Letcher and colleagues (1997) noted that species had a significant effect on ingestion and growth rates, especially as larva size increased and morphologies diverged.

While species effects were evident in some of these traits, and would surely be more important in some other species, these experiments confirmed the general importance of body size in governing the dynamics of fish early life history across species. These results from early life-history studies on Great Lakes fishes have provided an enduring framework for current and future investigations of larval fish ecology and population dynamics.

11.13   Environment matters too

Most of the examples presented so far have focused on the importance of biotic interactions, but environmental conditions also play an important role in determining recruitment for many Great Lakes species. Unlike bloaters, which deposit their eggs in relatively stable hypolimnetic conditions, many species spawn in shallower nearshore or riverine environments that are much more susceptible to changes in physical conditions.

In the early 1980s, William Taylor published several studies with his students and colleagues (for example, Taylor *et al.* 1987, Freeberg *et al.* 1990, Brown *et al.* 1993) that explored how the highly variable recruitment of lake whitefish is affected by the characteristics of the adult stock (mainly spawning biomass) as well as events during egg incubation and larva development. Lake whitefish spawn in late November, depositing most of their eggs in shallow water (<3 m deep), where they incubate for about 4 months. Taylor's group found that egg survival was nine times higher during harsh winters when the spawning grounds were covered by ice, than in warmer, ice-free years when the eggs were exposed to wind and wave action throughout the winter. Without the protection of the ice cover, mechanical destruction of the eggs and movement of eggs to poor incubation habitat (sand and muck) resulted in higher mortality. Following hatching, the importance of biotic factors increased. Larval survival was strongly dependent on the availability of intermediate sized copepods during the first 6 weeks of life.

Thus, final recruitment strength of lake whitefish depended on the integrated outcome of three main processes:

(1) the magnitude of egg deposition, which was largely determined by spawning biomass;
(2) egg survival, dictated by the extent of ice cover; and
(3) larval survival, which varied with the ratio of zooplankton densities to larval densities.

Recruitment models incorporating both abiotic and biotic factors explained about 60% of the variation in lake whitefish recruitment, compared to only about 10–35% for traditional stock–recruitment relationships. Although physical factors such as climatic conditions cannot be controlled, their impacts on year-class strength can be predicted, and used in making stock and harvest management decisions.

Joseph Mion and colleagues (1998) used the characteristics of survivors approach to evaluate the importance of river discharge in driving the recruitment variability of Lake Erie walleye. They hypothesized that high discharge in the rivers where walleye spawn would result in strong recruitment by reducing the residence time of larvae in the food-poor riverine environment and transporting them quickly to more productive nursery areas in the lake. They sampled newly hatched larvae at upstream sites to estimate daily larval production, and compared the temporal pattern of hatching to the hatching dates of surviving larvae (determined from otoliths) collected at mid- and downstream sites. The hatching-date distributions of surviving walleye larvae were markedly different from the temporal patterns of larval production; 75–84% of all survivors came from brief, discrete periods (4–7 days) of high survival during the hatching season. Contrary to their initial hypothesis, Mion and his colleagues found a dramatic and unexpected negative relationship between larval survival and river discharge. They attributed this negative effect to increased suspended sediment loads during high, turbulent discharge, which can suffocate and physically damage fragile, newly hatched larvae. Indeed, torn and abraded larvae were common in their upstream samples during high flows.

Low flow is necessary, but not sufficient, for larval walleye survival. The rate of seasonal warming and variable patterns of zooplankton availability combine to determine whether larvae starve or grow, and other fishes moving into the river to spawn create a predatory gauntlet for late hatching larvae. Thus, the timing of storms (high discharge events), water temperature changes, zooplankton blooms, and periods of high predation risk all interact to determine periods of high and low larval survival (Figure 11.21). Interestingly, the same suite of factors, including physical damage from wave action during storms, appears to govern survival of walleye spawned on open lake reefs, resulting in similar patterns of survival for river-spawned and lake-spawned walleye (Roseman *et al.* 1996). These results suggest that the increased storm frequency and intensity, an expected consequence of global warming, may cause a general decline in Lake Erie walleye recruitment. However, improving land use practices to reduce water runoff rates and sediment input can improve conditions for larval walleye survival (Mion *et al.* 1998).

Just as deforestation, dams, and other historical changes in the Great Lakes watershed had major abiotic impacts on spawning habitats in the lakes and their tributaries, current efforts to improve watersheds are having important effects on recruitment of fishes in the Great Lakes. For example, increasing numbers of young lake sturgeon (*Acipenser fulvescens*) are being observed in the Manistee and Muskegon Rivers in Michigan, after changes were implemented to stabilize flows on those rivers.

The effects of such changes are not always anticipated, however. Improvements in water quality due to better pollution controls, and reduced stream temperatures resulting from reforestation of riparian zones, have had the unintended consequence of turning many streams that had been unsuitable into excellent spawning habitat for lampreys. Due to the increase in the number of lamprey spawning streams, and the fact that some of them are too large to treat effectively with TFM, the job of lamprey control has become increasingly difficult. Similarly, stream improvements are making many Great Lakes tributaries suitable

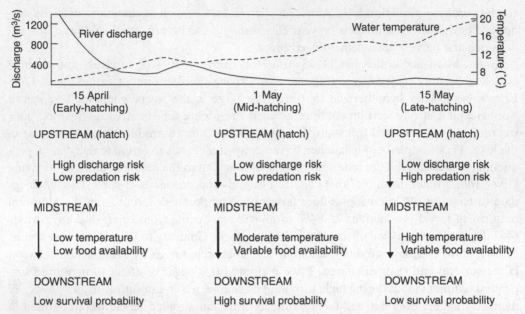

Figure 11.21 Conceptual model of generalized trends in river discharge and water temperature (plot shows data from the Maumee River, 1995) during the larval walleye hatch, coupled with the spatio-temporal influence of river discharge, predation, water temperature, and zooplankton density on survival of larval walleye during river out-migration. Interactions between time of hatching (early-, mid-, and late-hatching) and river location (up-, mid-, and downstream) largely determine which mechanisms are most important for determining larval survival. Arrows indicate movement between sites of larvae hatched during different periods (thicker arrows indicate greater relative probability of larval survival). Factors influencing larval survival at that location are listed next to each arrow. (From Mion *et al.* 1998.)

spawning habitat for the exotic salmonids that were stocked to control alewife populations. Initially, these predator populations were maintained exclusively by stocking. As flows have been stabilized at hydropower dams on many streams, particularly in Michigan, natural reproduction by chinook salmon has increased to the point that natural spawning accounts for more than 30% of the smolts entering the lake. While this change may substantially reduce hatchery costs, it is also greatly reducing the extent to which fishery managers can regulate salmonid populations. When reproduction was controlled artificially, managers could manipulate predator populations in response to changes in the prey base or the multi-billion-dollar sport fishery. This flexibility is being lost. Landscape-scale changes will continue to affect the Great Lakes fisheries and how they must be managed, and many of those effects will be mediated by dynamics that occur during the early life of fishes.

11.14   Summary

The biological communities of the Laurentian Great Lakes have undergone frequent and dramatic changes over the last 200 years, largely due to past and ongoing human impacts.

Over the last several decades, our knowledge of early life processes has become increasingly important to understanding and managing fish populations in these dynamic systems. Efforts to restore Great Lakes fisheries depended initially on finding a way to reduce the abundance of exotic sea lampreys. Knowledge of sea lamprey early life history allowed biologists to identify a vulnerable stage in their ontogeny and develop a chemical control method that eventually brought the lamprey population under control. Despite lamprey reductions and an aggressive stocking program, natural reproduction of lake trout remains very limited. This failure may be due to EMS, a vitamin deficiency that results in high larval mortality because of a lack of thiamine in alewives, the primary prey of lake trout. Other factors during early life stages, such as predation on eggs and larvae, may also be important.

The proliferation of alewives following the decimation of piscivore populations resulted in the dramatic decline and even extinction of several native planktivore species. Initially these impacts were attributed to competitive interactions with alewives, but later research revealed that alewife predation on pelagic eggs and larvae was also a major factor. Because the way species interact (through competition and predation) can change markedly as they grow, events occurring at all ontogenetic stages must be considered in order to fully understand population dynamics.

The discovery of daily growth increments in the otoliths of larval fishes made possible development of the characteristics of survivors approach, which uses the unique characteristics of surviving larvae to suggest which of many potential mechanisms are important in determining survival. Application of this approach to the decline of Lake Michigan bloaters in the 1970s suggested that size-dependent predation by alewives during the first few weeks of life was controlling bloater recruitment, and subsequent lab experiments supported this conclusion.

Our experience studying bloater recruitment dynamics suggested that body size might provide a general explanation for some of the differences among species in the mechanisms controlling survival of larval fishes. A review of published data supported this notion, and lab experiments with three Great Lakes species (bloater, alewife and yellow perch) demonstrated that, at least for these three species, body size explained much more variation in the outcome of most (but not all) processes affecting survival than did species.

While biotic interactions likely influence survival of most larval fishes, some species are strongly influenced by abiotic conditions. Recruitment of Great Lakes fishes that spawn in shallow, nearshore areas, such as lake whitefish, or in tributary streams, such as walleye, can be strongly affected by physical factors such as ice cover, frequency of storms, water temperatures, and suspended sediment, as well as by biological factors such as food availability and predation. Physical conditions in the Great Lakes and their tributaries are being modified by land use in the watershed, with substantial and often unanticipated consequences for fish populations. For every species, multiple mechanisms interact to determine survival and recruitment, and must be considered together, rather than in isolation, to understand and manage fish populations.

As the examples presented here illustrate, early life-history studies have become a prominent part of Great Lakes fishery research and management in recent decades. Fishery scientists have recognized that the dynamics of adult populations and communities are determined in large part by events and interactions that occur early in ontogeny.

This change in perspective is shaping Great Lakes management decisions from lamprey control to habitat and harvest management. At the same time, studies in the Great Lakes have contributed substantially to our understanding of the important processes affecting survival of fish early life stages in general.

Understanding Conservation Issues of the Danube River

Hubert Keckeis and Fritz Schiemer

11.15   The river and its fish fauna

11.15.1   Geomorphology and longitudinal zonation of the Danube River

The Danube River is the second largest river in Europe, with a drainage area of 805 000 km², a length of approximately 2850 km, and a discharge of 6450 m³ s⁻¹ at its mouth. From its source in Germany to its mouth in the Black Sea in Romania, it crosses nine countries (Germany, Austria, Slovakia, Hungary, Croatia, Serbia, Bulgaria, Romania, and Ukraine), representing a large variety of landscapes and climates.

Geomorphological conditions define three distinct sections of the river. The upper section (river 2850–1750 km) ranges from Germany to the border of Austria and Slovakia and has an average slope of 40 cm km⁻¹, with a high bedload-transport capacity. Before regulation, the morphology of the river in this section alternated between canyons with narrow riparian zones – where the river breaks through massive rocky layers – and braided alluvial sections with many side arms and backwaters in large floodplain areas. This was especially true in the plains in the eastern part of Austria. The middle section is characterized by a drastic reduction in the slope (6 cm km⁻¹) and lower bedload-transport capacity. This section is separated from the lower section (river 940–0 km) by a 100-km-long cataract (the "Iron Gate"), where the river cuts through the Carpathian Mountains. In the lower Danube River, the average slope is 3.9 cm km⁻¹ and the deposition of suspended solids increases significantly.

11.15.2   Biogeographical aspects

The fish fauna of the Danube River is the richest of any European river, with more than 100 species from 23 families (Busnita 1967, Bacalbaşa-Dobrovici 1989). The fish fauna is dominated by Cyprinidae (39 species), followed by Percidae (11), Gobiidae (11), Cobitidae (8), Salmonidae (7) and Acipenseridae (6). There is a clear succession of species associations along the longitudinal course of the river (Bacalbaşa-Dobrovici 1989). The species richness is due to the unique direction of flow of the Danube River. It flows from west to east, which leads to a significant role for it as a biocorridor, connecting the Ponto-Caspic and Central Asian areas in the east with the high mountain alpine regions in the west. The aquatic fauna is sharply delimited against that of Siberia and tropical Africa, but much less different from northeastern Europe and western Asia. Many of the aquatic organisms reveal a Ponto-Caspic distribution, indicating the role of the river as an important migration and recolonization route between and after the ice ages.

11.15.3   Hydrological characteristics and major human impacts in the Austrian section

The hydrological conditions along the Danube River's Austrian stretch are characterized by high and variable flow from the Alps. The long term monthly mean water levels are highest in June and lowest in November, with an amplitude of 2.5 m. Water-level fluctuations, however, are strong and unpredictable, and spates can occur throughout the year. High current velocities in the main channel and coarse grained substrates characterize the Austrian Danube as a hyporithral* river. Historically, the dynamic hydrology and the high sediment transport from the Alps produced large alluvial fans, especially in the tectonic basins below geomorphological constrictions. This was accompanied by a braided river course with extended floodplains prior to regulation (Figure 11.22).

Severe river regulation in the Austrian Danube was initiated in 1850 and continues to the present. The ecological effects, of course, were not taken into consideration. The main engineering approach was to create a single, straightened channel, stabilized by riverside embankments and ripraps. The former arms of the original braided system were cut off (Figure 11.22). The ecology of the Danube has been strongly affected by land use and changes in its catchment, by pollution, and most importantly by hydro-engineering. Over the past 50 years more than 90% of the upper Danube and its major tributaries have been dammed for hydropower production. The remaining few stretches have been severely affected by regulation. This also holds true for the remaining 50-km, free-flowing section downstream from Vienna. This reach, although considerably impacted by regulation, represents one of the last remnants of a river–floodplain system. Here, the hydrological dynamics, flood pulses, and bedload transport are partially operative and a high potential for re-establishing the hydrological regime remains. As the largest remnant of alluvial landscape in Europe, this stretch was declared a National Park in 1996.

The main features of the regulation scheme and its immediate and long term effects in this area were:

(1)   a general reduction of up to 80% of the former alluvial floodplain areas, backwaters, and side arms;
(2)   a loss of riverine inshore habitats, which had strong impacts on inshore retention characteristics and on the habitat value for rheophilic organisms; and
(3)   reduced hydrological connectivity, both of open surface-water connections and groundwater exchange between river and floodplains.

11.15.4   Status and ecological guilds of the fish fauna

Intensive surveys carried out during the last two decades revealed a large number of fish species. Compared to the historical records, most of the original fauna is still present.

---

* Mountain zone.

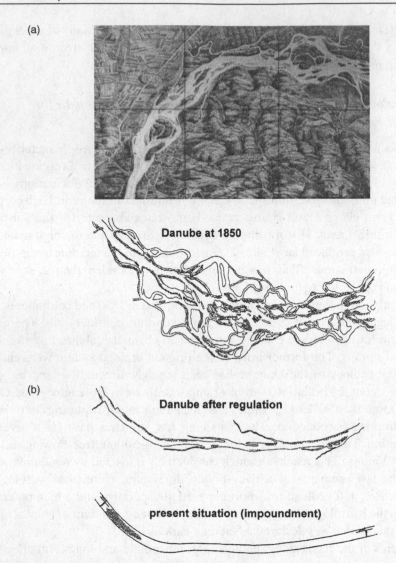

(a)

Danube at 1850

(b)

Danube after regulation

present situation (impoundment)

Figure 11.22   (a) Historical map of the Danube River near Vienna showing the character of a braided section before river regulation. (b) Impact of regulation schemes on river morphology and area of the Danube River near Vienna during the era of 1875 to present. (Reproduced with modifications from Zauner 1996 with permission of Hans Winkler.)

Only the large diadromous sturgeons (Acipenseridae: *Huso huso*, *Acipenser gueldenstädti*, *Acipenser stellatus*, and *Acipenser nudiventris*), which historically migrated from the Black Sea more than 2500 km upstream, have completely disappeared in the upper sections. Their migration route was blocked by the construction of large dams at the "Iron Gate" (Serbia, Romania). There are practically no records on the population structure prior to river regulation. The few quantitative data on the impact of river regulation and dam construction on fish abundance show that the populations of the two most characteristic fish species (Cyprinidae: *Chondrostoma nasus, Barbus barbus*) disappeared completely within a period of 2 years (1922–1924) after the construction of a dam in the Inn River, the largest tributary

Table 11.2   Categories for the ecological guilds of the Danube River fish fauna (Schiemer *et al.* 1994).

Category	Description
Rhitrale	Species that migrate at least during their reproductive period into oxygen-rich, cold water streams (upper or lower trout region)
Rheophilic A	Typical riverine species that spend their entire life cycle in the main stream. Larvae and juveniles require species-specific, ontogenetically changing inshore structures
Rheophilic B	Lithophilous species that must migrate into backwaters and side arms during certain seasons (summer-feeding and overwintering habitats)
Eurytopic	Generalist species without any significant preference, occurring either in the main channel, in side arms, or in backwaters. They reproduce mainly in backwaters (primarily phytolithophilous, lithophilous). Species-specific differences exist in diet and feeding habits, microhabitat selection, and preferred temperature
Stagnophilic/Limnophilic	Species that occur throughout their life cycle in abandoned backwaters with dense macrophytic vegetation. Some of these species are specifically adapted to extreme environmental conditions (for example, high water temperature, low oxygen context)
Exotic	Species that were introduced to provide a recreational fishery, escaped from aquaculture ponds, or were released from aquaria

of the Danube (Waidbacher & Haidvogel 1998). Most of the riverine species found in the Danube and in many other European rivers are on the "Red Lists" of endangered taxa (Lelek 1987; see Chapter 10, Box 10.1). Populations of several fish species have declined during the last decades, indicating that the conditions for sustaining a characteristic fish fauna in the Danube are disappearing. Evaluation of the present situation using qualitative data from numerous surveys and on the basis of species-specific ecological needs revealed six ecological guilds (Table 11.2). It is evident from habitat changes that rheophilic species must have declined in favor of the eurytopic and limnophilic groups (Figure 11.23).

The most significant ecological features for an endangered riverine fish fauna are the specific requirements for the larval and juvenile stages (see Chapter 7). Detailed studies on densities, distribution, and habitat requirements revealed that large numbers of larvae and juveniles of many fish species occur in the main channel of the river (Spindler 1988). A high diversity and a large number of endangered species are found among the populations of 0+ (young-of-the-year) fishes in small bays and in shallow sloped gravel banks of the main river. At artificial shores only a few individuals of eurytopic species are observed (Figure 11.24). Eurytopic forms dominate open backwaters in place of rheophilic and limnophilic species (Kurmayer *et al.* 1996). In disconnected backwaters with dense aquatic vegetation the larval fish associations are composed of eurytopic and limnophilic species.

Examples of clear habitat selection, even in the young larval stages, were also observed. During the first few months of their life, larvae of many riverine fish species undertake

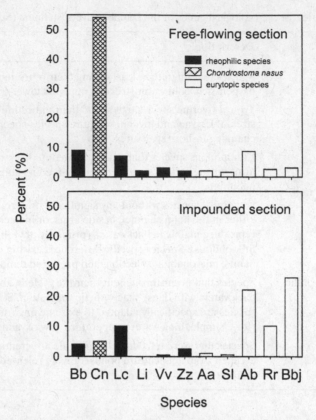

Figure 11.23 Comparison of the adult species composition in a free-flowing and in an impounded section of the Danube River in Austria. Bb, *Barbus barbus*; Cn, *Chondrostoma nasus*; Lc, *Leuciscus cephalus*; Li, *Leuciscus idus*; Vv, *Vimba vimba*; Zz, *Zingel zingel*; Aa, *Aspius aspius*; Sl, *Stizostedion lucioperca*; Ab, *Abramis brama*; Rr, *Rutilus rutilus*; Bbj, *Blicca björkna*. (Reproduced from Schiemer & Waidbacher 1992 with permission of John Wiley & Sons Limited.)

ontogenetic niche shifts with significant changes in habitat preferences (Figure 11.25; Schiemer & Spindler 1989). Characteristic 0+ species associations are significantly ordinated along different macrohabitats (lotic, lentic, artificial, parapotamic,* paleopotamic†), the explanatory abiotic variables are current velocity, substrate, depth, and slope of the individual sites (Wintersberger 1996a). A differential utilization of distinct areas by different size classes of the same species also exists. That is, rheophilic species occur in lotic habitats, and are also found in lentic areas, indicating a species-specific ontogenetic habitat shift and size-dependent spatial separation and resource utilization (Wintersberger 1996b).

A general relationship between number of species and shoreline sinuosity of the main river (Figure 11.26) demonstrates the importance of highly diverse inshore structures (gravel banks, bays) as nursery zones for rheophilic fishes. The ecological quality of inshore zones of large rivers can therefore be evaluated on the basis of the ecological requirements of fish

---

* Abandoned side arm.
† Oxbow lake.

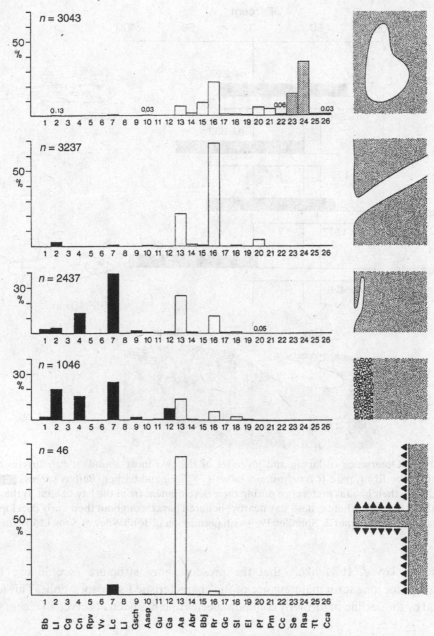

Figure 11.24 Occurrence and densities of species and ecological guilds of early life stages at different types of inshore habitats of the Danube River. Black bars, rheophilic species; white bars, eurytopic species; grey bars, limnophilic species. 1, *Barbus barbus*; 2, *Leuciscus leuciscus*; 3, *Cottus gobio*; 4, *Chondrostoma nasus*; 5, *Rutilus pigus virgo*; 6, *Vimba vimba*; 7, *Leuciscus cephalus*; 8, *Leuciscus idus*; 9, *Gymnocephalus schraetser*; 10, *Aspius aspius*; 11, *Gobio uranoscopus*; 12, *Gobio gobio*; 13, *Alburnus alburnus*; 14, *Abramis brama*; 15, *Blicca björkna*; 16, *Rutilus rutilus*; 17, *Gymnocephalus cernuus*; 18, *Stizostedion lucioperca*; 19, *Esox lucius*; 20, *Perca fluviatilis*; 21, *Proterorhinus marmoratus*; 22, *Cyprinus carpio*; 23, *Scardinius erythrophthalmus*; 24, *Rhodeus sericeus amarus*; 25, *Tinca tinca*; 26, *Carassius carassius*. (Reproduced from Schiemer & Spindler 1989 with permission of John Wiley & Sons Limited.)

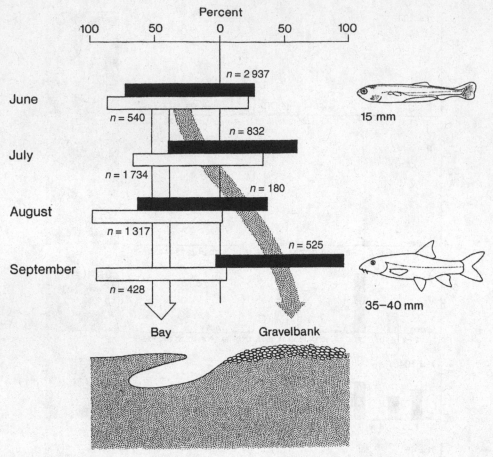

Figure 11.25 Occurrence of larvae and juveniles of the two most abundant fish species of the Austrian Danube River, nase (*Chondrostoma nasus* ☐) and barbel (*Barbus barbus* ■). Barbels change their habitat preference during their development from the bay habitat to the gravel banks in the main river, whereas nase stay nearby sheltered bays throughout their early development. (Reproduced from Schiemer & Spindler 1989 with permission of JohnWiley & Sons Ltd.)

embryos and larvae. It is likely that the present shore structure (see Figure 11.26) is inadequate for long term maintenance of the characteristic fish associations. This idea is supported by the decline of formerly common species that has been observed in recent years.

11.16   The nase: target species for river conservation

Nase (*Chondrostoma nasus*) were abundant in many European rivers, but their stocks have declined dramatically (Lusk & Halačka 1995, Kirchhofer 1996, Kappus *et al.* 1997). Factors that limit populations may change during life history, for example: ∕

(1)   Negative effects during the adult period may lead to reduced growth and reduced physiological condition, which can result in poor gamete quality (Trippel *et al.* 1997).

Figure 11.26 Relationship between number of species (age 0+, larvae and juveniles) and inshore sinuosity of the Danube River in Austria. The flowage line (that is, length of the shoreline) is almost straight at artificially constructed shores (left picture), whereas it is longer at natural gravel bars (right picture). (Graph is reproduced from Schiemer *et al.* 1991 with permission of E. Schweizerbart Verlagsbuchhandlungen, www.schweizerbart.de.)

(2) Reduced availability of adequate spawning areas (due to habitat fragmentation, or changes of the natural flow regime) may lead to decreased spawning success.

(3) Alterations of the flow regime as a result of damming and river regulation changes the size-distribution of transported particles, which may lead to silting of eggs with a concomitant reduction in the oxygen supply, which increases the mortality of the embryos (Marty *et al.* 1986, Keckeis *et al.* 1996, Bardonnet 2001).

(4) River channelization reduces the size, number, and quality of habitats available to 0+ fishes. It also alters the position and connectivity of habitats, prohibiting fish larvae and early juveniles from reaching the right place at the right time (that is, disrupted spatial and seasonal habitat relationship), thereby affecting survival of 0+ populations (Schiemer *et al.* 1991, Keckeis *et al.* 1997, Winkler *et al.* 1997). This principle is discussed further in Chapter 7.

The cumulative effect of these disturbances has important implications for the recruitment process.

Table 11.3 Summary of key habitat variables measured at about 80 spawning sites of *Chondrostoma nasus* over their entire distribution (Kamler & Keckeis 2000).

Characteristic	Average	Coefficient of variation (%)
Temperature (°C)	9.5	20
Current velocity (m s^{-1})	0.9	22
Oxygen (mg l^{-1})	10.9	25
Depth (cm)	39.9	37
Area (m^2)	89.5	61
Grain size (mm)	32.5	74

We therefore tried to identify the factors that are responsible for the present decline of *Chondrostoma nasus* populations in rivers. Examining extrinsic and intrinsic factors that influence the performance of the offspring enables a clear identification of the survival potential and helps in the development of successful management criteria based on the ecological demands of the organisms involved. A series of field and laboratory investigations was carried out, which proved to be very helpful for understanding development under natural conditions as well as population fluctuations.

11.16.1   Spawning habitat

Spawning of *Chondrostoma nasus* is tied to the inshore structure of the river itself or its tributaries. The spawning areas are characterized by a narrow range of specific characteristics, such as water depth, average current velocity, and spawning temperature, irrespective of the size and geographical position of the river (Table 11.3). These data may serve as a basis from which to predict or construct usable spawning sites in river restoration projects and to enhance spawning success in this endangered rheophilic cyprinid.

11.16.2   Spawning population, egg quality, and offspring viability

The size and age structure of the spawning population may influence recruitment, and in this respect the pathway from female attributes to egg properties and ultimately to offspring is very relevant. Younger and smaller females, as well as the oldest ones, produce small eggs, which leads to a reduction of offspring viability (Keckeis *et al.* 2000). The age and size structure of *Chondrostoma nasus* populations is affected by habitat modifications and pollution, with the effects being manifested in a heterogeneous size and age structure (Peňáz 1996). A 6-year observation of the structure of the spawning population in a tributary of the Danube revealed that larger individuals of both sexes dominated the population; new recruits of females rarely augmented the population (in only one out of 6 years). New males entered the population in intervals of 1–4 years (Kamler & Keckeis 2000). This record clearly demonstrates that many year classes are missing due to high mortality in early life.

11.16.3 Early development – the endogenous feeding period

Nase eggs and embryos possess many traits that indicate high survival and growth potential (Kamler *et al.* 1996, 1998). Embryos develop successfully over a large range of temperatures, from 9°C to 19°C. Within this range, survival rates are high and the duration of the incubation period ranges from 5.8 days at 19°C to 33.7 days at 10°C. The eggs are large, with a thick egg capsule that protects the embryos against mechanical damage, and yolk-sac larvae appear to be largely independent of minerals supplied from the water. Several traits that are crucial for survival help maximize the size that larvae attain at the end of the yolk-sac phase. Specifically, yolk size, the caloric value of egg dry matter, the efficiency of yolk utilization for growth, and embryonic growth rate are high compared to other species (Kamler & Keckeis 2000). As we have seen in other chapters of this book, the size attained by larvae at the transition from yolk to exogenous feeding is positively associated with survival. The total dry weight of *Chondrostoma nasus* at hatching (tissue + remaining yolk) is 1.485 mg (Kamler *et al.* 1998), which is high compared to the mean of 0.038 mg for marine fish and even other freshwater species (Table 1.1). Body size of *Chondrostoma nasus* at complete yolk absorption is also large, 1.08 mg dry weight, and is independent of temperature between 10°C and 19°C (Figure 11.27). Nase larvae, like bloater larvae (Section 11.11), can resist starvation for long periods (Keckeis *et al.* 2000), which gives them an extended window to initiate exogenous feeding. This is beneficial when environmental conditions are unpredictable or highly variable, which is typical of *Chondrostoma nasus* nursery grounds in rivers.

The relatively low metabolic expenditures and fast growth are due to high efficiencies of yolk energy utilization. Yolk-feeding nase have higher conversion efficiencies (K_1) than Atlantic salmon (*Salmo salar*), brown trout (*Salmo trutta*), and several other freshwater and marine species (averaging 57% in nase compared to 34–52% for others; Kamler *et al.* 1998). Conversion efficiencies continue to be high during exogenous feeding (Figure 11.28). During early ontogeny optimum temperatures increase from 8°C to 12°C for spawning, through 13–16°C for embryonic development, 15–18°C for yolk-feeding larvae, 19°C for exogenous feeding larvae, and 22°C for late larvae and early juveniles. These optimum temperatures for the physiological–biochemical processes in the individual fish closely parallel the spring rise in temperature in their nursery areas (Figure 11.29).

From these findings it can be concluded that nase embryos, larvae, and early juveniles have many intrinsic attributes that indicate high survival potential and help explain the previous success of this species in the Danube River. Low spawning success and low survival rates of young-of-the-year might therefore be attributed to unsuitable extrinsic factors.

11.16.4 Habitat requirements and refugia

As we have seen throughout this book, the success of a year class is primarily determined by reproductive success and mortality in the embryonic and larval periods. In rivers, the conditions along the inshore ecotones are decisive elements in the population dynamics of individual species. Habitat characteristics in rivers are dynamically controlled by hydrology, and, in particular, the patterns of current velocity, depth, temperature, and food availability that depend on the inshore relief and water level of the river (Figure 11.30). For larval

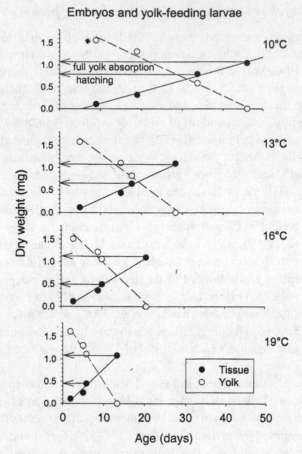

Figure 11.27 Dry weight of tissue and yolk from fertilization to full yolk absorption of *Chondrostoma nasus* embryos and larvae at different temperatures. The arrows indicate greater yolk in larvae hatching at higher temperatures. (Data from Kamler *et al.* 1998.)

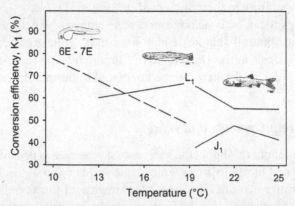

Figure 11.28 Shift of temperature-dependent conversion efficiencies (K_1; calculated as $P \times C^{-1} \times 100$) from embryonic stages 6–7, at larval stage 1; and at the first juvenile stage. (Data from Kamler *et al.* 1998, Keckeis *et al.* 2001.)

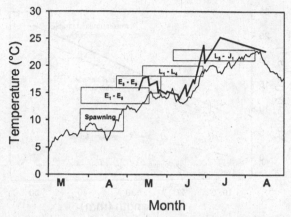

Figure 11.29 Shift in temperature requirements during *Chondrostoma nasus* spawning and early ontogeny in relation to water temperature in the field in 1994. Thin line shows temperatures in Danube River main channel, measured daily at 07:00 a.m. by the River Authority. Thick line shows mean daily temperatures taken in hourly intervals from three actual *Chondrostoma nasus* mesohabitats. Rectangles indicate the occurrence of optimum temperatures in the field. It is evident, that the development at optimal temperatures can only occur in inshore zones with higher temperatures. (Reproduced from Keckeis *et al.* 2001 with permission of Academic Press.)

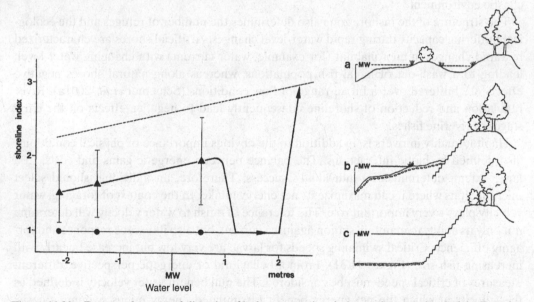

Figure 11.30 Degree of structural diversity of inshore zones in relation to the water level of the Danube River (left graph). The dashed line indicates a hypothetical situation in an undisturbed natural habitat. Inshore structure increases with increasing water level of the river. At the remaining richly structured gravel banks downstream of Vienna, the structural heterogeneity increases with increasing water level, but it is interrupted as the water level reaches the dyke (thick lines, triangles). When the water overflows the dyke, the heterogeneity increases again, as the floodplain is connected with the river. Habitat structure remains low at artificial shores (circles, thin line), so long as the water is higher than the dyke during high water. The inserts on the right side represent cross-sections of the (a) natural shore, (b) present situation at gravel bars of the free-flowing section, and (c) present situation at artificial shores. (Reproduced from Schiemer *et al.* 2001b with permission of E. Schweizerbart'sche Verlagsbuchhandlung.)

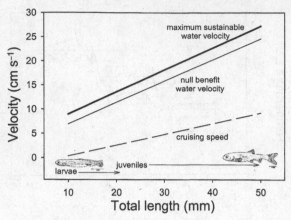

Figure 11.31   Cruising speed, null benefit water velocity, and maximum sustainable water velocity of *Chondrostoma nasus* larvae and juveniles. See text for explanation.

fishes, richly structured, small scale zones with low water currents, shallow depths, higher temperatures, finer substrate, and dense vegetation form islands in a highly fluctuating stochastic environment.

The structure of the inshore zone also determines the number of refuges and the ecological buffering capacity during rapid water-level changes. Artificial shores are characterized by rapid changes in microhabitat (for example, water current) with changing water level, leading to a wash-out of larval fish populations, whereas along natural shores negative effects are buffered over a broad range of flow conditions (Schiemer *et al.* 2001a). River regulation and reduction of shoreline consequently lead to negative effects on the early stages of riverine fishes.

Habitat quality in rivers is, in addition to the obvious importance of physical conditions, also defined by biotic interactions. The balance between energetic gains and costs, to a large extent, determines an individual's success. Therefore, an individual should select microhabitats where it can maximize its net energy intake. In the context of foraging, water velocity plays a very important role. The tolerance of a fish to water velocity will determine not only its ability to maintain station against a current, but also its escape response and foraging efficiency. Critical swimming speeds for larvae are very low but increase linearly with increasing fish size (Figure 11.31). From a behavioral or energetic perspective, different measures of critical speed must be considered. The null benefit water velocity is defined as the velocity at which the net energy benefit (assimilated energy minus swimming costs; Flore & Keckeis 1998) is zero, whereas the maximum sustainable swimming speed is the lowest water velocity at which a fish fails to maintain its position in the water column (Figure 11.31; Flore *et al.* 2001).

Below these critical values, water velocity influences fish feeding, growth, and metabolism. We found that the energy budget of nase larvae is very tight. That is, the ranges of current velocities and temperatures at which surplus energy was available for an individual fish are very narrow, especially in the early larval stages (Figure 11.32). For example, the water velocity at which a young larva obtains a positive energy balance at optimal food densities

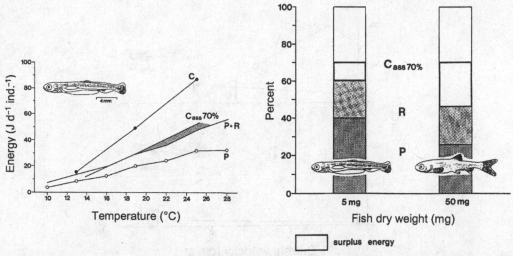

Figure 11.32 Upper: Elements of an energy budget for larval *Chondrostoma nasus* of 5 mg dry weight. Consumption over 24 h at three temperatures was measured at *ad libitum* food availability. Production rates were calculated from growth rates obtained at *ad libitum* food supply. Respiration values were based on routine metabolic rates (RMR) over a period of 24 h (Keckeis *et al.* 2001). Active metabolic rates were calculated by multiplying RMR by a factor of 2.5 to correct for food searching activity and food-induced thermogenesis. (From Schiemer *et al.* 2001a). Lower: Elements of an energy budget at 20°C compared for larvae (5 mg dry weight) and juveniles (50 mg dry weight) of *C. nasus*. The rates are given in percentage of the energy consumed. Production (*P*, dark stippled columns) and respiration (*R*, light stippled columns) as above. The strongly delineated part of the column indicates surplus energy, assuming an assimilation efficiency of 70%. (Reproduced from Schiemer *et al.* 2001b with permission of E. Schweizerbart'sche Verlagsbuchhandlung.)

ranges from 0 to $7 \, cm \, s^{-1}$. This window widens almost two-fold for a young juvenile (approximately $15 \, cm \, s^{-1}$; Figure 11.33). Another very important component of fitness is the ability to sustain swimming, since it affects habitat utilization, migration ability, and vulnerability to predation. The water-velocity tolerance of a fish will determine not only its maintenance of station against a current, but also its escape response and feeding efficiency. Measurements of critical water velocities are important for ecological management because they help to determine the limits of 0+ fishes for searching for food, consuming prey, and holding their position.

The transformation of natural shorelines to uniform, fast flowing channels has a potentially deleterious effect on the survival and distribution of the early life stages of fishes. Inshore areas with very low water velocities, which can be used as refuges, are thus decisive for the recruitment of 0+ rheophilic cyprinids. Combining the results of laboratory studies on critical swimming speeds with the ambient water current at nursery areas (Figure 11.34) and the densities of 0+ nase reveals a significant relationship. The population density drops drastically when the water current exceeds the critical swimming speeds of the fish. This is shown clearly in Figure 11.35, which compares the relative densities of *Chondrostoma nasus* larvae and juveniles at three different nursery areas on the Danube: a sheltered bay habitat (Bay) and two different gravel bank habitats (GB 1, GB 2). It is clear that young-of-the-year

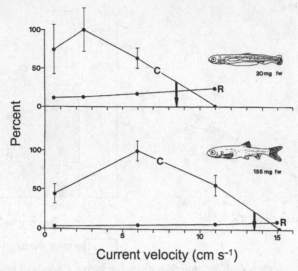

Figure 11.33   Energy acquisition (*C*) of two size classes of *Chondrostoma nasus* in different water currents and associated respiratory costs (*R*). The arrows indicate the null benefit water velocity. Feeding at higher currents results in a negative energy budget. (From Flore & Keckeis 1998.)

populations of nase are strongly controlled by the hydrological regime (water-level fluctuations) and the related structural and abiotic properties of the inshore zone of the river.

11.17   Consequences for conservation and future perspectives

The ecology of early life stages proves to be of the greatest significance for planning restoration programs for large rivers. The critical 0+ life stages are dependent on a broad array of conditions connected to the ecological integrity of river systems. They are dependent on the structural properties of the ecotonal inshore zone with high retention characteristics and potential to function as a refuge under fluctuating water levels. They are also dependent on the hydrological and habitat connectivity in a longitudinal sense, as well as the lateral integration between the river and its floodplains. In this respect the requirements of the early life stages are critical for formulating ecological goals for restoration as well as a monitoring tool to test the results of restoration programs.

Fish diversity, population dynamics, and production depend, to a large extent, on the complex and often contradictory effects of water-level fluctuations. The integration of autecological requirements of characteristic fish species helps to define natural conditions to improve structure, functioning, connectivity, and dynamics of the ecosystem. The results of habitat selection and early life history of nase, as one of the most common riverine fish species in central Europe, have been used, together with data for other groups, to design large restoration programs along the free-flowing section of the Danube (Schiemer 1999, Schiemer *et al.* 1999, Tockner *et al.* 1999, Ward *et al.* 1999) and along the main channel in the urban area of Vienna (Chovanec *et al.* 2000). The results of ongoing monitoring programs will hopefully show the extent to which this strategy achieved the planned ecological goals.

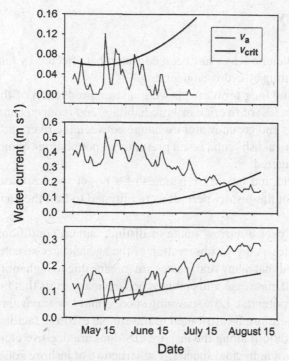

Figure 11.34 Thin line (v_a) represents the seasonal course of the water current at three different nursery areas of the Danube River. The thick line (v_{crit}) represents the critical swimming speed (Flore & Keckeis 1998) related to the size of *Chondrostoma nasus* larvae and juveniles in the habitat. Water current values above this line are unsuitable for holding position; water current values below the line represent suitable situations. The bay habitat (upper graph) was characterized by generally low water currents, the gravel bank (middle graph) had the highest water currents, with suitable conditions only during August, another gravel bank habitat (lower graph) provided suitable conditions at times when the two other nursery zones had high water current situations, thus acting as a refuge with a high inshore retention capacity during periods of floods.

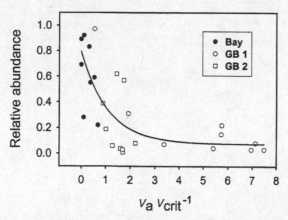

Figure 11.35 Relative density of *Chondrostoma nasus* larvae and juveniles at three inshore areas (one sheltered bay and two gravel banks) of the Danube River in relation to the ratio of the measured average current velocity at the respective habitat (v_a) and the critical swimming speed of the fish (v_{crit}; related to the average fish size at the corresponding site and sampling date). For details see Keckeis *et al.* (1997) and Winkler *et al.* (1997).

11.18   Summary

(1)   Since 1875 the Danube River has been considerably affected by land use, pollution, and most importantly by hydro-engineering.

(2)   The immediate and long term effects were a strong reduction of the former alluvial floodplain areas, a loss of riverine inshore habitats, and reduced connectivity of open-water connections and groundwater exchange between the river and its floodplains.

(3)   Most of the historical fish fauna is still present, but populations of most of the riverine species are endangered.

(4)   Autecological field and laboratory studies of a target species, nase (*Chondrostoma nasus*), at different life-history periods were initiated to find the factors that limit the populations.

(5)   Spawning occurs over a narrow range of distinct habitat conditions throughout the range of the species. A 6-year observation of the age and size structure of a spawning population showed that many year classes are missing due to high mortality in early life.

(6)   On the other hand, nase eggs and embryos possess many traits that indicate a high survival and growth potential. Low spawning success and low survival rates of young-of-the-year might therefore be attributed to unsuitable extrinsic factors.

(7)   In rivers, the conditions along the inshore ecotones are decisive elements in the population dynamics of individual species. The structure of inshore zones determines the ecological buffering capacity during water-level changes and the quality of refugees. By combining laboratory-derived data of critical swimming speeds with the ambient water current at nursery areas, a significant effect on the abundance of larvae and juveniles was observed. This finding shows that riverine young-of-the-year populations are strongly controlled by water-level changes and related structural properties of the inshore zone of the river.

(8)   The results of early life research on nase have been used to design large restoration programs for implementation along the free-flowing section of the Danube River.

Chapter 12
Methodological Resources

Robert G. Werner and Lee A. Fuiman

In this book, we have concentrated on the conceptual aspects of the contributions of early life stages to fishery science. Application of these concepts requires knowledge of appropriate techniques and methodologies, and access to a variety of specialized resources. Here, we identify some of the key references and other resources that would be useful for those working on the early life history of fishes.

12.1 General resources and bibliographies

12.1.1 Literature resources

Able, K.W. & Fahay, M.P. (1998) *The First Year in the Life of Estuarine Fishes in the Middle Atlantic Bight*. Rutgers University Press, New Brunswick, NJ, 342 pp.

Bagenal, T.B. & Braum, E. (1978) Eggs and early life history. In: *Methods for Assessment of Fish Production in Freshwater*, 3rd edn. (Ed. by T. Bagenal), pp. 165–201. International Biological Programme Handbook 3, Blackwell Scientific Publications, Oxford.

Blaxter, J.H.S. (1969) Development: eggs and larvae. In: *Fish Physiology*, Volume 3, *Development*. (Ed. by W.S. Hoar & D.J. Randall), pp. 178–252. Academic Press, New York.

Hempel, G. (1979) *Early Life History of Marine Fish: The Egg Stage*. Washington Sea Grant, Seattle, Washington, USA, 70 pp.

Hoar, W.S. & Randall, D.J. (1988) *Fish Physiology. Volume 11, The Physiology of Developing Fish. Part A, Eggs and Larvae*. Academic Press, San Diego.

Hoar, W.S. & Randall, D.J. (1988) *Fish Physiology. Volume 11, The Physiology of Developing Fish. Part B, Viviparity and Posthatching Juveniles*. Academic Press, San Diego.

Hoyt, R.D. (1988) *A Bibliography of the Early Life History of Fishes*. Western Kentucky University, Bowling Green, Kentucky, 2 volumes.

Kernehan, R.J. (1976) *A Bibliography of Early Life Stages of Fishes*. Bulletin No. 14, Ichthyological Associates, Ithaca, New York, 190 pp.

Lasker, R. (Ed.) (1981) *Marine Fish Larvae: Morphology, Ecology, and Relation to Fisheries*. University of Washington Press, Seattle, Washington, USA, 131 pp.

Moser, H.G., Richards, W.J., Cohen, D.M., Fahay, M.P., Kendall, A.W., Jr. & Richardson, S.L. (Eds.) (1984) *Ontogeny and Systematics of Fishes*. Special Publication No. 1. American Society of Ichthyologists and Herpetologists, Lawrence, Kansas, USA.

Murphy, B.R. & Willis D.W. (Eds) (1996) *Fisheries Techniques*, 2nd edn. American Fisheries Society, Bethesda, Maryland, USA.

Schreck, C.B. & Moyle, P.B. (Eds) (1990) *Methods for Fish Biology*. American Fisheries Society, Bethesda, Maryland, USA.

Snyder, D.E. (1983) Fish eggs and larvae. In: *Fisheries Techniques*. (Ed. by L.A. Nielsen & D.L. Johnson), pp. 165–197. American Fisheries Society, Bethesda, Maryland.

Wallus, R., Yeager, B.L. & Simon, T.P. (principal authors) (1990–1998) *Reproductive Biology and Early Life History of Fishes in the Ohio River Drainage*. Tennessee Valley Authority, Chattanooga, Tennessee, USA, 3 volumes.

Webb, J.F. (1999) Larvae in fish development and evolution. In: *The Origin and Evolution of Larval Forms*. (Ed. by B.K. Hall & M.H. Wake), pp. 109–158. Academic Press, San Diego.

12.1.2   Online resources

A Bibliography of the Early Life History of Fishes by Robert D. Hoyt:
  http://www.cnr.colostate.edu/~desnyder/elhsbib5.htm [February 20, 2002].
  http://scilib.ucsd.edu/sio/indexes/hoyt.html [February 20, 2002].
  gopher://biodiversity.bio.uno.edu:70/77/.indices/fishbib/bib [February 20, 2002].
Biodiversity and biological collections website:
  http://www.keil.ukans.edu [February 20, 2002].
FishBase: a global information system on fishes. Contains a continually expanding database on fishes:
  http://www.fishbase.org [February 20, 2002].
  http://ichtyonbl.mnhn.fr [February 20, 2002].
  http://filaman.uni-kiel.de [February 20, 2002].
Fish ecology mailing list: forum on fish and fisheries research and related topics, aimed to facilitate communication in the fish and fisheries research and academic communities:
  FISH-ECOLOGY@zeus.ccbb.ulpgc.es
  http://groups.yahoo.com/group/FISH-ECOLOGY [February 20, 2002].
LarvalBase: a global information system about fish larvae. Contains a continually expanding database on fish larvae:
  http://www.larvalbase.org [February 20, 2002].
Neodat II: a reference site for neotropical fishes with links to many museums and collections around the world:
  http://www.neodat.org [February 20, 2002].
NIA-NET: the official e-mail distribution list of the Neotropical Ichthyological Association (NIA). The NIA's mission is "to promote the study of Neotropical fishes":
  nia-net@inpa.gov.br
  http://www.pucrs.br/museu/nia/nianet.htm [February 20, 2002].
Plankton net mailing list: a forum for internet discussions and announcements among planktonologists researching on ecology, taxonomy, physiology, productivity, and monitoring of phytoplankton and zooplankton, in fresh, brackish, and sea water throughout the world:
  http://groups.yahoo.com/group/planktonnet [February 20, 2002].

12.1.3   Other resources

Early Life History Section of the American Fisheries Society. Membership open to all, with or without membership in the American Fisheries Society. Publishes a newsletter (*Stages*) and sponsors annual larval fish conferences:
  http://www2.ncsu.edu/elhs [February 20, 2002].
  http://www2.ncsu.edu/elhs/elhs-nl.html [February 20, 2002].
Larval fish conferences held annually and sponsored by the Early Life History Section of the American Fisheries Society (see http://www2.ncsu.edu/elhs).
Videotape: *Larval Fish Ecology: A Critical Management Concern*. Colorado State University Larval Fish Laboratory. A training videotape prepared for the U.S. Fish and Wildlife Service (1985, 24 minutes, available in English and Spanish):
  http://www.cnr.colostate.edu/~desnyder/lfevideo.htm [February 20, 2002].

12.2   Ichthyoplankton surveys and sampling techniques

Sampling ichthyoplankton is similar to plankton sampling except that questions of net avoidance and extrusion become more important with ichthyoplankton. Techniques also vary considerably between most freshwater environments and the open ocean.

12.2.1   Survey design and sampling gear

Literature resources

Anonymous (1984) A new larval fish light trap: the quatrefoil trap. *Progressive Fish-Culturist* **46**, 216–219.

Bagenal, T.B. & Nellen, W. (1980) Sampling eggs, larvae and juvenile fish. In: *Guidelines for Sampling Fish in Inland Waters*. (Ed. by T. Backiel & R.L. Welcomme), pp. 13–36. FAO of the United Nations, European Inland Fisheries Advisory Commission, Technical Paper 33, Rome, Italy.

Doherty, P.J. (1987) Light-traps: selective but useful devices for quantifying the distributions and abundances of larval fishes. *Bulletin of Marine Science* **41**, 423–431.

Harris, R., Wiebe, P., Lenz, J., Skjoldal, H.R. & Huntley, M. (Eds) (2000) *ICES Zooplankton Methodology Manual*. Academic Press, San Diego.

Hernandez, F.J., Jr. & Lindquist, D.G. (1999) A comparison of two light-trap designs for sampling larval and presettlement juvenile fish above a reef in Onslow Bay, North Carolina. *Bulletin of Marine Science* **64**, 173–184.

Kingsford, M.J. & Battershill, C.N. (1998) *Studying Temperate Marine Environments: A Handbook for Ecologists*. Canterbury University Press, Christchurch, New Zealand.

Lasker, R. (1985) An egg production method for estimating spawning biomass of pelagic fish: application to the northern anchovy (*Engraulis mordax*). *National Oceanic and Atmospheric Administration Technical Report* NMFS 36, Springfield, VA.

Olvera Limas, R. Ma., Padilla Garcia, M.A. & Ortuno Manzanarez, G. (1992) *Manual de Metodos para las Investigaciones Ictioplanctonicas del Instituto Nacional de la Pesca*, Secretaria de Pesca, Mexico City.

Pepin, P. & Shears, T.H. (1997) Variability and capture efficiency of bongo and Tucker trawl samplers in the collections of ichthyoplankton and other macrozooplankton. *Canadian Journal of Fisheries and Aquatic Sciences* **54**, 765–773.

Pham, T. & Greenwood, J. (1998) Push-net technique: a minimal interference system for sampling ichthyoplankton from near surface waters. *Proceedings of the Royal Society of Queensland* **107**, 51–55.

Ponton, D. (1994) Sampling neotropical young and small fishes in their microhabitats: an improvement of the quatrefoil light-trap. *Archiv für Hydrobiologie* **131**, 495–502.

Rivoirard, J., Simmonds, J., Foote, K., Fernandes, P. & Bez, N. (2000) *Geostatistics for Estimating Fish Abundance*. Blackwell Science, Oxford.

Smith, P.E. & Richardson, S.L. (1977) Standard techniques for pelagic fish egg and larva surveys. *FAO Fisheries Technical Paper* 175, 100 pp.

Smith, P.E. & Richardson, S.L. (1979) Selected bibliography on pelagic fish egg and larva surveys. *FAO Fisheries Circular* 706, 97 pp. (in English, French, and Spanish).

UNESCO (1968) Zooplankton sampling. *Monographs on Oceanic Methodology*. United Nations Educational, Scientific and Cultural Organization, Paris, 174 pp.

Werner, R.G. (1968) Addendum: methods of sampling fish larvae in nature. In: *Methods for Assessment of Fish Production in Freshwater*. (Ed. by W.E. Ricker), pp. 178–181. International Biological Programme Handbook 3, Blackwell Scientific Publications, Oxford.

12.2.2   Handling samples

Literature resources

Lavenberg, R.J., McGowen, G.E. & Woodsum, R.E. (1984) Preservation and curation. In: *Ontogeny and Systematics of Fishes*. (Ed. by H.G. Moser, W.J. Richards, D.M. Cohen, M.P. Fahay, A.W. Kendall, Jr. & S.L. Richardson), pp. 57–59. American Society of Ichthyologists and Herpetologists, Special Publication 1, Lawrence, Kansas, USA.

Olvera Limas, R. Ma., Padilla Garcia, M.A. & Ortuno Manzanarez, G. (1992) *Manual de Metodos para las Investigaciones Ictioplanctonicas del Instituto Nacional de la Pesca*. Secretaria de Pesca, Mexico City.

Snyder, D.E. (1983) Fish eggs and larvae. In: *Fisheries Techniques*. (Ed. by L.A. Nielsen & D.L. Johnson), pp. 165–197. American Fisheries Society, Bethesda, Maryland.

UNESCO. 1968. Zooplankton sampling. *Monographs on Oceanic Methodology*. United Nations Educational, Scientific and Cultural Organization, Paris, 174 pp.

12.3   Identification of species and stages

One of the key issues in the study of the early life stages of fishes is the problem of identification. The number of taxonomic characters available for making accurate identifications of eggs or larvae is considerably less than for juveniles or adults. Further, there are different types of characters and many of them change rapidly with development. In spite of these problems, many regional guides for identifying early life stages have been published. Some are based on laboratory-reared eggs and larvae from known parents, others are based on a series of field-caught specimens.

12.3.1   Taxonomic and examination methods

Literature resources

Balon, E.K. & Flegler-Balon, C. (1985) Microscopic techniques for studies of early ontogeny in fishes: problems and methods of composite descriptions. In: *Early Life Histories of Fishes. New Developmental, Ecological and Evolutionary Perspectives*. (Ed. by E.K. Balon), pp. 33–55. Dr. W. Junk Publishers, Dordrecht.

Boehlert, G.W. (1984) Scanning electron microscopy. In: *Ontogeny and Systematics of Fishes*. (Ed. by H.G. Moser, W.J. Richards, D.M. Cohen, M.P. Fahay, A.W. Kendall, Jr. & S.L. Richardson), pp. 43–48. American Society of Ichthyologists and Herpetologists, Special Publication 1, Lawrence, Kansas, USA.

Dunn, J.R. (1984) Developmental osteology. In: *Ontogeny and Systematics of Fishes*. (Ed. by H.G. Moser, W.J. Richards, D.M. Cohen, M.P. Fahay, A.W. Kendall, Jr. & S.L. Richardson), pp. 48–50. American Society of Ichthyologists and Herpetologists, Special Publication 1, Lawrence, Kansas, USA.

Galat, D.L. (1972) Preparing teleost embryos for study. *The Progressive Fish-Culturist* **34**, 43–48.

Govoni, J.J. (1984) Histology. In: *Ontogeny and Systematics of Fishes*. (Ed. by H.G. Moser, W.J. Richards, D.M. Cohen, M.P. Fahay, A.W. Kendall, Jr. & S.L. Richardson), pp. 40–42. American Society of Ichthyologists and Herpetologists, Special Publication 1, Lawrence, Kansas, USA.

Kendall, A.W., Jr., Ahlstrom, E.H. & Moser, H.G. (1984) Early life history stages of fishes and their characters. In: *Ontogeny and Systematics of Fishes*. (Ed. by H.G. Moser, W.J. Richards, D.M. Cohen, M.P. Fahay, A.W. Kendall, Jr. & S.L. Richardson), pp. 11–22. American Society of Ichthyologists and Herpetologists, Special Publication 1, Lawrence, Kansas, USA.

Matarese, A.C. & Sandknop, E.M. (1984) Identification of fish eggs. In: *Ontogeny and Systematics of Fishes*. (Ed. by H.G. Moser, W.J. Richards, D.M. Cohen, M.P. Fahay, A.W. Kendall, Jr. & S.L. Richardson), pp. 27–31. American Society of Ichthyologists and Herpetologists, Special Publication 1, Lawrence, Kansas, USA.

Potthoff, T. (1984) Clearing and staining techniques. In: *Ontogeny and Systematics of Fishes*. (Ed. by H.G. Moser, W.J. Richards, D.M. Cohen, M.P. Fahay, A.W. Kendall, Jr. & S.L. Richardson), pp. 35–37. American Society of Ichthyologists and Herpetologists, Special Publication 1, Lawrence, Kansas, USA.

Powles, H. & Markle, D.F. (1984) Identification of larvae. In: *Ontogeny and Systematics of Fishes*. (Ed. by H.G. Moser, W.J. Richards, D.M. Cohen, M.P. Fahay, A.W. Kendall, Jr. & S.L. Richardson), pp. 31–33. American Society of Ichthyologists and Herpetologists, Special Publication 1, Lawrence, Kansas, USA.

Sandknop, E.M., Sumida, B.Y. & Moser, H.G. (1984) Early life history descriptions. In: *Ontogeny and Systematics of Fishes*. (Ed. by H.G. Moser, W.J. Richards, D.M. Cohen, M.P. Fahay, A.W. Kendall, Jr. & S.L. Richardson), pp. 23–24. American Society of Ichthyologists and Herpetologists, Special Publication 1, Lawrence, Kansas, USA.

Sumida, B.Y., Washington, B.B. & Laroche, W.A. (1984) Illustrating fish eggs and larvae. In: *Ontogeny and Systematics of Fishes*. (Ed. by H.G. Moser, W.J. Richards, D.M. Cohen, M.P. Fahay, A.W. Kendall, Jr. & S.L. Richardson), pp. 33–35. American Society of Ichthyologists and Herpetologists, Special Publication 1, Lawrence, Kansas, USA.

Tucker, J.W., Jr. & Laroche, J.L. (1984) Radiographic techniques in studies of young fishes. In: *Ontogeny and Systematics of Fishes*. (Ed. by H.G. Moser, W.J. Richards, D.M. Cohen, M.P. Fahay, A.W. Kendall, Jr. & S.L. Richardson), pp. 37–39. American Society of Ichthyologists and Herpetologists, Special Publication 1, Lawrence, Kansas, USA.

Other resources

Armstrong, P.B. & Child, J.S. (1962) Stages in the development of *Ictalurus nebulosus*. Syracuse University Press. Syracuse, New York 8 pp., 16 plates. (folio) – available from the Early Life History Section of the American Fisheries Society (see http://www2.ncsu.edu/elhs).

12.3.2 Taxonomic guides

Literature resources

Auer, N.A. (Ed.) (1982) *Identification of Larval Fishes of the Great Lakes Basin with Emphasis on the Lake Michigan Drainage*. Special publication 82–3, Great Lakes Fishery Commission, Ann Arbor, USA.

Boltovskoy, D. (ed.). (1999) *South Atlantic Zooplankton*. Two Volumes. Backhuys Publishers, Leiden. 1,706 pp.

Brownell, C.L. (1979) Stages in the early development of 40 marine fish species with pelagic eggs from the Cape of Good Hope. Rhodes University, J.L.B. Smith Institute of Ichthyology. Bulletin 40.

Delsman, H.C. (1972) *Fish Eggs and Larvae from the Java Sea*. Linnaeus Press, Amsterdam.

Efremenko, F.N. (1985) *Illustrated Guide to Fish Larvae of the Southern Ocean*. BIOMASS Scientific Series, 5. Scott Polar Research Institute. Cambridge, England (published originally in *Cybium* 1983, 7(2) under the title: "Atlas of fish larvae of the Southern Ocean …").

Fahay, M.P. (1983) Guide to the early stages of marine fishes occurring in the western Atlantic Ocean, Cape Hatteras to the southern Scotian Shelf. *Journal of Northwest Atlantic Fishery* Science 4, 1–423.

Fish, M.P. (1932) Contributions to the early life histories of sixty-two species of fishes from Lake Erie and its tributary waters. *Bulletin of the Bureau of Fisheries, United States* 47, 293–398.

Garcia, A.M.A. & Moyano, P.D. (1990) *Estados juveniles de la ictiofauna en Los Canos de Las Salinas de La Bahi de Cadiz*. Instituto de Ciencias Marinas de Andalucia, Consejo Superior de Investigaciones Cientificas, Consejeria de Gobernacion. Junta de Andalucia, Cadiz, Spain, 163 pp.

Kellermann, A. (Ed.) (1989) *Identification Key and Catalogue of Larval Antarctic Fishes*. Biological Investigations of Marine Antarctic Systems and Stocks, Cambridge.

Kellermann, A. (1990) Identification key and catalogue of larval Antarctic fishes. *Berichte für Polarforschung* **67**, 1–136.

Koblitskaia, A.F. (1981) *Opredelitel' Molodi Presnovodnykh Ryb* (Identification of Young Freshwater Fishes). Legkaia I pishevaia promyshlennost', Moscow.

Leis, J.M. & Carson-Ewart, B.M. (Eds) (2000) *The Larvae of Indo-Pacific Coastal Fishes: An Identification Guide to Marine Fish Larvae*. Fauna Malesiana Handbooks, 2. Brill, Leiden.

Leis, J.M. & Trnski, T. (1989) *The Larvae of Indo-Pacific Shorefishes*. New South Wales University Press, Kensington, Australia.

Lo Bianco, S. (1931–1933) Uova, larve e stadi giovanili di Teleostei. *Fauna e Flora del Golfo di Napoli. Monografia della Stazione zoologica di Napoli* **38**, parts I and II. [Translated from Italian by Israel Program for Scientific Translations, Jerusalem, 1969; Available thru Smithsonian Institution Library or NTIS # 68-50346.]

Matarese, A.C., Kendall, A.W., Blood, D.M. & Vinter, M.V. (1989) Laboratory guide to early life history stages of Northeast Pacific fishes. *NOAA Technical Report* NMFS 80, 652 pp.

Miller, J.M., Watson, W. & Leis, J.M. (1979) An atlas of common nearshore marine fish larvae of the Hawaiian Islands. Univeristy of Hawaii Sea Grant College Program, *Miscellaneous Report* 80–02.

Mito, S. & Dotu, Y. (1961) Pelagic fish eggs and larvae from Japanese waters.– II. Lamprida, Zeida, Mugilina, Scombrina, Carngina, and Stromateina. *Scientific Bulletin of the Faculty of Agriculture, Kyushu University* **18**, 451–466.

Mito, S. & Dotu, Y.(1961) Pelagic fish eggs and larvae from Japanese waters.– I. Clupeina, Chanina, Stomiatina, Myctophida, Anguillida, Belonida, and Syngnathida. *Scientific Bulletin of the Faculty of Agriculture, Kyushu University* **18**, 285–310.

Mito, S. & Dotu, Y. (1962) Pelagic fish eggs from Japanese waters.–VI. Labrina. *Scientific Bulletin of the Faculty of Agriculture, Kyushu University* **19**, 493–502.

Mito, S. & Dotu, Y. (1963) Pelagic fish eggs from Japanese waters.–III. Percina. *Japanese Journal of Ichthyology* **11**, 39–64.

Moser, H.G. (Ed.) (1996) The early stages of fishes in the California Current Region. CalCOFI Report. Atlas No. 33, National Oceanic and Atmospheric Administration, National Marine Fisheries Service, La Jolla, CA, 1505 pp.

Neira, F.J., Miskiewicz, A.G. & Trnski, T. (1998) *Larvae of Temperate Australian Fishes: Laboratory Guide for Larval Fish Identification*. University of Western Australia Press, Nedlands, Western Australia.

Okiyama, M. (Ed.) (1988) *An Atlas of the Early Stage Fishes in Japan*. Tokai University Press, Tokyo, 1154 pp.

Olivar, M-P. & Fortuño, J-M. (1991) Guide to ichthyoplankton of the southeast Atlantic (Benguela Current Region). *Scientia Marina* **55**, 1–383.

Russell, F.S. (1976) *The Eggs and Planktonic Stages of British Marine Fishes*. Academic Press, London.

Santander B.H. & Castillo, O.S. de (1979) El ictioplancton de la costa peruana. *Boletin Instituto del Mar del Peru* **4**, 70–92.

Uchida, K., Imai, S., Mito, S., Fujita, S, Ueno, M., Shojima, Y., Senta, T., Tahuku, M. & Dotu, Y. (1958) *Studies on the Eggs, Larvae and Juveniles of Japanese Fishes*. Second Laboratory of Fisheries Biology, Fisheries Department, Faculty of Agriculture, Kyushu University, Fukuoka, Japan. 266 pp.

U.S. Fish and Wildlife Service (various authors) (1978) *Development of Fishes of the Mid-Atlantic Bight: An Atlas of Egg, Larval and Juvenile Stages*. U.S. Fish and Wildlife Service, Biological Services Program, FWS/OBS-78/12, 6 volumes.

Vatanachai, S. (1974) The identification of fish eggs and larvae obtained from the survey cruises in the South China Sea. *Proceedings of the Indo-Pacific Fisheries Council* **15**, 111–130.

Wang, J.C.S. (1981) *Taxonomy of the Early Life Stages of Fishes: Fishes of the Sacramento-San Joaquin Estuary and Moss Landing-Elkhorn Slough, California*. Ecological Analysist Inc., Concord, California, USA.

Watson, W. & Leis, J.M. (1974) *Ichthyoplankton of Kaneohe Bay, Hawaii: A One-Year Study of Fish Eggs and Larvae*. University of Hawaii, Honolulu.

Online resources

FishBase: a global information system on fishes. Contains a continually expanding database on fishes: http://www.fishbase.org [February 20, 2002].

Jayaseelan, M.J. (2000) *Fish eggs and larvae from Asian mangrove waters*. CD-ROM. Springer-Verlag.

LarvalBase: a global information system about fish larvae. Contains a continually expanding database on fish larvae:
http://www.larvalbase.org [February 20, 2002].

Larval fishes from Carrie Bow Cay, Belize – in color by David G. Smith: http://www.nmnh.si.edu/vert/larval [February 21,2002].

Preliminary guide to the identification of the early life history stages of ichthyoplankton of the western central Atlantic (draft edn) by William J. Richards:
http://www4.cookman.edu/noaa [February 20, 2002].

Others: Reference collections and archives of fish larvae

Albany Museum, Grahamstown, South Africa:
http://www.ru.ac.za/affiliates/am/ichthy.html [February 20, 2002].

Australian Museum, Sydney, Australia.

CSIRO Division of Fisheries Research, Hobart, Tasmania.

Larval Fish Laboratory, Colorado State University, USA:
http://www.cnr.colostate.edu/lfl [February 20, 2002].

Museum of Comparative Zoology, Harvard University, USA:
http://www.mcz.harvard.edu/fish [February 20, 2002].

Museum of Tropical Queensland, Townsville, Australia.

Museum of Victoria, Melbourne, Australia.

South Australian Museum, Adelaide, Australia.

University of Washington, USA:
http://artedi.fish.washington.edu [February 20, 2002].

12.4   Rearing eggs and larvae

In addition to the obvious value to aquaculture of being able to rear eggs and larvae, this ability is also fundamental in taxonomic, behavioral, and other types of experimental work. Culture techniques that are successful for one species may not be transferable to other species. But, all species require careful handling, high quality water, appropriate temperatures, and adequate food.

12.4.1   Literature resources

Bromage, N.R. & Roberts, R.J. (Eds) (1995) *Broodstock Management and Egg and Larval Quality*. Blackwell Science, Oxford.

Chamberlain, G.W., Midget, R.J. & Haby, M.B. (Eds) (1990) *Red Drum Aquaculture*. Texas A&M Sea Grant, College Program No. TAMU-SG-90-603.

Hunter, J.R. (1984) Synopsis of culture methods for marine fish larvae. In: *Ontogeny and Systematics of Fishes*. (Ed. by H.G. Moser, W.J. Richards, D.M. Cohen, M.P. Fahay, A.W. Kendall, Jr. & S.L. Richardson), pp. 24–27. American Society of Ichthyologists and Herpetologists, Special Publication 1, Lawrence, Kansas, USA.

Stickney, R.R. & Kohler, C.C. (1990) Maintaining fishes for research and teaching. In: *Methods for Fish Biology*. (Ed. by C.B. Schreck & P.B. Moyle), pp. 633–663. American Fisheries Society. Bethesda, Maryland, USA.

Stickney, R.R. (Ed.) (2000) *Encyclopedia of Aquaculture*. John Wiley & Sons, Inc., New York.

Suda, A. (1991) Recent progress in artificial propagation of marine species for Japanese sea-farming and aquaculture. In: *Marine Farming and Enhancement – Proceedings of the 15th US–Japan Meeting on Aquaculture, Kyoto, Japan October 22–23, 1986*. (Ed. by A.K. Spark), pp. 123–127. NOAA Technical Report NMFS 85.

Tucker, J.W. (1998) *Marine Fish Culture*. Kluwer Academic Publishers, Boston.

12.4.2   Online resources

Balyrut, E.A. (1989) *Aquaculture Systems and Practices: A Selected Review*. FAO United Nations Development Program, ADCP/REP/89/43:
http://www.fao.org/docrep/t8598e/t8598e00.htm [February 20, 2002].

12.5   Experimental protocols

More and more fishery scientists are turning to controlled experiments to understand the ecology of early life stages of fishes. All experiments are individual and distinct, but certain common elements are shared and these have been gathered together in the sources listed below. They will help anyone planning to establish an experimental approach to understanding early life stages of fishes.

12.5.1   Literature resources

Lalli, C.M. (Ed.) (1990) *Enclosed Experimental Marine Ecosystems : A Review and Recommendations*. Springer-Verlag, New York.

12.5.2   Online resources

Electronic Code of Federal Regulations. (2002) Title 40, Protection of environment. Chapter 1, Environmental Protection Agency. Part 797, *Environmental Effects Testing Guidelines*. 797.1600, Fish Early Life Stage Toxicity Test:
http://www.access.gpo.gov/nara/cfr/cfrhtml_00/Title_40/40cfr797_00.html [February 20, 2002].

12.6   Age, growth, and condition

A highly developed methodology has appeared over the last decade or so that allows fishery scientists to determine, to the day, the age and growth history of fish larvae. As pointed out in several chapters in this book, this is a very powerful tool for analyzing the population dynamics and ecology of the early life stages of fishes. In addition, several techniques have been developed to assess the nutritional condition of larvae.

12.6.1 Age and growth

Literature resources

Brothers, E.B. (1982) Aging reef fishes. In: *The Biological Bases for Reef Fishery Management*. (Ed. by G.R. Huntsman, W.R. Nicholson & W.W. Fox, Jr.), pp. 3–22. National Marine Fisheries Service, NOAA Technical Memorandum, SFC-80.

Brothers, E.B. (1984) Otolith studies. In: *Ontogeny and Systematics of Fishes*. (Ed. by H.G. Moser, W.J. Richards, D.M. Cohen, M.P. Fahay, A.W. Kendall, Jr. & S.L. Richardson), pp. 50–57. American Society of Ichthyologists and Herpetologists, Special Publication 1, Lawrence, Kansas, USA.

Brothers, E.B., Mathews, C.P. & Lasker, R. (1976) Daily growth increments in otoliths from larval and adult fishes. *Fishery Bulletin, U.S.* **74**, 1–8.

Chambers, R.C. & Miller, T.J. (1994) Evaluating fish growth by means of otolith incrment analysis: special properties of individual-level longitudinal data. In: *Recent Developments in Fish Otolith Research*. (Ed. by D.H. Secor, J.M. Dean & S.E. Campana), pp. 155–175. University of South Carolina, Columbia.

Secor, D.H., Dean, J.M. & Laban, E.H. (1991) *Manual for Otolith Removal and Preparation for Microstructural Examination*. Belle W. Baruch Institute for Marine Biology and Coastal Research, Technical Publication 1991–01.

Stevenson, D.K. & Campana, S.E. (Eds) (1992) Otolith microstructure examination and analysis. *Canadian Special Publication of Fisheries and Aquatic Sciences* 117, Department of Fisheries and Oceans, Ottawa.

Summerfelt, R.C. & Hall, G.E. (1987) *Age and Growth of Fish*. Iowa State University Press, Ames, Iowa, USA.

Online resources

Otolith Research Laboratory of Steven Campana, Department of Fisheries and Oceans, Canada (in English and French):
http://www.mar.dfo-mpo.gc.ca/science/mfd/otolith [February 20, 2002].

12.6.2 Otolith microchemistry

Literature resources

Campana, S.E. (1999) Chemistry and composition of fish otoliths: pathways, mechanisms and applications. *Marine Ecology Progress Series* **188**, 263–297.

Campana, S.E., Thorrold, S.R., Jones, C.M., Gunther, D., Tubrett, M., Longerich, H., Jackson, S., Halden, N., Kalish, J.M., Piccoli, P., Pontual, H. de, Troadec, H., Panfili, H., Secor, D.H., Severin, K.P., Sie, S.H., Thresher, R., Teesdale, W.J. & Cambell, J.L. (1997) Comparison of accuracy, precision and sensitivity in elemental assays of fish otoliths using the electron microprobe, PIXE and laser ablation ICPMS. *Canadian Journal of Fisheries and Aquatic Sciences* **54**, 2068–2079.

Gunn, J.S., Harrowfield, I.R., Proctor, C.H. & Thresher, R.E. (1992) Electron microanalysis of fish otoliths: evaluation of techniques for studying age and stock discrimination. *Journal of Experimental Marine Biology and Ecology* **158**, 1–36.

Milton, D.A. & Chenery, S.R. (1998) The effect of otolith storage methods on the concentration of elements detected by laser-ablation ICPMS. *Journal of Fish Biology* **53**, 785–794.

Proctor, C.H. & Thresher, R.E. (1998) Effects of specimen handling and otolith preparation on concentration of elements in fish otoliths. *Marine Biology* **131**, 681–694.

Thorrold, S.R., Campana, S.E., Jones, C.M. & Swart, P.K. (1997) Factors determining $\delta^{13}C$ and $\delta^{18}O$ fractionation in aragonitic otoliths of marine fish. *Geochimica et Cosmochimica Acta* **61**, 2909–2919.

Thresher, R.E. (1999) Elemental composition of otoliths as a stock delineator in fishes. *Fisheries Research* **43**, 165–204.

Online resources

Otolith Research Laboratory of Steven Campana, Department of Fisheries and Oceans, Canada (in English and French):
http://www.mar.dfo-mpo.gc.ca/science/mfd/otolith [February 20, 2002].

12.6.3   Biochemical measures of condition

Literature resources

Caldarone, E.M. & Buckley, L.J. (1991) Quantitation of DNA and RNA in crude tissue abstracts by flow injection analysis. *Analytical Biochemistry* **199**, 137–141.

Caldarone, E.M., Wagner, M., Onge-Burns, J. St. & Buckley, L.J. (2001) Protocol and guide for estimating nucleic acids in larval fish using a fluorescence microplate reader. *Northeast Fisheries Science Center Reference Document* 01-11, National Marine Fisheries Service, Woods Hole, Massachusetts, USA.

Canino, M.F. & Caldarone, E.M. (1995) Modification and comparison of two fluorometric techniques for determining nucleic acid contents of fish larvae. *Fishery Bulletin, U.S.* **93**, 158–165.

Buckley, L.J. & Bulow, F.J. (1987) Techniques for the estimation of RNA, DNA, and protein in fish. In: *Age and Growth of Fish*. (Ed. by R.C. Summerfelt & G.E. Hall), pp. 345–354. Iowa State University Press, Ames, Iowa.

Clemmesen, C. (1988) A RNA and DNA fluorescence technique to evaluate the nutritional condition of individual marine fish larvae. *Meeresforschung* **32**, 134–143.

Clemmesen, C. (1993) Improvements in the fluorometric determination of the RNA and DNA content in individual marine fish larvae. *Marine Ecology Progress Series* **100**, 177–183.

McGurk, M. & Kusser, W. (1992) Comparison of three methods of measuring rRNA and DNA concentrations of individual Pacific herring (*Clupea harengus*) larvae. *Canadian Journal of Fisheries and Aquatic Sciences* **49**, 967–974.

Theilacker, G.H. & Shen, W. (1993) Calibrating starvation-induced stress in larva fish using flow cytometry. *American Fisheries Society Symposium* **14**, 85–94.

12.7   Population dynamics

Understanding the population dynamics of early life stages is very difficult due in large part to the brief duration of the period. The basic principles are similar to those for more advanced stages, however. Those principles and techniques are outlined in the references below.

12.7.1   Literature resources

Bunn, N.A., Fox, C.J. & Webb, T. (2000) A literature review of studies on fish egg mortality: implications for the estimation of spawning stock biomass by the annual egg production method. Ministry of Agriculture, Fisheries and Food, Centre for Environment, Fisheries and Aquaculture Science, *Science Series Technical Report* 111, Lowestoft, UK, 37 pp.

Gulland, J.A. (1983) *Fish Stock Assessment: A Manual of Basic Methods*. Wiley-Interscience, Chichester.

Hilborn, R. & Walters, C.J. (1992) *Quantitative Fisheries Stock Assessment: Choice, Dynamics, and Uncertainty*. Chapman and Hall, New York.

Ricker, W.E. (1975) Computation and interpretation of biological statistics of fish populations. *Bulletin of the Fisheries Research Board of Canada* 191, Fisheries and Marine Service, Ottawa.

12.8 Modeling of early life stages

Mathematical modeling has been a very important part of the development of our understanding of early life stages of fishes. Models have been developed using many different techniques for a variety of purposes. Population models, bioenergetics models, age and growth models, individual-based models are some of the types that have been successfully applied to early life stages of fishes. Some of the following specific applications of models to the early life of fishes may be useful for learning modeling techniques. Other modeling applications are mentioned in individual chapters. We also mention software products that are designed for instruction.

12.8.1 Literature resources

Beyer, J.E. & Laurence, G.C. (1980) A stochastic model of larval growth. *Ecological Modelling* **8**, 109–132.

Beyer, J.E. & Laurence, G.C. (1981) Aspects of stochasticity in modelling growth and survival of clupeoid fish larvae. *Rapports et Procès-Verbaux des Réunions du Conseil International pour l'Exploration de la Mer* **178**, 17–23.

Breitburg, D.L., Rose, K.A. & Cowan, J.H., Jr. (1999) Linking water quality to larval survival: predation mortality of fish larvae in an oxygen-stratified water column. *Marine Ecology Progress Series* **178**, 39–54.

Carscadden, J.E., Frank, K.T. & Leggett, W.C. (2000) Evaluation of an environment–recruitment model for capelin (*Mallotus villosus*). *ICES Journal of Marine Science* **57**, 412–418.

Fiksen, Ø., Utne, A.C.W., Aksnes, D.L., Eiane, K., Helvik, J.V. & Sundby, S. (1998) Modelling the influence of light, turbidity and ontogeny on ingestion rates in larval cod and herring. *Fisheries Oceanography* **7**, 355–363.

Letcher, B.H. & Rice, J.A. (1997) Prey patchiness and larval fish growth and survival: inferences from an individual-based model. *Ecological Modeling* **95**, 29–43.

Letcher, B.H., Rice, J.A., Crowder, L.B. & Rose, K.A. (1996) Variability in survival of larval fish: disentangling components with a generalized individual-based model. *Canadian Journal of Fisheries and Aquatic Sciences* **53**, 787–801.

Vlymen, W.J. (1977) A mathematical model of the relationship between larval anchovy (*Engraulis mordax*) growth, prey microdistribution, and larval behavior. *Environmental Biology of Fishes* **2**, 211–233.

12.8.2 Others

EcoBeaker: Ecology Teaching Software. BeakerWare, Ithaca, New York, USA (see http://www. ecobeaker.com).

Populus: Simulations of Population Biology. D.N. Alstad, University of Minnesota (see http://www. cbs.umn.edu/populus).

12.9 Marking early life stages

12.9.1 Literature resources

Bergstedt, R.A., Eshenroder, R.L., Bowen, C., II, Seelye, J.G. & Locke, J.L. (1990) Mass-marking of otoliths of lake trout sac fry by temperature manipulation. *American Fisheries Society Symposium* **7**, 216–223.

Brothers, E.B. (1990) Otolith marking. *American Fisheries Society Symposium* **7**, 183–202.

Dabrowski, K. & Tsukamoto, K. (1986) Tetracycline tagging in coregonid embryos and larvae. *Journal of Fish Biology* **29**, 691–698.

Hettler, W.F. (1984) Marking otoliths by immersion of marine fish larvae in tetracycline. *Transactions of the American Fisheries Society* **113**, 370–373.

Nagiec, M. (1992) Persistence of tetracycline mark in otoliths of whitefish (*Coregonus lavaretus*). *Bulletin of the Sea Fisheries Institute* **3**, 77–80.

Pedersen, T. & Carlsen, B. (1991) Marking cod (*Gadus morhua* L.) juveniles with oxytetracycline incorporated into the feed. *Fisheries Research* **12**, 57–64.

Reinert, T.R., Wallin, J., Griffin, M.C., Conroy, M.J. & Van Den Avyle, M.J. (1998) Long term retention and detection of oxytetracycline marks applied to hatchery reared larval striped bass, *Morone saxatilis*. *Canadian Journal of Fisheries and Aquatic Sciences* **55**, 539–543.

Secor, D.H. & Houde, E.D. (1995) Larval mark–release experiments: potential for research on dynamics and recruitment in fish stocks. In: *Recent Developments in Fish Otolith Research*. (Ed. by D.H. Secor, S.E. Campana & J.M. Dean), pp. 423–444. University of South Carolina Press, Columbia.

Snyder, R.J., McKeown, B.A. & Colbow, R. (1992) Use of dissolved strontium in scale marking of juvenile salmonids: effects of concentration and exposure time. *Canadian Journal of Fisheries and Aquatic Sciences* **49**, 780–782.

Szedlmayer, S.T. & Howe, J.C. (1995). An evaluation of six marking methods for age-0 red drum, *Sciaenops ocellatus*. *Fishery Bulletin, U.S.* **93**, 191–195.

Volk, E.C., Schroder, S.L. & Fresh, K.L. (1990) Inducement of unique otolith banding patterns as a practical means to mass-marking juvenile Pacific salmon. *American Fisheries Society Symposium* **7**, 203–215.

Volk, E.C., Schroder, S.L., Grimm, J.J. & Ackley, H.S. (1994) Use of a bar code symbology to produce multiple thermally induced otolith marks. *Transactions of the American Fisheries Society* **123**, 811–816.

Wilson, C.A., Beckman, D.A. & Dean, J.M. (1987) Calcein as a fluorescent marker of otoliths of larval and juvenile fish. *Transactions of the American Fisheries Society* **116**, 668–670.

Appendix: List of Symbols

Symbol	Definition	Chapters
A	Total mortality (fraction dying per unit time)	3
AGR	Absolute growth rate in length (length increment per unit time)	2, 5
AGR$'$	Absolute growth rate in weight (weight increment per unit time)	2
a_i	Area represented by sampling station i	5
α	Parameter in power function relating either performance to size, length to weight, the decline in specific growth rate to time, or the shape of the survival curve	1, 2, 3
B	Biomass of stock (mass units)	5
B_s, B_{s-1}	Biomass of a population or cohort at respective stages (mass units)	3
β	Exponent in power function relating either performance to size, length to weight, or the decline in specific growth rate to time; also a constant in a Pareto model expressing the overall rate of decline in abundance of a cohort	1, 2, 3
c	Constant of integration or fraction of fishing mortality within a year before spawning	2, 9
$D_{\Delta L,i}$	Density of larvae of a selected length interval ΔL at sampling station i (number per unit area)	5
d	Fraction of natural mortality within a year before spawning	9
δ	"del value", $= [(R_{\text{sample}}/R_{\text{standard}}) - 1 \times 1000]$, where R_{sample} or R_{standard} is the ratio of a heavier isotope to a lighter (more abundant) isotope	6
E	Number of eggs per unit weight of female	5
E'	Number of eggs spawned per kilogram of female per batch during the period over which f is estimated	5
f	Fraction of females spawning during the time interval over which egg abundance is measured	5
fr_{ts}	Fraction mature at age t at time of spawning	9
F	Instantaneous fishing mortality rate (per unit time)	3, 9
F_{st}	Fixation index, heterozygosity of least inclusive group in relation to most inclusive group	6
g	Proportional growth rate in length or length-specific growth rate (per unit time)	2
g_0	Length-specific growth rate at hatching (per unit time)	2
g'	Proportional growth rate in weight or weight-specific growth rate (per unit time)	2
g'_0	Weight-specific growth rate at hatching (per unit time)	2

G	Instantaneous growth rate in length (per unit time)	2, 3, 7
G'	Instantaneous growth rate in weight (per unit time)	2, 3
H_s	Heterozygosity of subpopulation	6
H_t	Heterozygosity of total population	6
i	Index variable, identifies an interval of time, station, etc.	3
I_t	Larval abundance index	5
K	von Bertalanffy growth coefficient	2
K_1	Conversion coefficient	11
L	Length	1, 2
L_0	Length at hatching	2, 5
L_{juv}	Length at completion of metamorphosis	1
L_t	Length at age or time t	2, 5
L^∞	Mean asymptotic length	2
m	Proportion expected to die from fishing if no other causes of mortality present; conditional mortality	3
M	Instantaneous natural mortality rate (per unit time)	2, 3, 7
n	Proportion expected to die from natural mortality in the absence of fishing mortality; conditional mortality	3
N	Abundance	2, 3, 5
N_0	Abundance at beginning of stage or at time 0	5
N_s	Abundance at stage s	3
N_t	Abundance at age or time t	3, 5
N_{ts}	Number in cohort of age t alive at time of spawning	9
O_L	Ontogenetic index; fraction of the developmental period (embryonic and larval) that has already occurred, based on larva length	1
P_0	Number of offspring produced by population	5
P_t	Number of offspring of age t produced by the population, or fraction recruited to gear at age t	5, 9
R	Fraction of biomass of entire stock that is producing offspring or initial number in cohort	5, 9
R'	Fraction of females producing biomass	5
R_{sample}	Ratio of heavy to light isotope in a sample	6
$R_{standard}$	Ratio of heavy to light isotope in a standard	6
S	Survival rate; proportion surviving during a time period	3
SSB	Accumulated total weight of spawning stock over the lifetime of a cohort	9
t	Age, time interval, or survey period	2, 3, 5
t_0	Initial time or age	2
T	Temperature	1
W	Weight, mass, or dry mass	1, 2, 3
W_0	Weight at hatching	2
W_{its}	Average weight of an individual at age t at time of spawning	9
W_S, W_{S-1}	Weight of fish at respective stages	3
W_t	Weight at age or time t	2
W_{ts}	Weight of spawning segment of cohort at age t at time of spawning	9
X	Total number of stations sampled during a survey period	5
Z	Instantaneous total mortality rate (per unit time)	3, 5

Literature Cited

Ahlstrom, E.H., Amaoka, K., Hensley, D.A., Moser, H.G. & Sumida, B.Y. (1984) Pleuronectiformes: Development. In: *Ontogeny and Systematics of Fishes*. (Ed. by H.G. Moser, W.J. Richards, D.M. Cohen, M.P. Fahay, A.W. Kendall, Jr. & S.L. Richardson), pp. 640–670. American Society of Ichthyologists and Herpetologists, Special Publication 1, Lawrence, Kansas, USA.

Ahlstrom, E.H., Butler, J.L. & Sumida, B.Y. (1976) Pelagic stromateoid fishes (Pisces, Perciformes) of the eastern Pacific: kinds, distributions and early life histories and observations on five of these from the northwest Atlantic. *Bulletin of Marine Science* **26**, 285–402.

Armstrong, P.B. & Child, J.S. (1962) *Stages in the Development of* Ictalurus nebulosus. Syracuse University Press, Syracuse, New York. 8 pp., 16 plates. (folio).

Auer, N.A. (1982) Family Salmonidae, trouts. In: *Identification of Larval Fishes of the Great Lakes Basin with Emphasis on the Lake Michigan Drainage*. (Ed. by N.A. Auer), pp. 80–145. Great Lakes Fishery Commission, Special Publication 82–3, Ann Arbor, Michigan, USA.

Bacalbaşa-Dobrovici, N. (1989) The Danube River and its fisheries. *Canadian Special Publication of Fisheries and Aquatic Sciences* **106**, 455–468.

Bailey, K.M. (2000) Shifting control of recruitment of walleye pollock *Theragra chalcogramma* after a major climatic and ecosystem change. *Marine Ecology Progress Series* **198**, 215–224.

Bailey, K.M., Brodeur, R.D. & Hollowed, A.B. (1996) Cohort survival patterns of walleye pollock, *Theragra chalcogramma*, in Shelikof Strait, Alaska: a critical factor analysis. *Fisheries Oceanography* **5**(suppl. 1), 179–188.

Bailey, K.M. & Houde, E.D. (1989) Predation on eggs and larvae of marine fishes and the recruitment problem. *Advances in Marine Biology* **25**, 1–83.

Balon, E.K. (1975a) Reproductive guilds of fishes: a proposal and definition. *Journal of the Fisheries Research Board of Canada* **32**, 821–864.

Balon, E.K. (1975b) Terminology of intervals in fish development. *Journal of the Fisheries Research Board of Canada* **32**, 1663–1670.

Bardonnet, A. (2001) Spawning in swift water currents: implications for eggs and larvae. *Archiv für Hydrobiologie* **135**(suppl.), 271–291.

Beverton, R.J.H. & Holt, S.J. (1957) *On the Dynamics of Exploited Fish Populations*. Chapman and Hall, London.

Beyer, J.E. (1989) Recruitment stability and survival – size-specific theory with examples from the early life dynamics of marine fish. *Dana* **7**, 45–147.

Bigelow, H.B. & Welsh, W.W. (1925) Fishes of the Gulf of Maine. *Bulletin of the U.S. Bureau of Fisheries* **40**, 1–567.

Blaxter, J.H.S. (1976) Reared and wild fish – how do they compare? In: *Proceedings of the 10th European Symposium on Marine Biology*. (Ed. by G. Persoone & E. Jaspers), pp. 11–26. Universa Press, Wetteren, Belgium.

Blaxter, J.H.S. (1986) Development of sense organs and behavior of teleost larvae with special reference to feeding and predator avoidance. *Transactions of the American Fisheries Society* **115**, 98–114.

Blaxter, J.H.S. (1988) Pattern and variety in development. In: *Fish Physiology*, Volume 11A. (Ed. by W.S. Hoar & D.J. Randall), pp. 1–58. Academic Press, San Diego.

Blaxter, J.H.S. (2000) Enhancement of marine fisheries. *Advances in Marine Biology* **38**, 2–54.

Blaxter, J.H.S. & Batty, R.S. (1985) The development of startle responses in herring larvae. *Journal of the Marine Biological Association of the United Kingdom* **65**, 737–750.

Blaxter, J.H.S. & Hunter, J.R. (1982) The biology of the clupeoid fishes. *Advances in Marine Biology* **20**, 1–223.

Boehlert, G.W. & Mundy, B.C. (1993) Ichthyoplankton assemblages at seamounts and oceanic islands. *Bulletin of Marine Science* **53**, 336–361.

Bonetto, C.A. (1981) Contribucion al conocimiento limnologico del Rio Paraguay en su tramo inferior. *Esocur* **8**, 55–88.

Breitburg, D.L., Loher, T., Pacey, C.A. & Gerstein, A. (1997) Varying effects of low dissolved oxygen on trophic interactions in an estuarine food web. *Ecological Monographs* **67**, 489–507.

Brodeur, R.D. & Bailey, K.M. (1996) Predation on the early life stages of marine fish: a case study on walleye pollock in the Gulf of Alaska. In: *Survival Strategies in Early Life Stages of Marine Resources*. (Ed. by Y. Watanbe, Y. Yamshita & Y. Oozeki), pp. 245–259. Balkema, Rotterdam.

Brooks, J.L. & Dodson, S.I. (1965) Predation, body size, and composition of plankton. *Science* **150**, 28–35.

Brothers, E.B., Mathews, C.P. & Lasker, R. (1976) Daily growth increments in otoliths from larval and adult fishes. *Fishery Bulletin, U.S.* **74**, 1–8.

Brown, R.W., Taylor, W.W. & Assel, R.A. (1993) Factors affecting the recruitment of lake whitefish in two areas of northern Lake Michigan. *Journal of Great Lakes Research* **19**, 418–428.

Busnita, T. (1967) Die Ichthyofauna der Donau. In: *Limnologie der Donau*. (Ed. by R. Liepolt), pp. 198–224. Eine Monographische Darstellung, E. Schweizerbart'sche Verlagsbuchhandlung, Stuttgart.

Campana, S.E. (1996) Year-class strength and growth rate in young Atlantic cod *Gadus morhua*. *Marine Ecology Progress Series* **135**, 21–26.

Campana, S.E., Chouinard, G.A., Hanson, J.M. & Fréchet, A. (1999) Mixing and migration of overwintering Atlantic cod (*Gadus morhua*) stocks near the mouth of the Gulf of St. Lawrence. *Canadian Journal of Fisheries and Aquatic Sciences* **56**, 1873–1881.

Campana, S.E. & Jones, C.M. (1992) Analysis of otoliths microstructure data. In: *Otolith Microstructure Examination and Analysis*. (Ed. by D.K. Stevenson & S.E. Campana), pp. 73–100. *Canadian Special Publication of Fisheries and Aquatic Sciences* **117**.

Chapin, F.S., III, Zavaleta, E.S., Eviner, V.T., Naylor, R.L., Vitousek, P.M., Reynolds, H.L., Hooper, D.U., Lavorel, S., Sala, O.E., Hobbie, S.E., Mack, M.C. & Diaz, S. (2000) Consequences of changing biodiversity. *Nature* **405**, 234–242.

Chovanec, A., Schiemer, F., Cabela, A., Gressler, S., Grötzer, C., Pascher, K., Raab, R., Teufl, H. & Wimmer, R. (2000) Constructed inshore zones as river corridors through urban areas – the Danube in Vienna: preliminary results. *Regulated Rivers: Research and Management* **16**, 175–187.

Clemmesen, C. (1996) Importance and limits of RNA/DNA ratios as a measure of nutritional condition in fish larvae. In: *Survival Strategies in Early Life Stages of Marine Resources*. (Ed. by Y. Watanabe, Y. Yamashita & Y. Oozeki), pp. 67–82. A.A. Balkema, Rotterdam.

Collette, B.B., Pottthoff, T., Richards, W.J., Ueyanagi, S., Russo, J.L. & Nishikawa, Y. (1984) Scombroidei: development and relationships. In: *Ontogeny and Systematics of Fishes*. (Ed. by H.G. Moser, W.J. Richards, D.M. Cohen, M.P. Fahay, A.W. Kendall, Jr. & S.L. Richardson), pp. 591–620. American Society of Ichthyologists and Herpetologists, Special Publication 1, Lawrence, Kansas, USA.

Conner, J.V. (1979) Identification of larval sunfishes (Centrarchidae: Elassomatidae) from southeastern Louisiana. In: *Proceedings of the Third Symposium on Larval Fishes*. (Ed. by R.D. Hoyt), pp. 17–52. Western Kentucky University, Bowling Green, Kentucky, USA.

Cowan, J.H., Jr., Houde, E.D. & Rose, K.A. (1996) Size-dependent vulnerability of marine fish larvae to predation: an individual-based numerical experiment. *ICES Journal of Marine Science* **53**, 23–37.

Crecco, V. & Savoy, T. (1985) Effects of biotic and abiotic factors on growth and relative survival of young American shad in the Connecticut River. *Canadian Journal of Fisheries and Aquatic Sciences* **42**, 1640–1648.

Crowder, L.B. (1980) Alewife, rainbow smelt and native fishes in Lake Michigan: competition or predation? *Environmental Biology of Fishes* **5**, 225–233.

Crowder, L.B. & Binkowski, F.P. (1983) Foraging behaviors and the interaction of alewife, *Alosa pseudoharengus*, and bloater, *Coregonus hoyi*. *Environmental Biology of Fishes* **8**, 105–113.

Crowder, L.B. & Crawford, H.L. (1984) Ecological shifts in resource use by bloaters in Lake Michigan. *Transactions of the American Fisheries Society* **113**, 694–700.

Cury, P. & Roy, C. (1989) Optimal environmental window and pelagic fish recruitment success in upwelling areas. *Canadian Journal of Fisheries and Aquatic Sciences* **46**, 670–680.

Cushing, D.H. (1971) The dependence of recruitment on parent stock in different groups of fishes. *Journal du Conseil International pour l'Exploration de la Mer* **33**, 340–362.

Cushing, D.H. (1975) *Marine Ecology and Fisheries*. Cambridge University Press, London.

Cushing, D.H. (1977) The problems of stock and recruitment. In: *Fish Population Dynamics*. (Ed. by J.A. Gulland), pp. 116–133. John Wiley & Sons, Chichester.

Cushing, D.H. (1990) Plankton production and year-class strength in fish populations: an update of the match/mismatch hypothesis. *Advances in Marine Biology* **26**, 250–293.

Davis, T.J. & Blasco, D. (1997) *The Ramsar Convention Manual: A Guide to the Convention on Wetlands (Ramsar, Iran, 1971)*, 2nd edn. Ramsar Convention Bureau, Gland, Switzerland, 161 pp. (Online) Available: http://www.ramsar.org/lib_manual_1.htm [February 20, 2002].

Doherty, P.J., Planes, S. & Mather, P. (1995) Gene flow and larval duration in seven species of fish from the Great Barrier Reef. *Ecology* **76**, 2373–2391.

Doyle, M.J., Morse, W.W. & Kendall, A.W., Jr. (1993) A comparison of larval fish assemblages in the temperate zones of the northeast Pacific and northwest Atlantic Oceans. *Bulletin of Marine Science* **53**, 588–644.

Doyle, M.J., Rugen, W.C. & Brodeur, R.D. (1994) Neustonic ichthyoplankton in the western Gulf of Alaska during spring. *Fishery Bulletin, U.S.* **93**, 231–253.

Drinkwater, K. & Myers, R.A. (1987) Testing predictions of marine fish and shellfish landings from environmental variables. *Canadian Journal of Fisheries and Aquatic Sciences* **44**, 1568–1573.

Dunn, J.R. & Matarese, A.C. (1984) Gadidae: Development and relationships. In: *Ontogeny and Systematics of Fishes*. (Ed. by H.G. Moser, W.J. Richards, D.M. Cohen, M.P. Fahay, A.W. Kendall, Jr. & S.L. Richardson), pp. 283–299. American Society of Ichthyologists and Herpetologists, Special Publication 1, Lawrence, Kansas, USA.

Edmonds, J.S. & Fletcher, W.J. (1997) Stock discrimination of pilchards *Sardinops sagax* by stable isotope ratio analysis of otolith carbonate. *Marine Ecology Progress Series* **152**, 241–247.

Ehrenbaum, E. (1909) Eier und Larven von Fischen des nordischen Planktons, Teil II. *Verlag von Lipsius und Tischer, Keil und Leipzig* **1**, 1–217–413.

Faulk, C.K., Fuiman, L.A. & Thomas, P. (1999) Parental exposure to ortho, para-dichlorodiphenyl-trichloroethane impairs survival skills of Atlantic croaker (*Micropogonias undulatus*) larvae. *Environmental Toxicology and Chemistry* **18**, 254–262.

Fish, M.P. (1932) Contribution to the early life histories of sixty-two species of fishes from Lake Erie and its tributary waters. *Bulletin of the Bureau of Fisheries, Washington, D.C.* **47**, 293–398.

Fisher, R., Bellwood, D.R. & Job, S.D. (2000) Development of swimming abilities in reef fish larvae. *Marine Ecology Progress Series* **202**, 163–173.

Fisher, J.P., Fitzsimons, J.D., Combs, G.F. Jr. & Spitsbergen, J.M. (1996) Naturally occurring thiamine deficiency causing reproductive failure in Finger Lakes Atlantic salmon and Great Lakes lake trout. *Transactions of the American Fisheries Society* **125**, 167–178.

Fitzsimons, J.D., Brown, S.B., Honeyfield, D.C. & Hnath, J.G. (1999) A review of early mortality syndrome (EMS) in Great Lakes salmonids: relationship with thiamine deficiency. *Ambio* **28**, 9–15.

Flore, L. & Keckeis, H. (1998) The effects of water current on foraging behaviour of the rheophilic cyprinid, *Chondrostoma nasus*, during early ontogeny: evidence of a trade-off between energetic benefit and swimming costs. *Regulated Rivers: Research and Management* **14**, 141–154.

Flore, L., Keckeis, H. & Schiemer, F. (2001) Feeding, energetic benefit and swimming capabilities of 0+ nase (*Chondrostoma nasus* L.) in flowing water: an integrative laboratory approach. *Archiv für Hydrobiologie* **135**(suppl.), 409–424.

Fortier, L. & Leggett, W.C. (1985) A drift study of larval fish survival. *Marine Ecology Progress Series* **25**, 245–257.

Freeberg, M.H., Taylor, W.W. & Brown, R.W. (1990) Effect of egg and larval survival on the year-class strength of lake whitefish in Grand Travers Bay, Lake Michigan. *Transactions of the American Fisheries Society* **119**, 92–100.

Fry, B. & Arnold C.R. (1982) Rapid $^{13}C/^{12}C$ turnover during growth of brown shrimp (*Penaeus aztecus*). *Oecologia* **54**, 200–204.

Fuiman, L.A. (1989) Vulnerability of Atlantic herring larvae to predation by yearling herring. *Marine Ecology Progress Series* **51**, 291–299.

Fuiman, L.A. (1994) The interplay of ontogeny and scaling in the interactions of fish larvae and their predators. *Journal of Fish Biology* **45**(suppl. A), 55–79.

Fuiman, L.A. & Delbos, B.C. (1998) Developmental changes in visual sensitivity of red drum, *Sciaenops ocellatus*. *Copeia* **1998**, 936–943.

Fuiman, L.A. & Webb, P.W. (1988) Ontogeny of routine swimming activity and performance in zebra danios (*Teleostei: Cyprinidae*). *Animal Behaviour* **36**, 250–261.

Gabriel, W.L., Sissenwine, M.P. & Overholtz, W.J. (1989) Analysis of spawning stock biomass per recruit: an example for Georges Bank haddock. *North American Journal of Fisheries Management* **9**, 383–391.

Gartz, R.G., Miller, L.W., Fujimura, R.W. & Smith P.E. (1999) Measurement of larval striped bass (*Morone saxatilis*) net avoidance using evasion radius estimation to improve estimates of abundance and mortality. *Journal of Plankton Research* **21**, 561–580.

Gihr, M. (1957) Zur Entwicklung des Hechtes. *Revue Suisse De Zoologie* **64**, 355–474.

Govoni, J.J. (1993) Flux of larval fishes across frontal boundaries: examples from the Mississippi River plume front and the western Gulf Stream front in winter. *Bulletin of Marine Science* **53**, 538–566.

Grioche, A., Koubi, P. & Harlay, X. (1999) Spatial patterns of ichthyoplankton assemblages along the eastern English Channel French coast during spring 1995. *Estuarine, Coastal and Shelf Science* **49**, 141–152.

Gunn, J.S., Harrowfield, I.R., Proctor, C.H. & Thresher, R.E. (1992) Electron probe microanalysis of fish otoliths – evaluation of techniques for studying age and stock discrimination. *Journal of Experimental Marine Biology and Ecology* **158**, 1–36.

Hansson, S., Post, D.M., Kitchell, J.F. & McComish, T.S. (1997) Predation by yellow perch (*Perca flavescens*) in southern Lake Michigan – a model analysis. In: *Forage Fishes in Marine Ecosystems*, pp. 243–256. *Proceedings of the International Symposium on the Role of Forage Fishes in Marine Ecosystems*. Alaska Sea Grant College Program Report No. 97-01, University of Alaska, Fairbanks.

Harden Jones, F.R. (1968) *Fish Migrations*. Edward Arnold, London.

Harding, D., Nichols, J.H. & Tungate, D.S. (1978) The spawning of the plaice (*Pleuronectes platessa*) in the Southern North Sea and the English Channel. *Rapports et Procès-verbaux des Réunions, Conseil International pour l'Exploration de la Mer* **172**, 102–113.

Hardy, A. (1965) *The Open Sea: Its Natural History*. Houghton Mifflin, Boston.

Hardy, J.D., Jr. (1978) *Development of Fishes of the Mid-Atlantic Bight: An Atlas of Egg, Larval and Juvenile Stages. Volume 2. Anguillidae Through Syngnathidae*. U.S. Fish and Wildlife Service, Biological Services Program, FWS/OBS-78/12.

Hare, J.A. & Cowen, R.K. (1996) Transport mechanisms of larval and pelagic juvenile bluefish (*Pomatomus saltatrix*) from South Atlantic Bight spawning grounds to Middle Atlantic Bight nursery habitats. *Limnology and Oceanography* **41**, 1264–1280.

Hassler, T.J. (1970) Environmental influences on early development and year-class strength of northern pike in Lakes Oahe and Sharpe, South Dakota. *Transactions of American Fisheries Society* **99**, 369–375.

Heath, M. (1993) An evaluation and review of the ICES herring larval surveys in the North Sea and adjacent waters. *Bulletin of Marine Science* **53**, 795–817.

Heath, M., Zenitani, H., Watanabe, Y., Kimura, R. & Ishida, M. (1998) Modelling the dispersal of larval Japanese sardine (*Sardinops melanostictus*), by the Kuroshio Current in 1993 and 1994. *Fisheries Oceanography* **7**, 335–346.

Hedgecock, D. (1994) Does variance in reproductive success limit effective population sizes of marine organisms? In: *Genetics and Evolution of Aquatic Organisms*. (Ed. by A.R. Beaumont), pp. 122–132. Chapman and Hall, London.

Helbig, J.A. & Pepin, P. (1998) Partitioning the influence of physical processes on the estimation of ichthyoplankton mortality rates. I. Theory. *Canadian Journal of Fisheries and Aquatic Sciences* **55**, 2189–2205.

Herzka, S.Z. & Holt, G.J. (2000) Changes in isotopic composition of red drum (*Sciaenops ocellatus*) larvae in response to dietary shifts: potential applications to settlement studies. *Canadian Journal of Fisheries and Aquatic Sciences* **57**, 137–147.

Herzka, S.Z., Holt, S.A. & Holt, G.J. (2002) Characterization of settlement patterns of red drum (*Sciaenops ocellatus*) larvae to estuarine nursery habitat: a stable isotope approach. *Marine Ecology Progress Series* **226**, 143–156.

Hildebrand, S.F. & Cable, L. (1930) Development and life history of fourteen teleostean fishes at Beaufort, North Carolina. *Bulletin of the U.S. Bureau of Fisheries* **46**, 383–488.

Hjort, J. (1914) Fluctuations in the great fisheries of northern Europe viewed in the light of biological research. *Rapports et Procès-verbaux des Réunions, Conseil International pour l'Exploration de la Mer* **20**, 1–228.

Holt, J., Johnson, A.G., Arnold, C.R., Fable, W.A., Jr. & Williams, T.D. (1981) Description of eggs and larvae of laboratory reared red drum, *Sciaenops ocellata*. *Copeia* **1981**, 751–756.

Houde, E.D. (1987) Fish early life dynamics and recruitment variability. *American Fisheries Society Symposium* **2**, 17–29.

Houde, E.D. (1989) Comparative growth, mortality, and energetics of marine fish larvae: temperature and implied latitudinal effects. *Fishery Bulletin, U.S.* **87**, 471–495.

Houde, E.D. (1994) Differences between marine and freshwater fish larvae: implications for recruitment. *ICES Journal of Marine Science* **51**, 91–97.

Houde, E.D. (1996) Evaluating stage-specific survival during the early life of fish. In: *Survival Strategies in Early Life Stages of Marine Resources*. (Ed. by Y. Watanabe, Y. Yamashita & Y. Oozeki), pp. 51–66. Balkema, Rotterdam.

Houde, E.D. (1997) Patterns and trends in larval-stage growth and mortality of teleost fish. *Journal of Fish Biology* **51**(suppl. A), 52–83.

Hunter, J.R. & Leong, R. (1981) The spawning energetics of female northern anchovy (*Engraulis mordax*). *Fishery Bulletin, U.S.* **79**, 215–230.

Iles, T.D. & Sinclair. M. (1982) Atlantic herring: stock discreteness and abundance. *Science* **215**, 627–633.

Ishida, M. & Kikuchi, H. (1992) *Monthly Egg Production of the Japanese Sardine, Anchovy, and Mackerels Off the Southern Coast of Japan by Egg Censuses: January, 1989 through December 1990*. National Research Institute of Fisheries Science, Tokyo.

IUCN – World Conservation Union (2000) *IUCN Red List of Threatened Species*. (Online) Available: http://www.iucn.org/redlist/2000 [February 20, 2002].

Japan Fisheries Agency (1997) *Long-Term Forecast of Replacement of Dominant Species in the Small Pelagic Fish Community*. Japan Fisheries Agency, Tokyo.

Jones, G.P., Milicich, M.J., Emslie, M.J. & Lunow, C. (1999) Self-recruitment in a coral reef fish population. *Nature* **402**, 802–804.

Joseph, E.B., Massmann, W. H. & Norcross, J.J. (1964) The pelagic eggs and early larval stages of the black drum from Chesapeake Bay. *Copeia* **1964**, 425–434.

Jude, D.J. (1982) Family Acipenseridae, sturgeons. In: *Identification of Larval Fishes of the Great Lakes Basin with Emphasis on the Lake Michigan Drainage*. (Ed. by N.A. Auer), pp. 38–44. Great Lakes Fishery Commission, Special Publication 82–3, Ann Arbor, USA.

Kamler, E. & Keckeis, H. (2000) Reproduction and early life history of *Chondrostoma nasus*: implications for recruitment (a review). *Polski Archiv Hydrobiologie* **47**, 73–85.

Kamler, E., Keckeis, H. & Bauer-Nemeschkal, E. (1996) Egg energy content and partitioning in a rheophilic cyprinid, *Chondrostoma nasus* (L.). *Polski Archiv Hydrobiologie* **43**, 273–281.

Kamler, E., Keckeis, H. & Bauer-Nemeschkal, E. (1998) Temperature-induced changes of survival, development and yolk partitioning in *Chondrostoma nasus*. *Journal of Fish Biology* **53**, 658–682.

Kappus, B.M., Jansen, W., Böhmer, J. & Rahmann, H. (1997) Historical and present distribution and recent habitat use of nase, *Chondrostoma nasus*, in the lower Jagst River (Baden Würtemberg, Germany). *Folia Zoologica* **46**(suppl. 1), 51–60.

Keckeis, H., Bauer-Nemeschkal, E. & Kamler, E. (1996) Effects of reduced oxygen level on the mortality and hatching rate of *Chondrostoma nasus* embryos. *Journal of Fish Biology* **49**, 430–440.

Keckeis, H., Winkler, G., Flore, L., Reckendorfer, W. & Schiemer, F. (1997) Spatial and seasonal characteristics of 0+ fish nursery habitats of nase, *Chondrostoma nasus* in the river Danube, Austria. *Folia Zoologica* **46**, 133–150.

Keckeis, H., Bauer-Nemeschkal, E., Menshutkin, V.V., Nemeschkal, H.L. & Kamler, E. (2000) Effects of female attributes and egg properties on offspring viability in a rheophilic cyprinid, *Chondrostoma nasus*. *Canadian Journal of Fisheries and Aquatic Sciences* **57**, 789–796.

Keckeis, H., Kamler, E., Bauer-Nemeschkal, E. & Schneeweiss, K. (2001) Survival, development and food energy partitioning of nase larvae and early juveniles at different temperatures. *Journal of Fish Biology* **57**, 45–61.

Kellogg, L.L., Houde, E.D., Secor, D.H. & Gooch, J.W. (1996) Egg production and environmental factors influencing larval population dynamics in the Nanticoke River, 1992–1993. In: *Final Report to Maryland Department of Natural Resources.* (Ed. by E.D. Houde & D.H. Secor), pp. 1–142. Episodic water quality events and striped bass recruitment: larval mark–recapture experiments in the Nanticoke River. University of Maryland, Center for Estuarine and Environmental Science, Ref. No. (UMCEES)CBL 96-083.

Kendall, A.W. Jr. & Behnke, R.J. (1984) Salmonidae: development and relationships. In: *Ontogeny and Systematics of Fishes.* (Ed. by H.G. Moser, W.J. Richards, D.M. Cohen, M.P. Fahay, A.W. Kendall, Jr. & S.L. Richardson), pp. 142–149. American Society of Ichthyologists and Herpetologists, Special Publication 1, Lawrence, Kansas, USA.

Kikuchi, H. & Konishi, Y. (1990) *Monthly Egg Production of the Japanese Sardine, Anchovy, and Mackerels Off the Southern Coast of Japan by Egg Censuses: January, 1987 through December 1988.* National Research Institute of Fisheries Science, Tokyo.

Kimura, R., Kinoshita, T., Wada, T., Nakamura, M., Secor, D. & Piccoli, P. (2000a) Otolith Sr : Ca and growth history of juvenile sardines in the Kuroshio–Oyashio transition region. Abstracts for the *Meeting of the Japanese Society of Fisheries Science April 1–5, 2000*, p. 53.

Kimura, R., Watanabe, Y. & Zenitani, H. (2000b) Nutritional condition of first-feeding larvae of Japanese sardine in the coastal and oceanic waters along the Kuroshio Current. *ICES Journal of Marine Science* **57**, 240–248.

King, M. (1995) *Fisheries Biology: Assessment and Management.* Blackwell Science, Oxford.

King, T.L., Kalinowski, S.T., Schill, W.B., Spidle, A.P. & Lubinski, B.A. (2001) Population structure of Atlantic salmon (*Salmo salar* L.): a range-wide perspective from microsatellite DNA variation. *Molecular Ecology* **10**, 807–821.

Kirchhofer, A. (1996) Fish conservation in Switzerland – three case studies. In: *Conservation of Endangered Freshwater Fish in Europe.* (Ed. by A. Kirchofer & D. Hefti), pp. 135–145. Birkhaeuser-Verlag, Basel.

Krueger, C.C., Perkins, D.L., Mills, E.L. & Marsden, J.E. (1995) Predation by alewives on lake trout fry in Lake Ontario: Role of an exotic species in preventing restoration of a native species. *Journal Great Lakes Research*, **21**(suppl. 1), 458–469.

Kryzhanovskii, S.G. (1949) Eco-morphological principles and patterns of development among minnows, loaches and catfishes (Cyprinoidei and Siluroidei). Part II. Ecological groups of fishes and patterns of their distribution. Academy of Sciences of the USSR. *Study of the Institute of Animal Morphology* **1**, 237–331 (*Fisheries Research Board of Canada Translation* Series No. 2945, 1974).

Kubota, H., Oozeki, Y., Ishida, M., Konishi, Y., Goto, T., Zenitani, H. & Kimura, R. (1999). Distributions of eggs and larvae of Japanese sardine, Japanese anchovy, mackerels, round herring, jack mackerels and Japanese common squid in the waters around Japan, 1994 through 1996. *Resources Management Research Report Series* A-2.

Kurmayer, R., Keckeis, H., Schrutka, S. & Zweimüller, I. (1996) Macro- and microhabitats used by 0+ fish in a side-arm of the River Danube. *Archiv für Hydrobiologie* **113**(suppl.), 425–432.

Lasker, R. (1975) Field criteria for the survival of anchovy larvae: the relationship between inshore chlorophyll maximum layers and successful first feeding. *Fishery Bulletin, U.S.* **73**, 453–462.

Lasker, R. (1978) The relationship between oceanographic conditions and larval anchovy food in the California Current: identification of factors contributing to recruitment failure. *Rapports et Procès-verbaux des Réunions, Conseil International pour l'Exploration de la Mer* **173**, 212–230.

Lasker, R. (1985) An egg production method for estimating spawning biomass of pelagic fish: application to the northern anchovy (*Engraulis mordax*). *National Oceanic and Atmospheric Administration Technical Report* NMFS 36.

Leggett, W.C. & Deblois E. (1994) Recruitment in marine fishes: is it regulated by starvation and predation in the egg and larval stages? *Netherlands Journal of Sea Research* **32**, 119–134.

Lelek, A. (1987) *Threatened Fishes of Europe.* Aula Verlag, Wiesbaden.

Lemly, A.D. (1997) A teratogenic deformity index for evaluating impacts of selenium on fish populations. *Ecotoxicology and Environmental Safety* **37**, 259–266.

Letcher, B.H., Rice, J.A., Crowder, L.B. & Binkowski, F.P. (1997) Size- and species-dependent variability in consumption and growth rates of larvae and juveniles of three freshwater fishes. *Canadian Journal of Fisheries and Aquatic Sciences* **54**, 405–414.

Limburg, K.E. (1996) Growth and migration of 0-year American shad (*Alosa sapidissima*) in the Hudson River Estuary: otolith microstructural analysis. *Canadian Journal of Fisheries and Aquatic Sciences* **53**, 220–238.

Limburg, K.E. (2001) Through the gauntlet again: demographic restructuring of American shad by migration. *Ecology* **82**, 1584–1596.

Limburg, K.E., Pace, M.L. & Arend, K.K. (1999) Growth, mortality, and recruitment of larval (*Morone* spp.) in relation to food availability and temperature in the Hudson River. *Fishery Bulletin, U.S.* **97**, 80–91.

Limburg, K.E. & Schmidt, R.E. (1990) Patterns of fish spawning in Hudson River tributaries: response to an urban gradient? *Ecology* **71**, 1238–1245.

Lo, N.C.H. (1985) Egg production of the central stock of northern anchovy (*Engraulis mordax*), 1951–1982. *Fishery Bulletin, U.S.* **83**, 137–150.

Lo, N.C.H. (1986) Modeling life-stage-specific instantaneous mortality rates, and application to northern anchovy (*Engraulis mordax*), eggs and larvae. *Fishery Bulletin, U.S.* **84**, 395–407.

Lochmann, S.E., Taggart, C.T., Griffin, D.A., Thompson, K.R. & Maillet, G.L. (1997) Abundance and condition of larval cod (*Gadus morhua*) at a convergent front on Western Bank, Scotian Shelf. *Canadian Journal of Fisheries and Aquatic Sciences* **54**, 1461–1479.

Luecke, C., Rice, J.A., Crowder, L.B., Yeo, S.F. & Binkowski, F.P. (1990) Recruitment mechanisms of bloater in Lake Michigan: an analysis of the predatory gauntlet. *Canadian Journal of Fisheries and Aquatic Sciences* **47**, 524–532.

Lusk, S. & Halačka, K. (1995) Anglers catches as an indicator of population size of the nase, *Chondrostoma nasus. Folia Zoologica* **44**, 185–192.

MacKenzie, B.R. & Kiørboe, T. (2000) Larval fish feeding and turbulence: a case for the downside. *Limnology and Oceanography* **45**, 1–10.

MacKenzie, B.R. & Leggett, W.C. (1991) Quantifying the contribution of small-scale turbulence to the encounter rates between larval fish and their zooplankton prey: effects of wind and tide. *Marine Ecology Progress Series* **73**, 149–160.

Madenjian, C.P., Tyson, J.T., Knight, R.L., Kershner, M.W. & Hansen, M.J. (1996) First-year growth, recruitment, and maturity of walleyes in western Lake Erie. *Transactions of the American Fisheries Society* **125**, 821–830.

Mansueti, A.J. (1964) Early development of the yellow perch, *Perca flavescens. Chesapeake Science* **5**, 46–66.

Mansueti, A.J & Hardy, J.D. (1967) *Development of Fishes of the Chesapeake Bay Region: An Atlas of Egg, Larval, and Juvenile Stages.* Natural Resources Institute, University of Maryland, College Park, Maryland.

Mansueti, R.J. (1958) Eggs, larvae and young of the striped bass, *Roccus saxatilis. Chesapeake Biological Laboratory Contributions* **112**, 36.

Marshall, C.T., Yaragina, N.A., Lambert, Y. & Kjesbu, O.S. (1999) Total lipid energy as a proxy for total egg production by fish stocks. *Nature* **402**, 288–290.

Marty, C., Beall, E. & Parot, G. (1986) Influence de quelques paramètres du milieu d'incubation sur la survie d'alevins de saumon atlantique, *Salmo salar* L., en ruisseau experimental. *Internationale Revue der Gesamten Hydrobiologie* **71**, 349–361.

Mason, D.M. & Brandt, S.B. (1996) Effect of alewife predation on survival of larval yellow perch in an embayment of Lake Ontario. *Canadian Journal of Fisheries and Aquatic Sciences* **53**, 1609–1617.

Masuda, R. & Tsukamoto, K. (1998) Stock enhancement in Japan: review and perspective. *Bulletin of Marine Science* **63**, 337–358.

Matarese, A.C., Richardson, S.L. & Dunn, J.R. (1981) Larval development of Pacific tomcod, *Microgadus proximus*, in the northeast Pacific Ocean with comparative notes on larvae of walleye pollock, *Theragra chalcogramma*, and Pacific cod, *Gadus macrocephalus* (Gadidae). *Fishery Bulletin, U.S.* **78**, 923–940.

Matsuura, Y. & Hewitt R. (1995) Changes in the spatial patchiness of Pacific mackerel, *Scomber japonicus*, larvae with increasing age and size. *Fishery Bulletin, U.S.* **93**, 172–178.

McDonald, G., Fitzsimons, J.D. & Honeyfield, D.C. (Eds) (1998) Early stage mortality syndrome in fishes of the Great Lakes and Baltic Sea. *American Fisheries Society Symposium* **21**, Bethesda, Maryland.

McGurk, M.D. (1986) Natural mortality of marine pelagic fish eggs and larvae: role of spatial patchiness. *Marine Ecology Progress Series* **34**, 227–242.

Meekan, M.G. & Fortier, L. (1996) Selection for fast growth during the larval life of Atlantic cod *Gadus morhua* on the Scotian Shelf. *Marine Ecology Progress Series* **137**, 25–37.

Megrey, B.A., Hollowed, A.B., Hare, S.R., Macklin, S.A. & Stabeno, P.J. (1996) Contributions of FOCI research to forecasts of year-class strength of walleye pollock in Shelikof Strait, Alaska. *Fisheries Oceanography* **5**(suppl. 1), 189–203.

Methot, R.D., Jr. (1983) Seasonal variation in survival of larval *Engraulis mordax* estimated from the age distribution of juveniles. *Fishery Bulletin, U.S.* **81**, 741–750.

Miller, T.J., Crowder, L.B., Rice, J.A. & Marschall, E.A. (1988) Larval size and recruitment mechanisms in fishes: toward a conceptual framework. *Canadian Journal of Fisheries and Aquatic Sciences* **45**, 1657–1670.

Miller, T.J., Crowder, L.B. & Binkowski, F.P. (1990) Effects of changes in the zooplankton assemblage on growth of bloater and implications for recruitment success. *Transactions of the American Fisheries Society* **119**, 483–491.

Miller, T.J., Crowder, L.B., Rice, J.A. & Binkowski, F.P. (1992) Body size and the ontogeny of the functional response in fishes. *Canadian Journal of Fisheries and Aquatic Sciences* **49**, 805–812.

Miller, T.J., Crowder, L.B. & Rice, J.A. (1993) Ontogenetic changes in behavioural and histological measures of visual acuity in three species of fish. *Environmental Biology of Fishes* **37**, 1–8.

Miller, T.J., Herra, T. & Leggett, W.C. (1995) Characteristics of eggs and their subsequent newly-hatched larvae of Atlantic cod (*Gadus morhua*). *Canadian Journal of Fisheries and Aquatic Sciences* **52**, 1083–1093.

Mion, J.B., Stein, R.A. & Marschall, E.A. (1998) River discharge drives survival of larval walleye. *Ecological Applications* **8**, 88–103.

Mito, S. (1961) Pelagic fish eggs from Japanese waters. I. Clupeina, Chanina, Stomiatina, Myctophida, Anguillida, Belonida, and Syngnathida. *Scientific Bulletin of the Faculty of Agriculture, Kyushu University* **18**, 285–310.

Mori, K., Kuroda, K. & Konishi, Y. (1988) *Monthly Egg Production of the Japanese Sardine, Anchovy, and Mackerels Off the Southern Coast of Japan by Egg Censuses: January, 1978 through December 1986*. Tokai Regional Fisheries Research Laboratory, Tokyo.

Myers, R.A. & Barrowman, N.J. (1996) Is fish recruitment related to spawner abundance? *Fishery Bulletin, U.S.* **94**, 707–724.

Nakamura, M. (1969) *Cyprinid Fishes of Japan: Studies on the Life History of Cyprinid Fishes of Japan* (in Japanese). Research Institute for Natural Resources, Special Publication 4, Tokyo.

Nakata, K., Zenitani, H. & Inagake. D. (1995) Differences in food availability for Japanese sardine larvae between the frontal region and the waters on the offshore side of Kuroshio. *Fisheries Oceanography* **4**, 68–79.

National Research Institute of Fisheries Science (1992) *Long-Term Forecasting on the Distribution and Abundance of the Important Fishery Resources and the Related Oceanographic Conditions in the Sea of Tohoku and Chuo Blocs, No 88*. National Research Institute Fisheries Science, Tokyo.

National Research Institute of Fisheries Science (1994) *Long-Term Forecasting on the Distribution and Abundance of the Important Fishery Resources and the Related Oceanographic Conditions in the Sea of Chuo Blocs. No. 93, 31*. National Research Institute Fisheries Science, Tokyo.

Nelson, W.M., Ingham, M. & Schaaf, W. (1977) Larval transport and year class strength of Atlantic menhaden, (*Brevoortia tyrannus*). *Fishery Bulletin, U.S.* **75**, 23–41.

Okazaki, T., Kobayashi, T. & Uozumi, Y. (1996). Genetic relationship of pilchards (genus *Sardinops*) with anti-tropical distribution. *Marine Biology* **126**, 585–590.

Oozeki, Y. & Zenitani, H. (1996) Factors affecting the recent growth of Japanese sardine larvae (*Sardinops melanoscictus*) in the Kuroshio Current. In: *Survival Strategies in Early Life Stages of Marine Resources*. (Ed. by Y. Watanbe, Y. Yamshita & Y. Oozeki), pp. 95–104. Balkema, Rotterdam.

Pannella, G. (1971) Fish otoliths: daily growth layers and periodical patterns. *Science* **173**, 1124–1127.

Paradis, A.R., Pepin, P. & Brown, J.A. (1996) Vulnerability of fish eggs and larvae to predation: review of the influence of the relative size of prey and predator. *Canadian Journal of Fisheries and Aquatic Sciences* **53**, 1226–1235.

Pearl, R. (1928) *The Rate of Living*. Knopf, New York.

Pearson, J.C. (1929) Natural history and conservation of the redfish and other commercial sciaenids of the Texas coast. *Bulletin of the U.S. Bureau of Fisheries* **44**, 129–214.

Peňáz, M. (1996) *Chondrostoma nasus* – its reproduction strategy and possible reasons for a widely observed population decline – a review. In: *Conservation of Endangered Freshwater Fishes of Europe*. (Ed. by A. Kirchofer & D. Hefti), pp. 279–285. Birkhaeuser-Verlag, Basel.

Pepin, P., Orr, D.C. & Anderson, J.T. (1997) Time to hatch and larval size in relation to temperature and egg size in Atlantic cod (*Gadus morhua*). *Canadian Journal of Fisheries and Aquatic Sciences* **54**(suppl. 1), 2–10.

Pepin, P. & Shears, T.H. (1997) Variability and capture efficiency of bongo and Tucker trawl samplers in the collections of ichthyoplankton and other macrozooplankton. *Canadian Journal of Fisheries and Aquatic Sciences* **54**, 765–773.

Peterman, R.M. & Bradford, M.J. (1987) Wind speed and mortality rate of a marine fish, the northern anchovy (*Engraulis mordax*). *Science* **235**, 354–356.

Peterson, C.H. (2001) The "Exxon Valdez" oil spill in Alaska: acute, indirect and chronic effects on the ecosystem. *Advances in Marine Biology* **39**, 1–103.

Poling, K.R. & Fuiman, L.A. (1998) Sensory development and its relation to habitat change in three species of sciaenids. *Brain, Behavior and Evolution* **52**, 270–284.

Priede, I.G. & Watson, J.J. (1993) An evaluation of the daily egg production method for estimating biomass of Atlantic mackerel (*Scomber scombrus*). *Bulletin of Marine Science* **53**, 891–911.

Rice, J.A., Crowder, L.B. & Binkowski, F.P. (1985) Evaluating otolith analysis for bloater *Coregonus hoyi*: do otoliths ring true? *Transactions of the American Fisheries Society* **114**, 532–539.

Rice, J.A., Crowder, L.B. & Binkowski, F.P. (1987a) Evaluating potential sources of mortality for larval bloater: starvation vs. predation. *Canadian Journal of Fisheries and Aquatic Sciences* **44**, 467–472.

Rice, J.A., Crowder, L.B. & Holey, M.E. (1987b) Exploration of mechanisms regulating larval survival in Lake Michigan bloater: a recruitment analysis based on characteristics of individual larvae. *Transactions of the American Fisheries Society* **116**, 703–718.

Rice, J.A., Miller, T.J., Rose, K.A., Crowder, L.B., Marschall, E.A., Trebitz, A.S. & DeAngelis, D.L. (1993) Growth rate variation and larval survival: inferences from an individual-based size-dependent predation model. *Canadian Journal of Fisheries and Aquatic Sciences* **50**, 133–142.

Rice S.D, Spies, R.B., Wolfe, D.A. & Wright, B.A. (Eds) (1996) Proceedings of the Exxon Valdez Oil Spill Symposium. *American Fisheries Society Symposium* **18**, Bethesda, Maryland.

Richards, W.J., McGowan, M.F., Leming, T., Lamkin, J.T. & Kelley, S. (1993) Larval fish assemblages at the Loop current boundary in the Gulf of Mexico. *Bulletin of Marine Science* **53**, 475–537.

Ricker, W.E. (1954) Stock and recruitment. *Journal of the Fisheries Research Board of Canada* **11**, 559–623.

Ricker, W.E. (1969) Effects of size-selective mortality and sampling bias on estimates of growth, mortality, production and yield. *Journal of the Fisheries Research Board of Canada* **26**, 479–541.

Rilling, G.C. & Houde, E.D. (1999) Regional and temporal variability in growth and mortality of bay anchovy, (*Anchoa mitchilli*), larvae in Chesapeake Bay. *Fishery Bulletin, U.S.* **97**, 555–569.

Rose, K.A., Cowan, J.H., Jr., Houde, E.D. & Coutant, C.C. (1993) Individual-based model of environmental quality effects on early life stages of fish: a case study using striped bass. In: *Water Quality and the Early Life Stages of Fishes.* (Ed. by L.A. Fuiman), pp. 125–145. *American Fisheries Society Symposium*, volume 14, Bethesda, Maryland.

Roseman, E.F., Taylor, W.W., Hayes, D.B., Haas, R.C., Knight, R.L. & Paxton, K.O. (1996) Walleye egg deposition and survival on reefs in western Lake Erie. *Annals Zoologica Fennici* **33**, 341–351.

Rothschild, B.J. (1986) *Dynamics of Marine Fish Populations*. Harvard University Press, Cambridge. Massachusetts, USA.

Rutherford, E.S. & Houde, E.D. (1995) The influence of temperature on cohort-specific growth, survival and recruitment of striped bass (*Morone saxatilis*), larvae in Chesapeake Bay. *Fishery Bulletin, U.S.* **93**, 315–332.

Ruzzante, D.E., Taggart, C.T. & Cook, D. (1996) Spatial and temporal variation in the genetic composition of a larval cod (*Gadus morhua*) aggregation: cohort contribution and genetic stability. *Canadian Journal of Fisheries and Aquatic Sciences* **53**, 2695–2705.

Ruzzante, D.E., Taggart, C.T. & Cook, D. (1999) A review of the evidence for genetic structure of cod (*Gadus morhua*) populations in the NW Atlantic and population affinities of larval cod off Newfoundland and the Gulf of St. Lawrence. *Fisheries Research (Amsterdam)* **43**, 79–97.

Schaaf, W.E., Peters, D.S., Vaughan, D.S., Coston-Clements, L. & Krouse, C.W. (1987) Fish population responses to chronic and acute pollution: the influence of life history changes. *Estuaries* **10**, 267–275.

Schiemer, F. (1999) Conservation of biodiversity in floodplain rivers. *Archiv für Hydrobiologie* **115**(suppl.), 423–438.

Schiemer, F., Baumgartner, C. & Tockner, K. (1999) Restoration of floodplain rivers: the Danube Restoration Project. *Regulated Rivers: Research and Management* **15**, 231–244.

Schiemer, F., Jungwirth, M. & Imhof, G. (1994) *Die Fische der Donau-Gefährdung und Schutz. Grüne Reihe des Bundesministeriums für Umwelt, Jugend und Familie.* Styria Medienservice.

Schiemer, F., Keckeis, H., Reckendorfer, W. & Winkler, G. (2001a) The "inshore retention concept" and its significance for large rivers. *Archiv für Hydrobiologie* **135**(suppl.), 509–516.

Schiemer, F., Keckeis, H., Winkler, G. & Flore, L. (2001b) Large rivers: the relevance of ecotonal structure and hydrological properties for the fish fauna. *Archiv für Hydrobiologie* **135**(suppl.), 487–508.

Schiemer, F. & Spindler, T. (1989) Endangered fish species of the Danube River in Austria. *Regulated Rivers: Research and Management* **4**, 397–407.

Schiemer, F., Spindler, T., Wintersberger, H., Schneider, A. & Chovanec, A. (1991) Fish fry association: important indicators for the ecological status of large rivers. *Verhandlungen der Internationalen Vereinigung für Theoretische und Angewandte Limnologie* **24**, 2497–2500.

Schiemer, F. & Waidbacher, H. (1992) Strategies for Conservation of a Danubian fish fauna. In: *River Conservation and Management*. (Ed. by P.J. Boon, P. Calow & G.E. Petts), pp. 363–382. John Wiley and Sons Ltd, Chichester, UK.

Schmidt, J. (1905) The pelagic post-larval stages of the Atlantic species of *Gadus*. *Meddelelser fra Kommissionen for Havundersøgesler, Serie Fiskeri* **1**, 1–77.

Schmidt, J. (1916) On the early larval stages of the fresh-water eels (*Anguilla*) and some other North Atlantic muraenoids. *Meddelelser fra Kommissionen for Havundersøgesler, Serie Fiskeri* **5**, 1–20.

Schmidt, J. (1922) The breeding places of the eel. *Philosophical Transactions Royal Society B* **211**, 179–208.

Schwartzlose, R.A., Alheit, J., Bakun, A., Baumgartner, T.R., Cloete, R., Crawford, R.J.M., Fletcher, W.J., Green-Ruiz, Y., MacCall, A.D., Matsuura, Y., Nevarez-Martines, M.O., Parrish, R.H., Roy, C., Serra, R., Shust, K.V., Ward, M.N. & Zuzunaga, J.Z. (1999) Worldwide large-scale fluctuations of sardine and anchovy populations. *South African Journal Marine Science* **21**, 289–347.

Sclafani, M., Taggart, C.T. & Thompson, K.R. (1993) Condition, buoyancy, and the distribution of larval fish: implications for vertical migration and retention. *Journal of Plankton Research* **15**, 413–435.

Seber, G.A.F. & Wild, C.J. (1989) *Nonlinear Regression*. Series in Probability and Mathematical Statistics, John Wiley & Sons, New York.

Secor, D.H. (2000) Spawning in the nick of time? Effect of adult demographics on spawning behavior and recruitment in Chesapeake Bay striped bass. *ICES Journal of Marine Science* **57**, 403–411.

Secor, D.H. & Houde, E.D. (1995) Temperature effects on the timing of striped bass egg production, larval viability, and recruitment potential in the Patuxent River (Chesapeake Bay). *Estuaries* **18**, 527–544.

Secor, D.H. & Houde, E.D. (1998) Use of larval stocking in restoration of Chesapeake Bay striped bass. *ICES Journal of Marine Science* **55**, 228–239.

Selgeby, J.H., Bronte, C.R., Brown, E.H., Hansen, M.J., Holey, M.E., Vanamberg, J.P., Muth, K.M., Makauskas, D.B., McKeen, P., Anderson, D.M., Ferreri, C.P. & Schram, S.T. (1995) Proceedings of the 1994 international conference on restoration of lake trout in the Laurentian Great Lakes. *Journal Great Lakes Research* **21**(suppl. 1), 498–504.

Sinclair, M. (1988) *Marine Populations. An Essay on Population Regulation and Speciation*. University of Washington Press, Seattle.

Snyder, D.E. (1976) Terminologies of intervals of larval fish development. In: *Great Lakes Fish Egg and Larvae Identification: Proceedings of a Workshop*. (Ed. by J. Boreman), pp. 41–58. U.S. Fish and Wildlife Service, Biological Services Program, FWS/OBS-76-23.

Spencer, K., Shafer, D.J., Gauldie, R.W. & DeCarlo, E.H. (2000) Stable lead isotope ratios from distinct anthropogenic sources in fish otoliths: a potential nursery ground stock marker. *Comparative Biochemistry and Physiology Part A* **127**, 273–284.

Spindler, T. (1988) *Ecology of 0+ fish in the Danube near Vienna*. Ph.D. dissertation, University of Vienna (in German).

Stobutzki, I.C. & Bellwood, D.R. (1994) An analysis of the sustained swimming abilities of presettlement and postsettlement coral-reef fishes. *Journal Experimental Marine Biology and Ecology* **175**, 275–286.

Swearer, S.E., Caselle, J.E., Lea, D.W. & Warner, R.R. (1999) Larval retention and recruitment in an island population of a coral reef fish. *Nature* **402**, 799–802.

Taggart, C.T. & Frank, K.T. (1990) Perspectives on larval fish ecology and recruitment processes: probing the scales of relationships. In: *Large Marine Ecosystems: Patterns, Processes and Yields*. (Ed. by K. Sherman, L.M. Alexander & B.B. Gold), pp. 151–164. American Association for the Advancement of Science, Washington D.C.

Takahashi, M. (2001) *Growth and development of larval and juvenile Japanese anchovy (Engraulis japonicus) and their implications for recruitment to spawning population*. Ph.D. dissertation, University of Tokyo.

Taylor, W.W., Smale, M.A. & Freeberg, M.H. (1987) Biotic and abiotic determinants of lake whitefish (*Coregonus clupeaformis*) recruitment in northeastern Lake Michigan. *Canadian Journal of Fisheries and Aquatic Sciences* **44**(suppl.2), 313–323.

Thorrold, S.R., Jones, C.M., Campana, S.E., McLaren, J.W. & Lam, J.W.H. (1998) Trace element signatures in otoliths record natal river of juvenile American shad (*Alosa sapidissima*). *Limnology and Oceanography* **43**, 1826–1835.

Thorrold, S.R. & Williams, D. McB. (1996) Meso-scale distribution patterns of larval and pelagic juvenile fishes in the central Great Barrier Reef lagoon. *Marine Ecology Progress Series* **145**, 17–31.

Tockner, K., Schiemer, F., Baumgartner, C., Kum, G., Weigand, E., Zweimüller, I. & Ward, J.V. (1999) The Danube Restoration Project: species diversity patterns across connectivity gradients in the floodplain system. *Regulated Rivers: Research and Management* **15**, 245–258.

Townsend, D.W. (1992) Ecology of larval herring in relation to the oceanography of the Gulf of Maine. *Journal of Plankton Research* **14**, 467–493.

Trippel, E.A., Kjesbu, O.S. & Solemdal, P. (1997) Effect of adult age and size structure on reproductive output in marine fishes. In: *Early Life History and Recruitment in Fish Populations*. (Ed. by R.C. Chambers & E.A. Trippel), pp. 31–55. Fish and Fisheries Series 21. Chapman & Hall, London.

Turnpenny, A.W.H. & Williams, R. (1980) Effects of sedimentation on the gravels of an industrial river. *Journal Fish Biology* **17**, 681–693.

Van der Veer, H., Berghahn, R., Miller, J.D. & Rijnsdorp, A.D. (2000) Recruitment in flatfish, with special emphasis on North Atlantic species: progress made by the flatfish symposia. *ICES Journal of Marine Science* **57**, 202–215.

Wada, T. (1998) Migration range and growth rate in the Oyashio area. In: *Stock Fluctuations and Ecological Changes of the Japanese Sardine*. (Ed. by Y. Watanabe & T. Wada), pp. 27–34. Koseisha-Koseikaku, Tokyo.

Waidbacher, H. & Haidvogel, G. (1998) Fish migration and fish passage facilities in the Danube: past and present. In: *Fish Migration and Fish Bypasses*. (Ed. by M. Jungwirth, S. Schmutz & S. Weiss), pp. 85–98. Blackwell Science Ltd., Oxford.

Ward, J.V., Tockner, K. & Schiemer, F. (1999) Biodiversity of floodplain river ecosystems: ecotones and connectivity. *Regulated Rivers: Research and Management* **15**, 125–139.

Watanabe, Y. & Kuroki, T. (1997) Asymptotic growth trajectories of larval sardine (*Sardinops melanostictus*) in the coastal waters off western Japan. *Marine Biology* **127**, 369–378.

Watanabe, Y. & Saito, H. (1998) Feeding and growth of early juvenile Japanese sardines in the Pacific waters off central Japan. *Journal Fish Biology* **52**, 519–533.

Watanabe, Y., Yokouchi, K., Oozeki, Y. & Kikuchi, H. (1991) Preliminary report on larval growth of the Japanese sardine spawned in the dominant current area. *International Council for the Exploration of the Sea* C.M.-ICES 1991/L, 33 (mimeo).

Watanabe, Y., Zenitani, H. & Kimura, R. (1995) Population decline of the Japanese sardine (*Sardinops melanostictus*) owing to recruitment failures. *Canadian Journal of Fisheries and Aquatic Sciences* **52**, 1609–1616.

Watanabe, Y., Zenitani, H. & Kimura, R. (1996) Offshore expansion of spawning of the Japanese sardine, (*Sardinops melanostictus*), and its implication for egg and larval survival. *Canadian Journal of Fisheries and Aquatic Sciences* **53**, 55–61.

Watanabe, Y., Zenitani, H., Kimura, R., Sato, C., Okumura, Y., Sugisaki, H. & Oozeki, Y. (1998) Naupliar copepod concentrations in the spawning grounds of Japanese sardine (*Sardinops melanostictus*), along the Kuroshio Current. *Fisheries Oceanography* **7**, 101–109.

Watanabe, Y., Zenitani, H., Kimura, R., Watanabe, C., Okumura, Y., Sugisaki, H. & Oozeki, Y. (1999) Food availability for larval (*Sardinops melanostictus*). In: *The Kuroshio–Oyashio Transition Region Off Boso Peninsula*. (Ed. by M. Terazaki, K. Ohtani, T. Sugimoto & Y. Watanabe), pp. 194–205. Japan Marine Science Foundation, Tokyo.

Welcomme, R.L. (1998) Evaluation of stocking and introductions as management tools. In: *Stocking and Introduction of Fish*. (Ed. by I.G. Cowx), pp. 397–413. Fishing News Books, Oxford.

Wells, L. (1966) Seasonal and depth distribution of larval bloaters (*Coregonus hoyi*) in southeastern Lake Michigan. *Transactions of the American Fisheries Society* **95**, 388–396.

Wells, L. (1969) Effects of alewife predation on zooplankton populations in Lake Michigan. *Limnology and Oceanography* **14**, 556–565.

Werner, E.E. & Gilliam, J.F. (1984) The ontogenetic niche and species interactions in size-structured populations. *Annual Review of Ecology and Systematics* **15**, 393–425.

Werner, E.E. & Hall, D.J. (1988) Ontogenetic habitat shifts in bluegill: the foraging rate–predation risk trade-off. *Ecology* **69**, 1352–1366.

Werner, F.E., Page, F.H., Lynch, D.R., Loder, J.W., Lough, R.G., Perry, R.I., Greenberg, D.A. & Sinclair, M.M. (1993) Influences of mean advection and simple behavior on the distribution of cod and haddock early life stages on Georges Bank. *Fisheries Oceanography* **2**, 43–64.

Werner, R.G. (1967) Intralacustrine movements of bluegill fry in Crane Lake, Indiana. *Transactions of the American Fisheries Society* **96**, 416–420.

Werner, R.G. (1969) Ecology of limnetic bluegill fry in Crane Lake, Indiana. *American Midland Naturalist* **81**, 164–181.

Werner, R.G. & Lannoo, M.J. (1994) Development of the olfactory system of the white sucker (*Catostomus commersoni*), in relation to imprinting and homing: a comparison to the salmonid model. *Environmental Biology of Fishes* **40**, 125–140.

White, P., Robitaille, A.S. & Rasmussen, J.B. (1999) Heritable reproductive effects of benzo(a)pyrene on the fathead minnow (*Pimephales promelas*). *Environmental Toxicology and Chemistry* **18**, 1843–1847.

Williams, P.J., Brown, J.A., Gotceitas, V. & Pepin, P. (1996) Developmental changes in escape response performance of marine larval fish. *Canadian Journal of Fisheries and Aquatic Sciences* **53**, 1246–1253.

Winemiller, K.O. & Rose, K.A. (1992) Patterns of life-history diversification in North American fishes: implications for population regulation. *Canadian Journal of Fisheries and Aquatic Sciences* **49**, 2196–2218.

Winkler, G., Keckeis, H., Reckendorfer, W. & Schiemer, F. (1997) Temporal and spatial dynamics of 0+ *Chondrostoma nasus*, at the inshore zone of a large river. *Folia Zoologica* **46**, 151–168.

Wintersberger, H. (1996a) Species assemblages and habitat selection of larval and juvenile fishes in the River Danube. *Archiv für Hydrobiologie* **113**(suppl.), 497–505.

Wintersberger, H. (1996b) Spatial resource utilization and species assemblages of larval and juvenile fishes. *Archiv für Hydrobiologie* **115**(suppl.), 29–44.

Wurster, C.M., Patterson, W.P. & Cheatham, M.M. (1999) Advances in micromilling techniques: a new apparatus for acquiring high-resolution oxygen and carbon stable isotope values and major/minor elemental ratios from accretionary carbonate. *Computers and Geosciences* **25**, 1159–1166.

Youson, J.H. (1988) First metamorphosis. In: *Fish Physiology*, Volume 11B. (Ed. by W.S. Hoar & D.J. Randall), pp. 135–196. Academic Press, San Diego.

Zauner, G. (1996) *Ecological Study on Percids of the Upper Danube* (Ed. by. W. Morawetz & H. Winkler), Biosystematics and Ecology Series 9. Austrian Academy of Sciences Press (in German).

Zaunreiter, M., Junger, H. & Kotrschal, K. (1991) Retinal morphology of cyprinid fishes: a quantitative histological study of ontogenetic changes and interspecific variation. *Vision Research* **31**, 383–394.

Zenitani, H., Ishida, M., Konishi, Y., Goto, T., Watanabe, Y. & Kimura, R. (1995) Distributions of eggs and larvae of Japanese sardine, Japanese anchovy, mackerels, round herring, Japanese horse mackerel and Japanese common squid in the waters around Japan, 1991 through 1993. *Resources Management Research Report Series* **A-1**.

Zenitani, H., Nakata, K. & Inagake, D. (1996) Survival and growth of sardine larvae in the offshore side of the Kuroshio. *Fisheries Oceanography* **5**, 56–62.

Subject Index

Taxonomic Index